FISHERIES MANAGEMENT AND CONSERVATION

FISHERIES MANAGEMENT AND CONSERVATION

William Hunter III

Researcher, National Science Foundation, U.S.A.

AAP APPLE ACADEMIC PRESS

Fisheries Management and Conservation

© Copyright 2011*
Apple Academic Press Inc.

First Published in the Canada, 2011
Apple Academic Press Inc.

1265 Goldenrod Circle, NE,
Palm Bay, FL 32905 USA

4164 Lakeshore Road, Burlington,
ON, L7L 1A4 Canada
E-mail: info@appleacademicpress.com
www.appleacademicpress.com

First issued in paperback 2021

ISBN 978-1-926692-66-1 (hardbound)
ISBN: 978-1-46656-200-4 (ebook)
ISBN: 978-1-77463-244-4 (paperback)

William Hunter III

Printed at Thomson Press (I) Ltd.
9 8 7 6 5 4 3 2 1

Cover Design: Psqua

Library and Archives Canada Cataloguing in Publication Data
CIP Data on file with the Library and Archives Canada

CONTENTS

INTRODUCTION

The current state of the world's marine ecology calls for direct action on an international scale. At its heart, fisheries science strives to find ways to protect natural resources so that sustainable exploitation of fish and marine life is possible. Many management strategies have been used in an attempt to find this balance, making fisheries science an important field for today's ever-threatened ecosystems.

In this collection of articles, the focus is on human interaction with the marine world and ways to make these interactions more sustainable. The driving force behind this research is, of course, changing policy at an international level to create a mutually beneficial outcome for both marine life and humans.

The most prevalent management practice is setting up fisheries, or preserves, where harvest numbers are monitored in an effort to maintain a fish population in a given area. Fisheries have been created to address specific problems such as the potential for stock population structures to lose diversity and resilience to environmental fluctuations; overexploitation and therefore a loss in significant potential yield; ecosystems with economic infrastructures that will cycle between collapse and recovery with progressively less productive cycles; and potential negative changes in the trophic balance.

The efficacy of this practice is being researched in depth because fish accounts for 15% of the protein consumed and used by humans. Camilo Mora et al. consider these questions in "Management and Effectiveness of the World's Marine Fisheries" by surveying the strategies at fisheries around the world. Their research gives evidence that there is a stark contrast between countries and groups that support initiatives and actual implementation of them. Additionally, when scientific advice is used in shaping transparent policy, the resulting systems are more sustainable and better enforced.

Richard Cudney-Bueno et al. discuss the possibilities of marine reserves re-populating nearby areas that have been overfished in "Rapid Effects of Marine Reserves via Larval Dispersal." Whether or not reserves have the potential to help recover affected areas is more dependent on dispersal pattern than density-dependent adult spillover and other oceanographic effects.

The management strategies at fisheries worldwide are diverse, but one of the most effective techniques is discussed in "Lack of Cross-Scale Linkages Reduces Robustness of Community-Based Fisheries Management." Richard Cudney-Bueno and Xavier Basurto find that rapid increase in fish populations is potentially seen when local groups form to protect resources. While small community-based management systems where stakeholders monitor and enforce cooperation are effective, larger government must become involved to maintain incentives and prevent over-exploitation from outsiders.

One of the biggest problems with fisheries today is that there is little monitoring of fish harvests that is without bias. "A Step Towards Seascape Scale Conservation: Using Vessel Monitoring Systems (VMS) to Map Fishing Activity" discusses the potential for using independent VMS equipment to monitor boat patterns. This will hopefully lead to reduced bycatch (where other fish are caught along with the target fish) and reduced destruction to other parts of the marine ecosystem. Matthew J. Witt and Brendan J. Godley believe that international, unbiased reporting on fishing activity will also lead to better management policies worldwide.

If fisheries are not the most sustainable option for recovery, more research is required, such as that done by Natalie C. Jan and Amanda C.J. Vincent. They discuss their theory on spatial areas where fishing is allowed in "Beyond Marine Reserves: Exploring the Approach of Selecting Areas where Fishing is Permitted, Rather than Prohibited." Currently only one percent of the ocean is set aside as fish preserves. They propose areas where fishing is permitted to certain quotas rather than totally banned. This will potentially minimize losses to both fishermen and wildlife but also requires less effort.

Fisheries influence the natural world as well as the human world. Einar Árnason et al. look at the way in which directional natural selection of the Atlantic cod (*Gadus morphua*) is pushed along because of fishery pressures. "Intense Habitat-Specific Fisheries-Induced Selection at the Molecular Pan I: Locus Predicts Imminent Collapse of a Major Cod Fishery" finds that Pan I alleles, correlated with the depth at which fish are found, have been selected against shallow water fish.

This represents a change in a predator/prey relationship where humans are the predators. This research is important because they predict a collapse of the Iceland fishery as evidenced by similar clues of decreasing length and age of maturation before a previous collapse in Newfoundland. Outside of fisheries, the natural world can also be affected by human activity.

In "Fish Communities in Coastal Freshwater Ecosystems: The Role of the Physical and Chemical Setting," Kristin K. Arend and Mark B. Bain discuss the effects of nearby human activity on species abundance, biomass, total phosphorous load, embayment area, and submerged vegetation. Energy transfer and nutrient cycling potentially change the local ecosystems. In general, more vegetation leads to higher populations of smaller fish, potentially changing the trophic levels of an area, which has implications for all ecosystems affected by human activity.

The policies surrounding the fishing community worldwide vary greatly in strategy and effectiveness. One of the most positive techniques is incorporating local culture and knowledge into management practices. In "Analysing Ethnobotanical and Fishery-Related Importance of Mangroves of the East-Godava Delta (Andhra Pradesh, India) for Conservation and Management Purposes," F. Dahdouh-Guebas et al. discuss the potential for educating local cultures that utilize mangroves. Mangroves in this region are endangered due to their use and destruction in wood fuel, construction, shrimp aquaculture, and pollution. By enacting policy that takes cultural norms into perspective and the need for these individuals to make a living, more sustainable harvesting can be utilized.

In the same vein, Ierecê M. L. Rosa et al. look at the importance of using local knowledge in preserving longsnout seahorses (*Hippocampus reidi)* in "Fishers' Knowledge and Seahorse Conservation in Brazil." Because local fishermen know breeding patterns and good habitat conditions and appreciate the future economic value of maintaining the seahorse population, their help goes a long way in creating conservation strategies.

The negative impacts of human activity or inaction are becoming more commonplace around the globe. Christopher D. Stallings discusses the direct results of artisanal fishing on large predator fish in "Fishery-Independent Data Reveal Negative Effect of Human Population Density on Caribbean Predatory Fish Communities." Not only do the actual numbers of fish decrease, the species richness for an area diminishes as well. Fabien Leprieur et al. consider the three hypotheses on non-native species invasions in their article "Fish Invasions in the World's River Systems: When Natural Processes are Blurred by Human Activities." Human activity, biotic resistance (a species-rich area will prevent non-native species), and biotic acceptance (a suitable habitat for natives is a suitable habitat for non-natives) are all put to the test in a series of fish invasions. On a global scale, they find that human activity vastly accounts for invasive fish species presence in an area and no evidence for the other two theories. Finally, in "Estimating the Worldwide Extent of Illegal Fishing," David J. Agnew et al. attempt to find a correlation between governance and levels of illegal or underreported fishing. These missing fish cause overexploitation of fish stocks, slow ecological recovery, and unsustainable harvesting. They find that poorer countries are more susceptible to illegal fishing by their own people and to licensing to outside nations. To solve this dilemma, increased cooperation and international community involvement is needed. Despite much of the negative findings and media, there have been some success stories.

In "Integrated Ecosystem Assessment: Lake Ontario Water Management," Mark B. Bain et al. find that after the community assessed different management strategies for the least environmental impact, the current one was actually the most effective. In another finding, Mark B. Bain et al. describe the success story of the shortnose sturgeon (*Zcipenser brevirostrum*) in "Recovery of a U.S. Endangered Fish." The Hudson River in New York has recovered the sturgeon population by over 400% since the 1970s by protecting the species and its habitat. This is an area with high human activity, and the authors want to prove that the U.S. Endangered Species Act can be considered successful if used appropriately.

The research underpinning fisheries science is expanding at a rapid rate with new technology, global efforts, and increasing concern about the natural world. Effective policy requires cooperation on an international level to maintain fish species as both a food source and a vital part of the marine ecosystem. Therefore, it is not enough to change management practices in the short term; they must reflect the long-term conservation impacts as well.

— William Hunter III

Chapter 1

Beyond Marine Reserves: Exploring the Approach of Selecting Areas where Fishing is Permitted, Rather than Prohibited

Natalie C. Ban and Amanda C. J. Vincent

ABSTRACT

Background

Marine populations have been declining at a worrying rate, due in large part to fishing pressures. The challenge is to secure a future for marine life while minimizing impacts on fishers and fishing communities.

Methods and Principal Findings

Rather than selecting areas where fishing is banned – as is usually the case with spatial management – we assess the concept of designating areas where

Originally published as Ban NC, Vincent ACJ (2009) Beyond Marine Reserves: Exploring the Approach of Selecting Areas where Fishing Is Permitted, Rather than Prohibited. PLoS ONE 4(7): e6258. https://doi.org/10.1371/journal.pone.0006258. © 2009 Ban, Vincent. https://creativecommons.org/licenses/by/4.0/

fishing is permitted. We use spatial catch statistics for thirteen commercial fisheries on Canada's west coast to determine the minimum area that would be needed to maintain a pre-ascribed target percentage of current catches. We found that small reductions in fisheries yields, if strategically allocated, could result in large unfished areas that are representative of biophysical regions and habitat types, and have the potential to achieve remarkable conservation gains.

Conclusions

Our approach of selecting fishing areas instead of reserves could help redirect debate about the relative values that society places on conservation and extraction, in a framework that could gain much by losing little. Our ideas are intended to promote discussions about the current status quo in fisheries management, rather than providing a definitive solution.

Introduction

The oceans have suffered declines in faunal biomass and biodiversity [1], [2], [3], with fisheries constituting the single biggest human-induced pressure on marine life [4]. Marine reserves (no-fishing zones) have been widely hailed as providing one powerful tool – but not a panacea – for halting the decline of overexploited fish and invertebrate populations [5], [6], [7], [8], [9]. The evidence that they increase biomass, abundance, and average size of exploited organisms within their boundaries [5], [6] has prompted international commitments to establish marine protected areas (including reserves) under the Convention on Biological Diversity and at the World Summit on Sustainable Development [10], [11]. Nevertheless, and despite this accord on the value of marine reserves, they are being implemented far too slowly to meet agreed targets for marine protection [10]. Their impacts on fisheries remain uncertain, and the extent of fisheries benefits, if any, varies [9], [12].

Given the slow accumulation of marine reserves relative to international targets, we turn the problem on its head [e.g., 13]. We embrace the challenge of presuming that the entire ocean is initially protected from fishing rather than open to fishing [14], [15], [16]. At present, fisheries exploitation is specifically excluded (i.e., areas are protected) in less than 1% of the oceans [10]. Given biodiversity concerns and the challenging task of managing fisheries with limited data, it is increasingly vital to explore ways to restrict fisheries spatially while respecting their socioeconomic and nutritional contributions. Such restrictions should, ideally, also meet systematic conservation planning criteria of representation and persistence [17].

Conceptually the approach of selecting fishing areas is very similar to using fisheries as a "cost" to represent economic losses to fisheries in marine reserve selection [18], [19]. "Cost" in this context refers to the socio-economic or political cost of adding an area to a marine reserve [20]. The typical approach in marine reserve selection is to ensure representation of biodiversity features whilst minimize the cost to fisheries. By treating fisheries as a cost, marine reserve selection tools require that they be summarized in one layer – this often involves adding or averaging the catches, effort, or catch-per-unit-effort of fisheries for each area [20]. By targeting fisheries instead of treating them as a cost, we select the most productive fishing regions while minimizing the area fished. The main advantage of this approach is that each fishery can be selected for individually, thereby ensuring that all fisheries would be able to continue.

The goal of our research was to explore the possible conservation gains that might accrue from different hypothetical levels of restriction on fisheries. We use data from the Pacific coast of British Columbia, Canada (approximately 49°N to 54°N latitude) for this initial foray. Our purpose is not to provide a definitive answer to fisheries management or reserve selection. Rather, we seek to promote discussion about the current status quo of our ocean management approaches.

Results

Our analyses show that very small reductions in fisheries yields – if allocated in a strategic manner across space – can offer promising conservation benefits in both space and composition. For example, catch reductions of only 2%–5% could result in no-fishing areas constituting 20% or 30% of previously fished areas (Fig. 1). Every subsequent reduction in target catches yielded yet larger no-fishing areas (Fig. 1 and Fig. 2). Moreover, for each scenario, the multiple solutions that released the greatest area from fishing (Fig. 2) described no-fishing areas that included representation from all twelve ecosections in British Columbia (Table 1); these ecosections delineate marine regions based on physical criteria. Maintaining catches at 95% of recent levels (or more, depending on the fishery) resulted in no-fishing areas that protected at least 17%, and an average of 55%, of each physical and habitat feature (Table 2). In this scenario the total area protected would be 30% in exchange for a mean 4.6% reduction in catches (Table 3).

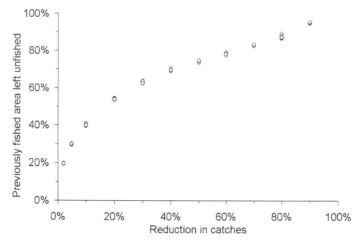

Figure 1. Decreases in areas fished resulting from reductions of catches for 13 commercial marine fisheries (British Columbia, Canada). Each of 11 scenarios was repeated 10 times, with 100 runs of one million iterations each (11,000 runs). The result requiring the least area of each of the 10 repetitions per scenarios is graphed (i.e., there are 10 data points per scenario; some overlap closely and appear as one).

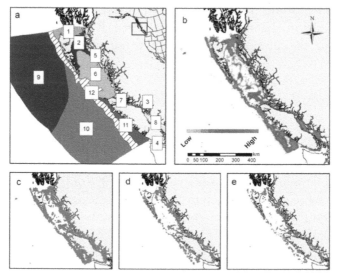

Figure 2. Marine ecosections in British Columbia and selected permitted fishing area solutions. The marine ecosections (a) are 1 = Dixon Entrance; 2 = Hecate Strait; 3 = Johnstone Strait; 4 = Juan de Fuca Strait; 5 = North Coast Fjords; 6 = Queen Charlotte Sound; 7 = Queen Charlotte Strait; 8 = Strait of Georgia; 9 = Subarctic Pacific; 10 = Transitional Pacific; 11 = Vancouver Island Shelf; 12 = Continental Slope. The selection frequency map (b) shows the importance of areas to commercial fisheries based pm the summed solution results from Marxan. The permitted fishing area solutions (in blue) are for a sample of the scenarios that minimize the area fished with the corresponding percent reduction in commercial fishing catches: (c) 5%, (d) 20%, (e) 40%.

Table 1. Gap analysis by ecosection for the most spatially limited result for each scenario.

| | Area of ecosection (ha * 1000) | % of area fished | Percent reduction in catches (italics), resulting in percent protected (plain, in %) | | | | | | | | | | |
			2%	*5%*	*10%*	*20%*	*30%*	*40%*	*50%*	*60%*	*70%*	*80%*	*90%*
Continental Slope	3,330	53.8	55.3	60.0	64.8	74.5	78.6	79.2	84.5	88.5	91.9	89.6	97.2
Dixon Entrance	1,089	55.4	57.8	64.8	72.0	79.1	83.0	90.8	86.9	89.9	95.3	93.6	98.1
Hecate Strait	1,280	77.0	36.5	43.0	50.0	57.6	64.9	70.9	74.5	82.7	91.7	85.5	95.5
Johnstone Strait	239	98.0	11.4	19.2	24.1	46.3	58.1	65.1	98.0	73.0	81.7	81.8	96.1
Juan de Fuca Strait	150	90.8	15.5	27.0	67.4	56.7	74.6	72.8	69.5	96.6	100	97.7	100
North Coast Fjords	958	91.9	23.6	33.9	46.9	59.6	68.7	72.8	80.5	83.3	91.3	85.2	96.4
Queen Charlotte Sound	3,642	55.7	60.7	68.1	75.9	82.5	87.5	89.1	89.8	90.1	97.4	94.2	99.2
Queen Charlotte Strait	220	94.5	7.9	18.1	33.2	46.2	45.6	69.4	66.4	56.8	93.2	74.6	86.0
Strait of Georgia	815	94.8	7.9	63.0	14.6	27.3	31.2	51.7	50.7	64.5	79.4	77.8	92.7
Subarctic Pacific	17,098	0.3	99.8	99.9	99.9	99.9	99.9	100	99.6	100	100	100	100
Transitional Pacific	14,850	0.1	100	100	100	100	100	100	100	100	100	100	100
Vancouver Island Shelf	1,670	89.2	17.8	24.4	30.9	42.2	56.3	66.8	70.7	76.4	87.9	75.5	93.8

Table 2. Detailed analysis of the result of the 5% catch reduction scenario that produced the greatest area unfished, indicating ecosystem components that would be protected.

		Total area (ha) of each ecological feature	% outside permitted fishing areas
Depth	Shallow (0–20 m)	743,853	40.6
	Photic (20–50 m)	1,521,555	42.9
	Mid-depth (50–200 m)	60,400,258	94.3
	Deep (200–1000 m)	3,469,678	43.9
	Abyssal (>1000 m)	33,627,695	99.7
Temperature (summer at seabed bottom)	Warm (9–15˚C)	2,438,557	32.9
	Cool (<9˚C)	42,820,022	88.0
Slope	Flat (0–5%)	40,556,889	87.2
	Sloping (5–20%)	4,737,411	67.4
	Steep (>20%)	42,749	43.0
Current	High (>3 knots)	212,713	39.0
	Low (<3 knots)	45,162,974	85.3
Substrate	Mud	2,295,529	27.6
	Sand	4,852,577	47.7
	Hard	3,631,788	53.5
Exposure	High	42,616,399	89.0
	Moderate	1,287,192	17.4
	Low	1,470,964	30.4
Relief	High	206,158	17.6
	Moderate	20,839,047	93.9
	Low	43,040,993	87.7
Salinity (annual average at surface)	Mesohaline (5–18ppt)	147,957	22.1
	Polyhaline (18–28 ppt)	11,279,517	91.7
	Euhaline (28–33 ppt)	43,945,636	87.0
Stratification	Mixed	4,931,996	36.8
	Weakly-mixed	2,083,666	42.3
	Stratified	37,823,783	94.4
Kelp		79,806	19.5
Eelgrass		10,449	28.7
Clam		18,978	22.5
Herring spawn		99,737	22.3
Sponge reefs		69,733	85.0

Table 3. Actual catch reductions and estimated direct financial impact for each fishery under the scenario that (a) reduced overall catch by 2%, 5% and 10% and (b) produced the greatest area unfished at that level.

Commercial fishery	Actual catch reduction (%)			Estimated direct impact (US$)[*]		
	2% catch reduction scenario	5% catch reduction scenario	10% catch reduction scenario	2% catch reduction scenario	5% catch reduction scenario	10% catch reduction scenario
Crab	2.0	5.0	10.0	665,624	1,664,103	3,327,551
Geoduck	2.0	5.0	10.0	113,128	282,906	565,828
Green urchin	2.0	5.0	10.0	5,021	12,512	25,122
Groundfish trawl	2.0	5.0	10.0	470,332	1,175,818	2,351,668
Krill	1.9	4.7	9.9	NA	NA	NA
Prawn	2.0	5.0	10.0	234,238	585,538	1,171,154
Red urchin	2.0	5.0	10.0	186,231	465,469	932,242
Sablefish longline	1.2	1.5	3.4	35,061	42,496	95,713
Sablefish trap	2.0	5.0	10.0	266,145	665,345	1,330,782
Schedule two	2.0	5.0	10.0	19,455	48,632	97,280
Sea cucumber	2.0	4.7	10.0	35,661	83,147	178,338
Shrimp trawl	2.0	4.2	7.2	207,272	430,335	748,438
ZN catch	2.0	5.0	10.0	17,223	43,057	86,118
Total				2,255,391	5,499,359	10,910,233

[*]Ex-vessel data obtained from the Sea Around Us online database (www.seaaroundus.org) [21]. We used mean annual prices from 2000 to 2004 in the reporting units of US$. At the time of writing the Canadian and US currencies were about par.
NA = ex-vessel data not available.

Estimates of the direct losses for the 2%, 5%, and 10% reduction scenarios based on ex-vessel prices [21] revealed that the combined cost to fisheries ranges from US$2.3 million per year to US$11 million per year for the above scenarios (Table 3). While this is less than one percent of British Columbia's seafood industry – which is valued at $1.4 billion annually – and an even smaller portion of British Columbia's oceans economy – valued at $11.4 billion annually [22] – it could be significant for some fisheries. In addition, the approach we cite would result in spin-off losses (e.g., job losses in the seafood processing industry).

The approach we employed for selecting permitted fishing areas used catches averaged over multiple years as the input, yet the result of the 5% reduction scenario also performed well when analyzed using documented annual catches for geoduck, green urchin, red urchin and sea cucumber fisheries (Table 4). As expected, we found some inherent spatial and temporal variability in the proportion of catches that would fall within the permitted fishing areas each year. The greatest range for a target of 95% of catches retained across all fisheries was a 2–12% reduction in sea cucumber catches, depending on the year.

Table 4. Proportion of annual commercial fisheries catches that fall within the permitted fishing area result of the 95% target scenario.

Fishery	Annual data	Average	Standard deviation	Minimum	Maximum
Geoduck	2002–2004	95.04%	1.91%	92.96%	96.73%
Green urchin	1998–2003	94.29%	2.89%	90.49%	97.05%
Red urchin	1997–2003	95.07%	0.39%	94.72%	95.82%
Sea cucumber	1997–2004	94.81%	3.07%	88.25%	97.97%

Discussion

Potential Conservation and Fisheries Benefits of Permitted Fishing Areas

The practical approach used in this study allows for explicit analyses of trade-offs between small reductions in fisheries – in a spatially strategic manner – and large gains for marine conservation through spatial protection. Managing marine environments by selecting permitted fishing areas rather than marine reserves would represent a much-needed paradigm shift in areas where little headway is being made in marine reserve establishment. Instead of debating the merit of each potential marine reserve, the discourse could focus on analyses of the ecological benefits of small reductions in fishing.

This approach has the potential to offer real conservation benefits. At a minimum, the approach outlined here would protect the same proportion of fished populations as the target reduction in catches, assuming even catchability in space. Because we suggest reducing quotas by the target percentage used for the spatial selection of permitted fishing areas, effort within these areas would not increase. Therefore any alterations to source-sink dynamics would already have occurred with previous fishing patterns. Even small marine reserves that protect only a fraction of populations have been shown to increase the size, number, and diversity of fish within their boundaries [5], [23], [24]. Given the usually exponential increase in fecundity of fishes that grow larger within protected areas, protecting even a small proportion of the population could greatly enhance numbers in areas that continue to be fished. By changing the reserve to fished ratio, recruitment effects could be significantly greater than anything seen to date. For larger species, fisheries yields may respond to no-fishing areas with a response that exceeds the results of conventional fisheries management by 60 % [25]. However, fecundity has not increased in all closed areas [26], and hence the effects of marine reserves can be unpredictable.

Even though ecological goals were not included a priori in the designation of the permitted fishing areas, the areas that fell outside permitted fishing areas included good representativeness across ecosections [27] (Table 1). Further detailed analysis of the scenario with 5% catch reduction showed that the areas outside the permitted fishing area represented key physical and habitat features (Table 2).

Even while protecting large (and representative) tracts of ocean, the proposed approach of designating permitted fishing areas could reasonably be expected also to strengthen fisheries in three ways if the experience in some marine reserves holds [e.g.], [6], [23,28]. First, the removal of destructive fishing gear from the areas outside the permitted fishing areas should promote improved habitat quality

[29], while also reducing bycatch [30]. Second, given the benefits of even small reserves for population recovery [28], the areas outside the permitted fishing areas could be enhanced by fish populations within permitted fishing areas [6], [31]. Third, many fisheries around the world are operating unsustainably [32], such that reductions in catch while setting permitted fishing areas could also move these fisheries closer to desired biological reference points for sound management [33]. Such changes might well offset catch reductions over the long run.

We are well aware that the actual effects of permitted fishing areas are untested and hence uncertain. Clearly and importantly, our estimates of direct losses to fisheries need refinement to achieve greater realism. Our approach is simplistic in that we currently use ex-vessel prices and production, and hence assume that productivity is proportional to profitability. Models of fishing behaviour, and analyses of current allocation rights and dependencies, should be developed to provide better estimates of the potential costs to fisheries, such that they can be incorporated into the analysis of conservation benefits. A more advanced version of our analyses would also incorporate other commercial fisheries, recreational fisheries, timing of fishing effort, and more detail on ecologically important areas. Ironically, launching the assessment process we propose – in a consultative fashion – might be a particularly effective way of eliciting or prompting the collection of just such important data, which are seldom available (or at least publicly accessible) in even the best resourced management jurisdictions.

The flexibility of the approach used here could help to enhance societal acceptance of and compliance with spatial planning, particularly among fishers. The decision support tool we used – Marxan – facilitates decision-making, without making decisions. Indeed, because it offers multiple solutions that may differ only slightly in their efficiency, the exact choice of permitted fishing areas can be adjusted for social acceptability and ecological viability [34]. Fishers' input will be important in setting commercial catch targets by fishery, verifying formal data [35], mapping and scaling fisheries that lack formal spatial data, and in agreeing to the permitted fishing areas. By using Marxan to set catches as targets instead of as a cost – the common approach in reserve selection – each fishery is targeted the same way, and therefore (under the assumptions of this approach) would incur a loss proportional to that fishery. In contrast, when all fisheries are combined into one cost, some fisheries may be disproportionately affected.

The approach of selecting permitted fishing areas would be expected to yield useful results in other geographic areas. Gear types used in British Columbia are typical of commercial fisheries elsewhere – trawl, hook and line, gillnet, seine, trap, and dive – and bioeconomic models suggest consistency in fisher behavior across locations [36]. Moreover, modeling has previously shown that optimal harvesting strategies always include marine reserves for populations with sedentary adults, even before consequent improvements in habitat recovery are considered [37]. Trials of this approach must, however, be taken elsewhere to determine whether, for example, the resultant no-fishing zones are generally ecologically representative.

Assessing the Costs and Benefits

While optimistic about the potential of our approach, we are well aware that many challenges remain to be resolved. First, we would benefit from knowing more about the connectedness of the fished and unfished area over space and time, in order to understand responses to spatial management; this is also true of MPA design and conventional management tools. Second, some fisheries in the portfolio that already operate sustainably might gain few benefits from the spatial management we propose, and would essentially be making concessions for other fisheries and/or for broader conservation principles. Third, our approach focuses only on fisheries (and only commercial fisheries in this trial study), whereas other marine and terrestrial uses also significantly affect the ocean. Fishing, however, is the main threat, and hence a tangible starting place for making conservation gains. Fourth, the large no-fishing zones arising from our approach, might lead to claims that no further areas need be protected, whatever their claims for conservation. Fifth, it remains to be determined whether the areas protected through this approach would provide the same conservation benefits conventional marine reserve selection. We do, however, know that both approaches tend to lead to protection for areas that are less valuable economically.

As ever, no single management measure will achieve all goals. The effectiveness of our approach in terms in accelerating protection will depend, in large measure, on the extent to which fishers gain yields in proportion to the benefits they cede in the no-fishing zones. Some conflict is still likely if, for example, the best fishing grounds – and hence the areas most likely to be included in permitted fishing areas – are also (a) the most sensitive habitats with the highest fish densities or (b) the most sensitive habitats even if they do not have high fish densities. These areas would ideally be protected in no-fishing zones, to the annoyance of fishers. Worse, however, would be to leave them in the permitted fishing areas, where they might come under more concentrated fishing unless quotas were reduced commensurate with the spatial contraction of the fishery. In terms of spatial management, the best approach is likely to combine the selection of permitted fishing areas with the identification and protection of sensitive habitats, whilst respecting the needs of fishers with limited ranges, such as small-scale fishing fleets or anglers. In addition, conventional fisheries management approaches will continue to be essential to managing fisheries within fished areas, especially to reduce the race to fish, realign incentives to reduce bycatch and habitat damage, and promote profitable fisheries; these are some of the roots of fisheries management failures [9].

The available of fishing data will influence the completeness of the selection of permitted fishing areas. In particular, historical fishing data are rarely available. In some regions, the areas receiving less fishing effort – areas more likely to fall outside of permitted fishing areas – might be depleted through past fishing. Some marine reserves sited in degraded and depleted areas have shown remarkable recovery [38], and therefore such areas likely have the capacity to recover. In the case of British Columbia, areas that have been fished the most historically – those close to population centers – are still among the most heavily fished, and hence sequential exhaustion of past fishing sites is less likely to be an issue than in other regions. In regions where spatial fishing data are not available – such as most small-scale fisheries – interviews with fishers could be used to collect such data.

Approaches such as ours are unlikely to succeed without strong community and political support. Given that selecting permitted fishing areas would restrict fishers' flexibility in where to fish, support for the concept is not guaranteed. However, many fisheries are continuing to decline, and some scientists are calling for large reductions in catches [39]. By specifying targets for each fishery based on sustainability estimates and permitting all fisheries in the same areas, ecological benefits may be greater than by managing each fishery independently.

The designation of permitted fishing areas will face many of the same obstacles as in the selection of marine reserves. First, there are data availability issues and knowledge gaps. In both cases, we usually lack spatial data for at least some fisheries, biological and range data for at least some species, and an appropriate understanding of dispersal and connectivity [40]. Given this lack of spatially structured data, specific modeling of the anticipated increases and decreases in species and ecosystem dynamics may be a challenge. Second, similar implementation and management issues might arise for permitted fishing areas and marine reserves. Enforcement would still be a challenge, and the political will to proceed with establishment has to exist for advances to be made.

Conclusion

Given the dismal state of many fisheries, time is ripe to debate alternative approaches. We have little to lose – and much to gain – in trying a new approach in areas where marine conservation advances have been inadequate. It appears, ab initio, that large areas that are representative of ecoregions and habitats might be protected at a small cost to fisheries (although particularly sensitive areas may have to be included a priori). Moreover, the dependency of the approach on explicit commercial catch targets for each fishery forces us to define the trade-offs we are willing to make to ensure a healthy ocean. The alternative to the approach described here seems to be the continuation of the status quo, which has resulted in the sequential collapse of fisheries [1], [41] with only a small proportion of the ocean protected by marine reserves. At a minimum, a debate about management assumptions is much needed, if innovative approaches are to emerge.

Methods

Selection of Permitted Fishing Areas: The Decision Support Program Used

We applied Marxan [42], [43], a decision support tool that has commonly been used to plan reserves, to spatial catch statistics for 13 commercial marine fisheries in British Columbia, Canada. Marxan tries to find the least expensive solution to the following objective function, using a simulated annealing algorithm [44]:

Total score = $\sum$ planning unit cost + (boundary length modifier * $\sum$ boundary cost) + feature penalty

We created a grid of 2 km by 2 km cells (with each cell considered to be one planning unit), to cover the study area, populated it with the spatial catch data, and then ran scenarios to select fishing areas. As spatial catch data were not available for recreational fisheries, our analysis is limited to commercial fisheries. We set the boundary length modifier – which controls the boundary to area ratio of the Marxan output – high enough so that the results were spatially compact. We pre-specified targets for commercial catches (kg) for each fishery, then set the penalty factor high enough to ensure that were met. Marxan provides a good approximation to an optimal solution by incorporating a random component to adding and removing planning units. Rather than settling on a single outcome, Marxan produces many solutions for any target that is proposed. The frequency with which particular planning units are chosen across different solutions is a measure of how important those planning units are for meeting the commercial catch targets efficiently.

The Data

We obtained spatial catch data from Fisheries and Ocean Canada for 13 commercial fisheries in British Columbia, Canada. The scale of analysis – the province of British Columbia – matches the scale of management. For confidentiality reasons, eight sets of data had been summarized in 4 km by 4 km grids: ZN fishery (hook and line inshore rockfish), shrimp trawl, schedule 2 (hook and line, other species), sablefish trap, sablefish longline, prawn trap, groundfish trawl, and crab. Similarly, the other five sets of data had been grouped into 10 km by 10 km grids: sea cucumber, red urchin, krill, green urchin, and geoduck. Many of the commercial fisheries have a high percentage of observer coverage and have vessel monitoring systems; we therefore believe that the data are as reliable as catch data can be. If fishing data are to be used for the purpose of selecting fishing areas, workshops with managers and fishers could be used to assess the reliability of the data. Also, if the fishing industry knows that the data could be used for such a purpose, reliability of the data may improve.

We normalized all data to the average annual catch (kg) for each planning unit. Data were averaged per annum over the temporal duration of spatial data collection, which extended between 3 and 12 years for any given fishery (1993–2004). We used 2 km by 2 km planning units that assumed even spatial distribution of catches for each original 4 km by 4 km or 10 km by 10 km grid.

The Scenarios and Analyses

We set each of 11 scenarios to maintain a particular target level of recent mean annual commercial catches (kg), from 98% to 10%; the targeted yield reductions thus ranged from 2% to 90%. We repeated each scenario ten times, with 100 runs of one million iterations each. The results for each scenario integrated all 13 fisheries, with each fishery maintaining at least that target catch.

We carried out a detailed assessment of the run with a target of a 5% reduction in catches (in all fisheries) by examining the proportion of different habitat types or surrogates that fell within the areas where fishing was allowed to continue (permitted fishing areas). Our intention was to determine which features would be protected by selecting permitted fishing areas, and which would remain unprotected. The habitat types were described by depth, exposure, relief, slope, current, temperature, substrate, salinity and stratification. In addition, limited spatial information was available for the distribution of kelp, eelgrass, herring spawn areas, and clam beds.

We further assessed the performance of the run with targeted 5% reduction in catches on annual spatial catch data for the four commercial fisheries for which we had annual data: geoduck, green urchin, red urchin and sea cucumber. Furthermore, we took the scenarios for the 2%, 5% and 10% reduction that resulted in the least area fished, and assessed the predicted reduction in catches, and the expected losses based on ex-vessel prices, for all 13 fisheries.

Acknowledgements

This is a contribution from Project Seahorse. We thank I. Cote, S. Palumbi, L. Kaufman, C. Roberts, and an anonymous reviewer for comments.

Author Contributions

Conceived and designed the experiments: NCB ACV. Performed the experiments: NCB. Analyzed the data: NCB. Contributed reagents/materials/analysis tools: NCB ACV. Wrote the paper: NCB ACV.

References

1. Myers RA, Worm B (2003) Rapid worldwide depletion of predatory fish communities. Nature 423: 280–283.

2. Worm B, Barbier EB, Beaumont N, Duffy JE, Folke C, et al.. (2006) Impacts of biodiversity loss on ocean ecosystem services. Science 314: 787–790.

3. Sibert J, Hampton J, Kleiber P, Maunder M (2006) Biomass, size, and trophic status of top predators in the pacific ocean. Science 314: 1773–1776.

4. Jackson JBC, Kirby MX, Berger WH, Bjorndal KA, Botsford LW, et al.. (2001) Historical overfishing and the recent collapse of coastal ecosystems. Science 293: 629–637.

5. Halpern BS, Warner RR (2002) Marine reserves have rapid and lasting effects. Ecology Letters 5: 361–366.

6. Roberts CM, Bohnsack JA, Gell F, Hawkins JP, Goodridge R (2001) Effects of marine reserves on adjacent fisheries. Science 294: 1920–1923.

7. Sala E, Aburto-Oropeza O, Paredes G, Parra I, Barrera JC, et al.. (2002) A general model for designing networks of marine reserves. Science 298: 1991–1993.

8. Conover DO, Munch SB (2002) Sustaining fisheries yields over evolutionary time scales. Science 297: 94–96.

9. Hilborn R, Stokes K, Maguire J-J, Smith T, Botsford LW, et al.. (2004) When can marine reserves improve fisheries management? Ocean & Coastal Management 47: 197–205.

10. Wood LJ, Fish L, Laughren J, Pauly D (2007) Assessing progress towards global marine protection targets: Shortfalls in information and action. UBC Fisheries Centre Working Paper Series #2007-03: 1–39.

11. Mora C, Andréfouët S, Costello MJ, Kranenburg C, Rollo A, et al.. (2006) Coral reefs and the global network of marine protected areas. Science 312: 1750–1751.

12. Sale PF, Cowen RK, Danilowicz BS, Jones GP, Kritzer JP, et al.. (2005) Critical science gaps impede use of no-take fishery reserves. Trends in Ecology & Evolution 20: 74–80.

13. Coase RH (1960) The problem of social cost. The Journal of Law and Economics 3: 1.

14. Walters C (2000) Impacts of dispersal, ecological interactions, and fishing effort dynamics on efficacy of marine protected areas: How large should protected areas be? Bulletin of marine science 66: 745–757.

15. Walters C (1998) Designing fisheries management systems that do not depend upon accurate stock assessment. In: Pitcher TJ, Hart PJB, Pauly D, editors. Reinventing fisheries management. London: Kluwer Academic Publishers. pp. 279–288.

16. Dayton PK (1998) Reversal of the burden of proof in fisheries management. Science 279: 821–822.

17. Margules CR, Pressey RL (2000) Systematic conservation planning. Nature 405: 243–253.

18. Stewart R, Possingham H (2005) Efficiency, costs and trade-offs in marine reserve system design. Environmental Modeling and Assessment 10: 203–213.

19. Stewart RR, Noyce T, Possingham HP (2003) Opportunity cost of ad hoc marine reserve design decisions: An example from south australia. Marine Ecology Progress Series 253: 25–38.

20. Game ET, Grantham HS (2008) Marxan user manual: For marxan version 1.8.10. St. Lucia, Queensland, Australia and Vancouver, British Columbia, Canada: University of Queensland and Pacific Marine Analysis and Research Association. pp. 1–127.

21. Sumaila UR, Marsden AD, Watson R, Pauly D (2007) A global ex-vessel fish price database: Construction and applications. Journal of Bioeconomics 9: 39–51.

22. Government of British Columbia (2007) Economic contribution of the oceans sector in british columbia. Vancouver: Prepared by GSGislason & Associates Ltd. http://wwwenvgovbcca/omfd/reports/oceansector-economicspdf .

23. Tetreault I, Ambrose RF (2007) Temperate marine reserves enhance targeted but not untargeted fishes in multiple no-take mpas. Ecological Applications 17: 2251–2267.

24. Halpern BS (2003) The impact of marine reserves: Do reserves work and does reserve size matter? Ecological Applications 13: S117–S137.

25. Gaylord B, Gaines S, Siegel D, Carr M (2005) Marine reserves exploit population structure and life history in potentially improving fisheries yields. Ecological Applications 15: 2180–2191.

26. Lindholm JB, Auster PJ, Ruth M, Kaufman L (2001) Modeling the effects of fishing and implications for the design of marine protected areas: Juvenile fish responses to variations in seafloor habitat. Conservation Biology 15: 424–437.

27. Zacharias MA, Howes DE, Harper JR, Wainwright P (1998) The british columbia marine ecosystem classification: Rationale, development, and verification. Coastal Management 26: 105–124.

28. Russ GR, Alcala AC (1996) Marine reserves: Rates and patterns of recovery and decline of large predatory fish. Ecological Applications 6: 947–961.

29. Collie JS, Hall SJ, Kaiser MJ, Poiner IR (2000) A quantitative analysis of fishing impacts on shelf-sea benthos. Journal of Animal Ecology 69: 785–798.

30. Roberts CM, Hawkins JP, Gell FR (2005) The role of marine reserves in achieving sustainable fisheries. Philosophical Transactions of the Royal Society 360: 123–132.

31. Polacheck T (1990) Year around closed areas as a management tool. Natural Resource Modeling 4: 327–354.

32. Pauly D, Christensen V, Guenette S, Pitcher TJ, Sumaila UR, et al.. (2002) Towards sustainability in world fisheries. Nature 418: 689–695.

33. Collie J, Gislason H (2001) Biological reference points for fish stocks in a multispecies context. Canadian Journal of Fisheries and Aquatic Sciences 58: 2167–2176.

34. Fernandes L, Day J, Lewis A, Slegers S, Kerrigan B, et al.. (2005) Establishing representative no-take areas in the great barrier reef: Large-scale implementation of theory on marine protected areas. Conservation Biology. 19: 1733–1744. doi:10.1111/j.1523-1739.2005.00302.X.

35. Johannes REF, M.M.R. Hamilton RJ (2000) Ignore fishers' knowledge and miss the boat. Fish & Fisheries 1: 257.

36. Walters CJ, Martell SJD (2004) Fisheries ecology and management. New Jersey: Princeton University Press. 399 p.

37. Hastings A, Botsford LW (1999) Equivalence in yield from marine reserves and traditional fisheries management. Science 284: 1537–1538.

38. Samoilys MA, Martin-Smith KM, Giles BG, Cabrera B, Anticamara JA, et al.. (2007) Effectiveness of five small philippines' coral reef reserves for fish

populations depends on site-specific factors, particularly enforcement history. Biological Conservation 136: 584–601.

39. Schrank WE (2007) Is there any hope for fisheries management? Marine Policy 31: 299–307.

40. Palumbi SR (2004) Marine reserves and ocean neighborhoods: The spatial scale of marine populations and their management. Annual Review of Environment and Resources 29: 31–68.

41. Pauly D, Christensen V, Dalsgaard J, Froese R, Torres F Jr. (1998) Fishing down marine food webs. Science 279: 860–863.

42. Ball IR, Possingham H (2000) Marxan (v1.8.2): Marine reserve design using spatially explicit annealing, a manual.

43. Possingham HP, Ball IR, Andelman S (2000) Mathematical methods for identifying representative reserve networks. In: Ferson S, Burgman M, editors. Quantitative methods for conservation biology. New York: Springer-Verlag. pp. 291–305.

44. Kirkpatrick S, Gelatt CD, Vecchi MP (1983) Optimization by simulated annealing. Science 220: 671–680.

Chapter 2

Estimating the Worldwide Extent of Illegal Fishing

David J. Agnew, John Pearce, Ganapathiraju Pramod,
Tom Peatman, Reg Watson, John R. Beddington and
Tony J. Pitcher

ABSTRACT

Illegal and unreported fishing contributes to overexploitation of fish stocks and is a hindrance to the recovery of fish populations and ecosystems. This study is the first to undertake a world-wide analysis of illegal and unreported fishing. Reviewing the situation in 54 countries and on the high seas, we estimate that lower and upper estimates of the total value of current illegal and unreported fishing losses worldwide are between $10 bn and $23.5 bn annually, representing between 11 and 26 million tonnes. Our data are of sufficient resolution to detect regional differences in the level and trend of illegal fishing over the last 20 years, and we can report a significant correlation between governance and the level of illegal fishing. Developing countries are most at risk from illegal fishing, with total estimated catches in West Africa being 40% higher than reported catches. Such levels of exploitation

Originally published as Agnew DJ, Pearce J, Pramod G, Peatman T, Watson R, Beddington JR, et al. (2009) Estimating the Worldwide Extent of Illegal Fishing. PLoS ONE 4(2): e4570. https://doi.org/10.1371/journal.pone.0004570.   © 2009 Agnew et al. https://creativecommons.org/licenses/by/4.0/

severely hamper the sustainable management of marine ecosystems. Although there have been some successes in reducing the level of illegal fishing in some areas, these developments are relatively recent and follow growing international focus on the problem. This article provides the baseline against which successful action to curb illegal fishing can be judged.

Introduction

It is widely accepted that there is a severe problem with future global food security. Driven by substantial world population growth, demand for fish protein continues to increase, but a large number of the world's fish stocks are currently depleted and therefore not capable of producing their maximum sustainable yield [1]. Illegal and unreported fishing (in this article taken to include illegal and unreported catches but to exclude discards and artisanal unregulated catches) prejudices the managed recovery of the world's oceans from severe fish depletions [2]–[4]. It is reported to lead to a loss of many billions of dollars of annual economic benefits [5], [6], creates significant environmental damage through the use of unsustainable fishing practices [7] and has wider consequences for food supply [8].

Estimating the level of illegal fishing is, by its very nature, extremely difficult and has not previously been attempted on a global scale. Fishing vessels, especially those fishing in high seas waters and under third party access agreements to EEZ waters (Exclusive Economic Zones, which can extend up to 200 nm from the coast), are highly mobile. Although there are a number of studies of the level of IUU (Illegal, Unreported and Unregulated) fishing in individual fisheries (both EEZs and high seas) [3], [9]–[16], only two studies have attempted to estimate the impacts of IUU over a whole region [5], [6]. In this article we set out, for the first time, a detailed study which arrives at global estimates of current and historical illegal and unreported catches.

Results

The term "Illegal, Unreported and Unregulated" (IUU) fishing can cover a wide range of issues [see discussion in 13], [17]–[19]. We confined our analysis to illegal and unreported catches (IU), namely those taken within an EEZ which are both illegal and retained, and which are usually unreported, and all unreported catches taken in high seas waters subject to a Regional Fisheries Management Organisation's (RFMO) jurisdiction. We excluded discards and unregulated artisanal catches, which will be analysed in a future publication. With illegal and unreported catches rents are captured by illegal fishermen but lost to legitimate fishermen and management authorities. Note that the word "landings" is often used to distinguish catches that are retained from catches that are discarded. For simplicity, and to avoid confusion with the suggestion that fish are necessarily landed in the country in whose waters they are caught, we use the word catches here to mean catches that are retained and discards to mean catches that are discarded.

In total, 54 EEZs and 15 high seas regions were analysed, providing an estimate of global illegal and unreported catch for 292 case study fisheries which comprise 46% of the reported total world marine fish catch. All data sources were combined to provide upper and lower estimates of IU for each fishery. The total catch of case study and non-case study fish from the EEZs and high seas regions analysed comprises 75% of global catch.

There were significant differences in the level of illegal and unreported catch and the trends in those catches between regions. The level of IU was highest in the Eastern Central Atlantic (Area 34) and lowest in the Southwest Pacific (Area 81) (Table 1). Since the 1990s we estimate that the level of IU has declined in 11 areas and increased in 5 (Table 2). We estimate that the overall loss from our studied fisheries is 13–31% (lower and upper estimates) with a mean of 18%, and that this was worth some $5-11 bn in 2003.

Table 1. Summary of regional estimates of illegal fishing, averaged over 2000–2003.

Region	Reported catch of case study species	Catch of case study species as a percentage of total regional catch	Lower estimate of Illegal catch (t)	Upper estimate of Illegal catch (t)	Lower estimate of value (US$m)	Upper estimate of value (US$m)
Northwest Atlantic	557,147	25%	22,325	82,266	20	74
Northeast Atlantic	6,677,607	60%	364.908	842.467	328	758
Western Central Atlantic	390,942	22%	21,745	58,514	20	53
Eastern Central Atlantic	1,154,586	32%	294,089	562,169	265	506
Southwest Atlantic	1,403,601	65%	227,865	673,712	205	606
Southeast Atlantic	1,351,635	79%	52,972	139,392	48	125
Western Indian	2,165,792	52%	229,285	559,942	206	504
Eastern Indian	2,263,158	44%	467,865	970,589	421	874
Northwest Pacific	7,358,470	32%	1,325,763	3,505,600	1,193	3,155
Northeast Pacific	196,587	7%	2,326	8,449	2	8
Western Central Pacific	3,740,192	36%	785,897	1,729,588	707	1,557
Eastern Central Pacific	1,374,062	73%	129,772	278,450	117	251
Southwest Pacific	451,677	61%	5,227	32,848	5	30
Southeast Pacific	9,799,047	73%	1,197,547	2,567,890	1,078	2,311
Antarctic	136654	100%	9593	9593	9	9
Total	**39,021,155**	**46%**	**5,140,928**	**12,040,052**	**4,627**	**10,836**

Regional trends reveal issues related to the quality of fishery management. In the Northeast Atlantic, reasonable estimates of the level of illegal fishing are available from various reports and assessments conducted by the International Council for the Exploration of the Sea, ICES [20]. These indicate that as pressure on stocks increased following the end of the 'gadoid outburst' (exceptional recruitment from cod family fish between the mid 1970s to late 1980s) the level of illegal and unreported catch increased, and has only recently improved. The decline in IU that we show in the Western Central Atlantic is due to a reduction in the upper bound of uncertainty over unreported tuna catches. The introduction by the International Commission for the Conservation of Atlantic Tunas (IC-CAT) of statistical document schemes required for trade in tuna has significantly decreased the amount of unreported tuna catch in the Central Atlantic [9]. In the Eastern Central Atlantic there appears to have been a steady increase in illegal fishing, which is at a much higher level than in the western central Atlantic. This is a large area, covering many states with a wide variety of fisheries and governance (Morocco to Angola), some of which, such as Guinea, Sierra Leone and Liberia suffered increasing illegal catches as a result of internal strife in the 1990s. We have increasing uncertainty about the level of illegal fishing in the Soutwest Atlantic from the mid-1990s, but overall the proportion of illegal catch appears to have increased at this time, once again in response to declining resource status. In contrast, the exclusion of foreign vessels from Exclusive Economic Zones in the Southeast Atlantic, and the imposition of national control in Southeast Atlantic coastal states from the late 1980s, led to a marked reduction in illegal fishing at that time.

Table 2. Trends in regional estimates of illegal fishing, averaged over 5 year periods 1980–2003.

Region	1980–1984	1985–1989	1990–1994	1995–1999	2000–2003
Northwest Atlantic	26%	19%	39%	15%	9%
Northeast Atlantic	10%	10%	12%	11%	9%
Western Central Atlantic	16%	14%	14%	11%	10%
Eastern Central Atlantic	31%	38%	40%	34%	37%
Southwest Atlantic	15%	18%	24%	34%	32%
Southeast Atlantic	21%	25%	12%	10%	7%
Western Indian	31%	24%	27%	25%	18%
Eastern Indian	24%	29%	30%	33%	32%
Northwest Pacific	16%	15%	23%	27%	33%
Northeast Pacific	39%	39%	7%	3%	3%
Western Central Pacific	38%	37%	37%	36%	34%
Eastern Central Pacific	20%	17%	13%	14%	15%
Southwest Pacific	10%	9%	7%	7%	4%
Southeast Pacific	22%	21%	24%	23%	19%
Antarctic	0%	0%	2%	15%	7%
Average	**21%**	**21%**	**21%**	**20%**	**18%**

The figure given is the mid-point between the lower and upper estimates of illegal and unreported catch in the case study species, expressed as a percentage of reported catch of case study species.

The decline of illegal fishing in the Western Indian Ocean reflects gradually increasing control over time by coastal states, particularly those in the extreme north and countries of the Southern African Development Community, and a reduction in the unreported catch estimated by the Indian Ocean Tuna Commission (IOTC). The increase in estimated illegal fishing in the Northwest Pacific is almost entirely due to the influence of China and Russia, since estimates of illegal catch in other states in the area are relatively small. However, the confidence in this estimate is not as good as for other estimates in this analysis, which is reflected in an increase in uncertainty in this region. Northeast Pacific illegal catch is currently estimated to be low and to have steadily declined over recent years, but, surprisingly, we were unable to obtain good estimates from the USA. Western Central Pacific data include coastal states of the western Pacific seaboard, where the information available to us suggests that a relatively high level of illegal and unreported catch has been present with little change over the years. For instance, in Indonesia a huge amount of unreported catch (over 1.5 million tonnes annually) has recently been revealed by an FAO study of the Arafura Sea, much of this illegal [21]. In the Eastern and Southeastern Pacific a similar situation of low change exists, but with a much lower estimated proportion of illegal fishing. In the Southwest Pacific increasing control by coastal states has led to a significant reduction in illegal fishing over the last 20 years.

Finally, in the Antarctic, the only illegal fishing issue is unregulated and unreported fishing for toothfish, which peaked in 1996 and has since significantly reduced.

As would be expected, the highest levels of illegal fishing are associated with high value demersal fish, lobsters and shrimps/prawns (Figure 1). It is somewhat surprising at first glance that the proportion of illegal catch is low for tunas. The reason for this is that most tuna catches are taken within the areas of RFMOs where the small amounts of unreported fishing are generally associated with large volume catches (for instance of yellowfin and bigeye tuna) and in some regions (e.g., the Inter-American Tropical Tuna Commission and the IOTC) unreported catches of tunas are now very small.

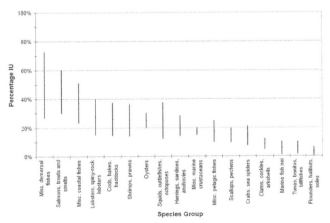

Figure 1. Illegal and unreported catch, expressed as a percentage of reported catch, by species group 2000–2003. Upper and lower bounds are given.

Taking the total estimated value of illegal catch losses and raising by the proportion of the total world catch analysed in this paper, lower and upper estimates of the total value of current illegal and unreported fishing losses worldwide are between $10 bn and $23.5 bn annually, representing between 11 and 26 million tonnes. The estimates previously made by MRAG [6] ($9 bn) and Pauly et al.. [5] ($25 bn) fall at either end of this range. Estimates of losses from illegal logging are of the same magnitude, roughly 10% of world timber trade with illegal products worth at least $15 bn a year [22].

Since there are strong economic drivers for illegal fishing [3], [4] and it occurs in situations of poor fisheries management and control [2], [6], we might expect that the level of illegal fishing would be related to fish price, governance and indicators of the control problem, such as the area of a country's EEZ and the number of patrol vessels at its disposal. We found no significant relationship between illegal fishing and the price of fish or the size of the EEZ or the fishery in our study, but we did find a significant relationship with World Bank governance indicators measured in 2003 [23], which was strongest with the log of illegal fishing level. This relationship was significant for the whole dataset (Figure 2) (R2 0.400, p<0.001, n = 54), for Africa, Europe and Asia separately (R2 0.393, 0.375, 0.429, p<0.01, 0.05 and 0.01 respectively, n = 16 in each case), and with different indicators of governance such as the Corruption Perceptions index [24] (R2 0.371, p<0.001, n = 50).

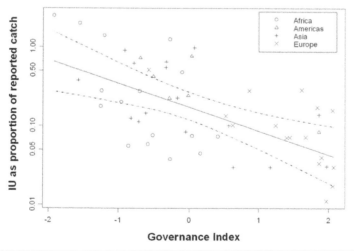

Figure 2. Relationship between the amount of illegal fishing (expressed as a proportion of the reported catch that is additionally taken as illegal and unreported catch) and an average of four World Bank indices of governance (Government Effectiveness, Regulatory Quality, Rule of Law and Control of Corruption, measured in 2003; 23). Although there is a significant linear relationship between governance and the proportion of IU, the log-linear relationship shown above is a better fit to the data and has R^2 = 0.4081, p<0.001 with 53 degrees of freedom. The broken lines are 95% confidence intervals.

Discussion

This is the first time, to our knowledge, that a significant relationship has been demonstrated on a global scale between the level of illegal and unreported fishing and indices of governance, and it points to the benefits of improving governance. This is not to say that developing countries with poor governance records are necessarily to blame for illegal fishing, but that they are more vulnerable to illegal activities, conducted by both their own fishers and vessels from distant water fishing nations. In Africa, for instance, many coastal states licence vessels from distant water fishing nations such as China, Taiwan, Korea, the EU and Russia to fish in their waters, and there is a significant illegal fishing problem from many of these vessels [6]. This represents a failure of control on behalf of the flag state as well as the coastal state. Furthermore, many vessels engaged in IUU activities are registered with so-called 'flag of convenience' states, and whilst these are mostly developing countries the vessels themselves are usually owned and operated by developed country companies [25].

On a world scale, poor performance in the control of illegal fishing is pervasive. In a recent review [26] over half of the countries (30/53 top fishing countries) assessed for compliance with illegal and unreported fishing in the FAO (UN) Code of Conduct for Responsible Fisheries [27], [28; Article 6.10; Articles 7.6.1, 7.7.1, 7.7.5, and 7.8.1] were awarded fail grades (less than 4/10). Only a quarter (16/53) were rated as 'passable' (6/10 or more). Moreover, implementation of ecosystem-based management requires control of illegal fishing, and here again almost half of the countries surveyed failed (16/33), while only two received a 'good' rating [29].

Illegal and unreported fishing can have very significant effects on stocks. For instance, unreported catches of bluefin tuna from the Mediterranean (estimated by the International Commission for the Conservation of Atlantic Tunas to have been 19,400 t in 2006 and 28,600 in 2007; 30) have significantly contributed to the rapid decline in the stock, and a failure by the European Union to control unreported catches led to a failure to generate any recovery in North Sea cod until very recently [31]. There is a correspondence between our regional estimates of illegal and unreported fishing and the number of depleted stocks in those regions. For instance out of 53 demersal stocks recognised in the eastern central Atlantic, 32 of which could be assessed, 60% were overexploited in the early 2000s [32] compared to 30% of EU stocks and 15% of New Zealand stocks [2]. Thus out of these three areas those with the highest and lowest proportion of depleted stocks also had the highest and lowest levels of illegal fishing (Table 1). This may be both because illegal and unreported fishing is contributing to overexploitation of stocks, and because the general management of stocks (including the quality of research, for instance) is likely to be better in areas of higher quality governance. Illegal fishing in regions with poor governance has often been linked to organised crime [25], [33], but where fish have a high value, this can be an issue even in countries with good governance [34].

Illegal fishing creates significant collateral damage to ecosystems. Illegal fishing, by its very nature, does not respect national and international actions designed to reduce bycatch and mitigate the incidental mortality of marine animals such as sharks, turtles, birds and mammals. Such practices are common: examples are illegal fishing in marine reserves in west Africa [6], [7] and the bycatch of albatross in illegal and unreported longline and gillnet operations in the Antarctic [11], [35]. Only a solution of the illegal fishing problem will generate the compliance with these wider ecosystem management measures. Moreover, as part of the move to explore ecosystem-based management, estimates of unreported catches have proved to be necessary to balance ecosystem models [36]. Where unreported extraction of fish from major stocks is not included this can bias both single species stock assessment and ecosystem-based analyses in a dangerous direction of allowing more fishing than would otherwise be thought sustainable.

Clearly some progress has been made in some areas over the last decade; our study identifies reductions in illegal fishing in 11 areas since the early 1990s and indeed this trend has continued in the years since 2003. The worst period for illegal and unreported fishing world wide appears to have been the mid-1990s, driven by a combination of factors: a growing world demand for fish and significant overcapacity of the world's fishing fleet set against increasing limitation of access to distant water fishing nations and a lack of new or alternative fishing opportunities [1], [37].

The solutions most often proposed to eliminate illegal fishing are associated with increased governance and the rule of law - increased cooperation between regional management authorities in management and control activities, increased capacity to undertake surveillance and enforcement of port state control [38], and other means of reducing the economic incentives to engage in IUU fishing, such as increased sanctions and trade measures [4], [39]. Recent successes emphasise this. There has been a significant reduction in illegal and unreported catches of cod from 50% to 20% of the reported catch in the Barents Sea following cooperative port state controls implemented by the states party to the Northeast Atlantic Fisheries Commission [40] and in the Antarctic IUU catches have been reduced from 33,700 t in 1996/97 to 3,600 t in 2006/07 through cooperative international and state action [36].

These activities are encouraging, but set in the context of burgeoning demand for food and particularly protein, there will continue to be enormous pressure on fish stocks over the next 50 years and it is essential that the international community address effectively the large illegal and unreported catch of fish reported in this paper. Given the recent change in political will to tackle the issue of illegal fishing, [e.g. 38], further improvements might be expected to come from legally mandating compliance with the FAO (UN) Code of Conduct for Responsible Fisheries [25], which would provide countries with an international legal basis for economic and other sanctions that discourage illegal fishing. Some countries already include some of the provisions of the Code of Conduct in their national legislation (for example in Australia, South Africa, Norway, Namibia, Malaysia). Others, such as the EU, are now proposing to implement much stronger import controls and sanctions to restrict trade in IUU fish. This paper provides the baseline against which action to curb illegal fishing can be judged.

Methods

We used the "anchor points and influence table" approach of Pitcher et al.. [13] which employs detailed reports (from published scientific literature and in-country specialist studies) to establish point estimates and upper and lower bounds of the level of illegal fishing in different fisheries, and identifies changes to these levels over time based on historical data or likely trends based on known changes in management regime. In the source studies a number of different methods have been used to estimate the level of illegal fishing, including surveillance data, trade data, stock assessments based on fishery-independent (survey) data and expert opinion [7]. Some of these methods deliver a point estimate of the level of illegal fishing, some deliver statistical estimates with confidence intervals, and some deliver upper and lower bounds. We took the approach that, when trying to integrate the results of these various estimation methods with their differing levels of reliability, using the extreme upper and lower limits produced less variation than trying to make a point estimate. Where it was available the point estimate was used to set initial bounds for a percentage figure, but we used expert opinion to guide the upper and lower limits, and did not treat them as two point estimates.

Countries were selected based on the volume (tonnage) of catches reported to have been taken in their EEZ in order of magnitude (i.e., their importance as fishing areas). A few additional countries with smaller catches (5 in all, 1.4% of world catch) were included because of their importance in understanding the distribution of illegal fishing. All Regional Fishery Management Organisations (RFMO) were examined. Because data were required by EEZ rather than FAO area, and in order to keep catches consistent with FAO totals, catch data for each EEZ selected and for each high seas FAO region were extracted from the Sea Around Us project database of estimated catches (41; www.seaaroundus.org). The Sea Around Us project has attributed FAO catches reported to EEZs and High Seas by means of a geospatial algorithm [41]. Within each study, catches and IUU fishing (discarded and illegal assembled separately) of the four highest volume (i.e., tonnage caught) species were estimated from the source studies (seven countries only had three species listed). This led to 292 separate fishery estimations per year. Percentage trends for IUU in each five year block were multiplied by the annual reported catch data to form overall annual estimates, separated into illegal catches and discards (including other unreported fish catch such as recreational and legal but unreported artisanal catches).

EEZs selected for the main time-trend analysis were: Angola, Argentina, Australia, Bangladesh, Brazil, Canada, China, Denmark, Ecuador, Faeroe Islands, France, Germany, Ghana, Iceland, India, Indonesia, Iran, Ireland, Italy, Japan, Republic of Korea, Latvia, Malaysia, Mexico, Morocco, Mozambique, Myanmar, Namibia, Netherlands, New Zealand, Nigeria, Norway, Pakistan, Peru, Philippines, Poland, Portugal, Russia, Senegal, South Africa, Spain, Sri Lanka, Sweden, Thailand, United Kingdom, Tanzania, Yemen. RFMOs analysed for time trends of illegal and unreported catches were: Northwest Atlantic Fisheries organisation (NAFO), Northeast Atlantic Fisheries Commission (NEAFC), the Commission for the Conservation of Antarctic Marine Living Resources (CCAMLR), the Inter American Tropical Tuna Commission (IATTC), the International Commission for the Conservation of Atlantic Tuna (ICCAT), the Indian Ocean Tuna Commission (IOTC), the Western and Central Pacific Fisheries Commission (WCPFC) and the Commission for the Conservation of Southern Bluefin Tuna (CCSBT). Our primary data sources were several key composite studies [6], [20], [42]–[44], supplemented by country-specific studies [15], [19], [33], [45]–[86].

In order to estimate the global level of illegal and unreported catches (IU) a single estimate for the price of a tonne of fish each year was used. The price data used were those reported by FAO [87].

For some countries a historical time series of estimates of IU could not be derived from available data sources, although data were available from the early 2000s from MRAG [6] (Guinea, Kenya, Liberia, Papa New Guinea, Seychelles, Sierra Leone, Somalia). While they have been included in the analysis of governance relating to illegal fishing, these data do not contribute to the table showing the trends in illegal catch over time by region.

For each case study and species the analysis generated the following:

T_{cy} total reported tonnage of all wild fish caught in the case study EEZ/RFMO area c in year y

t_{csy} reported tonnage of fishery s in case study c

U_{csy} upper bound estimate of illegal catch

L_{csy} lower bound estimate of illegal catch

The estimate of illegal catch as a proportion of reported catch for a case study and year was calculated as $P_{yc} = \dfrac{U_{cy} + L_{cy}}{2t_{cy}}$ where $U_{cy} = \sum\limits_{s=1}^{4} U_{csy}$, and so on.

Regional estimates were developed by combining the high seas estimates along with EEZ estimates within that region. Where an EEZ was covered by a number of different FAO regions, these EEZs were where possible divided into two separate estimates (e.g., the estimate for the Russian EEZ was broken down by for the Atlantic and Pacific catches, and Canada and Mexico for west and east coasts). If this was not possible, the data reported by FAO area and recorded in the FAO FISHSTAT database were used to determine the approximate percentage of catches taken in each area and the estimates distributed uniformly with reported catches (e.g., South Africa, Australia and USA).

The confidence intervals shown in Figure 2 were created by estimating the confidence intervals for 1000 simulated datasets where, for each country, the level of IU was sampled from a uniform distribution defined by our upper and lower estimates of IU, and governance was sampled from a gaussian distribution with mean and standard deviation as presented in Lambsdorff [24]. The confidence intervals plotted in the paper are the maximum upper 95% and minimum lower 95% limits from the 1000 simulations.

Acknowledgements

We are grateful to numerous colleagues who commented on earlier drafts and reports, in particular Darius Campbell from the UK Department for Environment, Food and Rural Affairs and Tim Bostock from the UK Department for International Development who provided support throughout the project.

Author Contributions

Conceived and designed the experiments: DA. Performed the experiments: DA JP GP. Analyzed the data: DA JP GP TP TP. Contributed reagents/materials/analysis tools: DA GP RW JB TP. Wrote the paper: DA JP TP TP.

References

1. FAO (2007) The state of world fisheries and aquaculture 2006. Rome: FAO.

2. Beddington JR, Agnew DJ, Clark CW (2007) Current Problems in the Management of Marine Fisheries. Science 316: 1713–1716.

3. Sumaila UR, Alder J, Keith H (2006) Global scope and economics of illegal fishing. Marine Policy 30: 696–703.

4. Agnew DJ, Barnes C (2004) Economic Aspects and Drivers of IUU Fishing: Building a Framework. In: Gray K, Legg F, Andrews-Chouicha E, editors. Fish Piracy: combating illegal, unreported and unregulated fishing. Paris: OECD Publishing. pp. 169–200.

5. Pauly D, Christensen V, Guénette S, Pitcher TJ, Sumaila UR, et al.. (2002) Towards sustainability in world fisheries. Nature 418: 689–695.

6. MRAG (2005) Review of Impacts of Illegal, Unreported and Unregulated Fishing on Developing Countries. London: MRAG. Available: http://www.dfid.gov. uk/pubs/files/illegal-fishing-mrag-report.pdf . Accessed 20 September 2008.

7. MRAG (2005) IUU fishing on the high seas: Impacts on Ecosystems and Future Science Needs. London: MRAG. Available: http://www.dfid.gov.uk/pubs/files/ illegal-fishing-mrag-impacts.pdf. Accessed 20 September 2008.

8. Brashares J, Arcese P, Sam MK, Coppolillo PB, Sinclair ARE, et al.. (2004) Bushmeat Hunting, Wildlife Declines, and Fish Supply in West Africa. Science 306: 1180–1183.

9. Restrepo V (2004) Estimation of unreported catches by ICCAT. In: Gray K, Legg F, Andrews-Chouicha E, editors. Fish Piracy: combating illegal, unreported and unregulated fishing. Paris: OECD Publishing. pp. 155–157.

10. Anganuzzi A (2004) Gathering data on unreported activities in Indian Ocean tuna fisheries. In: Gray K, Legg F, Andrews-Chouicha E, editors. Fish Piracy: combating illegal, unreported and unregulated fishing. Paris: OECD Publishing. pp. 147–154.

11. Agnew DJ, Kirkwood GP (2005) A statistical method for estimating the level of IUU fishing: application to CCAMLR Subarea 48.3. CCAMLR Science 12: 119–141.

12. Ainsworth CH, Pitcher TJ (2005) Estimating Illegal, Unreported and Unregulated catch in British Columbia's Marine Fisheries. Fish Res 75: 40–55.

13. Pitcher TJ, Watson R, Forrest R, Valtýsson H, Guénette S (2002) Estimating Illegal and Unreported Catches From Marine Ecosystems: A Basis For Change. Fish and Fisheries 3: 317–339 (2002).

14. Tesfamichael D, Pitcher TJ (2007) Estimating the unreported catch of Eritrean Red Sea fisheries. African Journal of Marine Science 29: 55–63.

15. Varkey DA, Ainsworth CH, Pitcher TJ, Johanes G (2008) Estimating illegal and unreported catches in Raja Ampat Regency, Indonesia. In: Kalikoski D, Pitcher TJ, editors. Assessing Illegal, Unreported and Unregulated Fishery Catches

(IUU): Some case studies. Vancouver: University of British Columbia. Fisheries Centre Research Reports. In press.

16. Bailey M, Rotinsulu C, Sumaila UR (2008) The migrant anchovy fishery in Kabui Bay, Raja Ampat, Indonesia: Catch, profitability, and income distribution. Mar Pol 32: 483–488.

17. Bray K (2000) A global review of illegal, unreported and unregulated (IUU) fishing. Rome: FAO paper AUS:IUU/2000/6. Available: ftp://ftp.fao.org/fi/document/Ec-OpenRegistries/BRAY_AUS-IUU-2000-6.pdf . Accessed 24 November 2008.

18. Agnew DJ (2000) The illegal and unregulated fishery for toothfish in the Southern Ocean, and the CCAMLR Catch Documentation Scheme. Marine Policy 24: 361–374.

19. Ainsworth CH, Pitcher TJ (2005) Estimating illegal, unreported and unregulated catch in British Columbia's marine fisheries. Fish Res 75: 40–55.

20. ICES (2007) Report of the ICES Advisory Committee on Fishery Management, Advisory Committee on the Marine Environment and Advisory Committee on Ecosystems, 2007. Denmark: ICES.

21. Nurhakim S, Nikijuluw VPH, Badrudin M, Pitcher TJ, Wagey GA (2008) A Study Of Illegal, Unreported and Unregulated (IUU) Fishing In The Arafura Sea, Indonesia. Rome: FAO.

22. Brack D (2006) Illegal Logging. London: Chatham House. Available: www.illegal-logging.info/papers. Accessed July 2008.

23. Kaufmann D, Kraay A, Mastruzzi M (2007) Governance Matters VI: Governance Indicators for 1996–2006. World Bank Policy Research Working Paper No. 4280.

24. Lambsdorff JG (2003) Background Paper to the 2003 Corruption Perceptions Index. Transparency International and University of Passau. Available: http://www.icgg.org/downloads/FD_CPI_2003.pdf. Accessed July 2008.

25. Gianni M, Simpson W (2005) The Changing Nature of High Seas Fishing: how flags of convenience provide cover for illegal, unreported and unregulated fishing. Australian Department of Agriculture, Fisheries and Forestry, International Transport Workers Federation, and WWF International. Available: http://www.daff.gov.au/__data/assets/pdf_file/0008/5858/iuu_flags_of_convenience.pdf . Accessed September 2008.

26. Pitcher TJ, Pramod G, Kalikoski D, Short K (2008) Safe Conduct? Twelve Years Fishing under the UN Code. Gland: WWF.

27. FAO (1995) Code of conduct for Responsible Fisheries. Rome: FAO.

28. FAO (1997) FAO Technical guidelines for Responsible Fisheries 4. Rome: FAO.

29. Pitcher TJ, Kalikoski D, Short K, Varkey D, Pramod G (2008) An evaluation of progress in implementing ecosystem-based management of fisheries in 33 countries. Mar Pol: in press. Available: http://www.sciencedirect.com/science?_ob =ArticleURL&_udi=B6VCD-4T24FVX-1&_user=10&_rdoc=1&_fmt=&_ orig=search&_sort=d&view=c&_version=1&_urlVersion=0&_userid=10&m d5=9d09bb192635f23593a02115ea24cc4c#FCANote. Accessed September 2008.

30. ICCAT (2008) Report of the Standing Committee on Research and Statistics (SCRS). Madrid: ICCAT. Available: http://www.iccat.int/Documents/Meetings/Docs/2008_SCRS_ENG.pdf . Accessed November 2008.

31. WWF (2007) Mid-Term Review of the EU Common Fisheries Policy. Brussels (Belgium): World Wide Fund for Nature.

32. FAO (2006) Report of the FAO/CECAF working group on the assessment of demersal resources Conakry, Guinea, 19–23 September 2003. Rome: FAO.

33. Vaisman A (2001) Trawling in the mist. Industrial fisheries in the Russian part of the Bering Sea. Cambridge: TRAFFIC Network Report.

34. Putt J, Anderson K (2007) A national study of crime in the Australian fishing industry. Canberra: Research and Public Policy Series 76, Australian Institute of Criminology.

35. Scientific Committee for the Conservation of Antarctic Marine Living Resources (2007) Report of the Twenty-Sixth Meeting of the Scientific Committee. Hobart (Australia): CCAMLR.

36. Ainsworth CH, Pitcher TJ, Heymans JJ, Vasconcellos M (2008) Reconstructing historical marine ecosystems using food web models: Northern British Columbia from Pre-European contact to present. Ecol Model 216: 354–368.

37. Watson R, Pauly D (2001) Systematic distortions in world fisheries catch trends. Nature 44: 534–536.

38. High Seas Task Force (2006) Closing the net: Stopping illegal fishing on the high seas. Governments of Australia, Canada, Chile, Namibia, New Zealand, and the United Kingdom, World Wildlife Fund, World Conservation Union, and the Earth Institute at Columbia University, 2006. Available: http://www. high-seas.org/. Accessed 24 November 2008.

39. Commission of the European Communities (2007) Establishing a Community system to prevent, deter and eliminate illegal, unreported and unregulated fishing. EC COM(2007) 602 final. Available: http://eur-lex.europa.eu/

LexUriServ/LexUriServ.do?=uriCOM:2007:0602:FIN:EN:PDF. Accessed September 2008.

40. WWF (2008) Illegal fishing in Arctic waters. Oslo: WWF International Arctic Programme. Available: http://assets.panda.org/downloads/iuu_report_version_1_3_30apr08.pdf . Accessed September 2008.

41. Watson R, Kitchingman A, Gelchu A, Pauly D (2004) Mapping global fisheries: sharpening our focus. Fish Fisheries 5: 168–177.

42. Pitcher TJ, Kalikoski D, Pramod G, editors. (2006) Evaluations of Compliance with the FAO (UN) Code of Conduct for Responsible Fisheries. 14(2)Vancouver: University of British Columbia. Fisheries Centre Research Reports.

43. Pramod G, Pitcher TJ, Agnew D, Pearce J (2008) Sources of information supporting estimates of unreported fishery catches (IUU) for 59 countries and the high seas. 16(4)Vancouver: University of British Columbia. Fisheries Centre Research Reports.

44. Kalikoski D, Pitcher TJ, editors. (2008) Assessing Illegal, Unreported And Unregulated Fishery Catches (IUU): Some case studies. Vancouver: University of British Columbia. Fisheries Centre Research Reports. In press.

45. CEDEPESCA (2003) La pesqueria de calamar en Argentina. Buenos Aires: CEDEPESCA. Comunidad Pesquera no 9.

46. DAFF (2005) Effective Export Controls For Illegally Harvested Abalone Discussion Paper. Canberra: Australian Government Department of Agriculture, Fisheries and Forestry. Available: http://www.daff.gov.au/__data/assets/pdf_file/0006/5856/abalone_discussion_paper.pdf. Accessed September 2008.

47. Flewwelling P (2001) Fisheries Management and MCS in South Asia. Rome: FAO. GCP/INT/648/NOR: Field Report C-6.

48. Chimanovitch M (2001) Zona maritime. Available: www.terra.com.br/istoe/1633/brasil/1633_zona_maritima.htm .

49. Weidner DM, Hall DL (1993) World Fishing Fleets: An Analysis of Distant Water Fleet Operations, Past-Present-Future: Volume 4. Silver Spring: NMFS.

50. Bernal P, Oliva D, Aliaga B, Morales C (1999) New regulations in Chilean Fisheries and Aquaculture: ITQ's and Territorial Users Rights. Ocean Coast Man 42: 119–142.

51. Zuleta A (2004) The management of the small pelagic fishery in Chile. Rome: FAO. FAO Fisheries Report No. 700.

52. Patterson KR, Pitcher TJ, Stokes TK (1993) A stock collapse in a fluctuating environment: the chub mackerel Scomber japonicus (Houttuyn) in the eastern central Pacific. Fish Res 18: 199–218.

53. Hariri KI, Nichols P, Krupp F, Mishrigi S, Barrania A, et al.. (2002) Status of the Living Marine Resources in the Red Sea and Gulf of Aden Region and their Management. Washington: The World Bank. Available: http://www.persga.org/UI/English/Download/Vol3bStatusofLMRinRSGA.pdf). Accessed September 2008.

54. Falaye A (2008) Illegal unreported unregulated (IUU) fishing in West Africa (Nigeria & Ghana). London: MRAG. Available: http://www.mrag.co.uk/Documents/IUU_WestAfrica.pdf. Accessed November 2008.

55. Rajan PT (2003) A field guide to marine food fishes of Andaman and Nicobar Islands. Kolkata (India): Zoological Survey of India.

56. Willoughby N, Monintja D, Badrudin M (1997) Do Fisheries Statistics Give the Full Picture? Indonesia's Non-Recorded Fish Problem. Report of the Regional Workshop on the Precautionary Approach to Fishery Management. Rome: FAO. Report BOPB/REP/82. Available: ftp://ftp.fao.org/docrep/fao/007/ad914e/ad914e02.pdf), pp. 163–172. Accessed September 2008.

57. Nurhakim S, Nikijuluw VPH, Badrudin M, Pitcher TJ, Wagey GA (2008) A Study Of Illegal, Unreported and Unregulated (IUU) Fishing In The Arafura Sea, Indonesia. Rome: FAO.

58. Palma MA, Tsamenyi M (2008) Case study on the impacts of Illegal, Unregulated and Unreported fishing in the Sulawesi Sea. Singapore: APEC. Available: http://www.apec.org/apec/publications/all_publications/fisheries_working.MedialibDownload.v1.html?url=/etc/medialib/apec_media_library/downloads/workinggroups/fwg/pubs/2008.Par.0001.File.v1.1. Accessed 24 November 2008.

59. Taghavi SA (1999) Fisheries MCS in the Islamic Republic of Iran. Report of a regional workshop on fisheries monitoring, control and surveillance. Rome: FAO. pp. 67–74. GCP/INT/648/NOR Field Report C-3 (En).

60. Long R, Grehan A (2002) Marine Habitat Protection in Sea Areas under the Jurisdiction of a Coastal Member State of the European Union: The Case of Deep-Water Coral Conservation in Ireland. Int J Mar Coast Law 17: 235–261.

61. Clarke S (2008) Illegal Fishing in the Exclusive Economic Zone of Japan. London: MRAG. Available: http://www.mrag.co.uk/Documents/IUU_Japan.pdf. Accessed November 2008.

62. Lozano-Montes HM, Pitcher TJ, Haggan N (2008) Shifting environmental and cognitive baselines in the upper Gulf of California. Frontiers Ecol Envir 6: 75–80.

63. Tudela S, Kai AK, Maynou F, El Andalossi M, Guglielmi P (2005) Driftnet fishing and biodiversity conservation: the case study of the large-scale Moroccan

driftnet fleet operating in the Alboran Sea (SW Mediterranean). Biol Cons 121: 65–78.

64. Baddyr M, Guénette S (2002) The Fisheries off the Atlantic Coast of Morocco 1950–1997. In: Zeller D, Watson R, Pauly D, editors. Fisheries Impacts on North Atlantic Ecosystems: Catch, Effort and National/Regional Data Sets. vol. 9(3)Vancouver: Univ. British Columbia. pp. 191–205. Fisheries Centre Research Reports.

65. Guénette S, Balguerías E, Santamaría MTG (2001) Spanish fishing activities along the Saharan and Moroccan coasts. In: Zeller D, Watson R, Pauly D, editors. Fisheries Impacts on North Atlantic Ecosystems: Catch, Effort and National/Regional Data Sets. vol. 9(3)Vancouver: Univ. British Columbia. pp. 206–213. Fisheries Centre Research Reports.

66. Melnychuk M, Guénette S, Martín-Sosa P, Balguerías E (2001) Fisheries in the Canary Islands, Spain. In: Zeller D, Watson R, Pauly D, editors. Fisheries Impacts on North Atlantic Ecosystems: Catch, Effort and National/Regional Data Sets. vol. 9(3)Vancouver: Univ. British Columbia. pp. 221–224. Fisheries Centre Research Reports.

67. Kelleher K, Rottingen A, Limmitada MGA (2002) Planning cost-effective fisheries. Monitoring, Control and Surveillance in Mozambique. Oslo: NORAD. Available: http://www.nfh.uit.no/norad/reports/mcs20mozambique20kelleher. doc. Accessed September 2008.

68. MRAG, CAPFISH (2008) Study and Analysis of the Status of IUU Fishing in the SADC Region and an Estimate of the Economic, Social and Biological Impacts, Volume 2 – Main Report. London: MRAG. Available: http://www.stopillegalfishing.com/docs/study_of_the_status_of_IUUfishing_in_sadcregion_n_estimate_ESBI_vol2_eng.pdf. Accessed September 2008.

69. Butcher JG (2002) Getting into Trouble: The Diaspora of Thai Trawlers, 1965–2002. Int J Maritime History 14: 85–121.

70. Butcher JG (2004) The Closing of the Frontier. A History of the Marine Fisheries of Southeast Asia c. 1850–2000. Singapore: Institute of Southeast Asian Studies.

71. Pe M (2004) National Report Of Myanmar on the Sustainable Management of The Bay of Bengal Large Marine Ecosystem (BOBLME). Rome: FAO.GCP/RAS/179/WBG. Available: http://www.fao.org/fi/oldsite/BOBLME/website/reports.htm. Accessed July 2008.

72. Pearse NI, Folson WB (1979) Fisheries of Cameroon, 1973. NMFS Foreign Fish Leaflet (74–12).

73. Esmark M, Jensen N (2004) The Barents sea cod - Last of the large cod stocks. Oslo: WWF. WWF Norway Report 4/2004.

74. Castillo S, Mendo J (1987) Estimation of Unregistered Peruvian Anchoveta (Engraulis ringens) in Official Catch Statistics, 1951 to 1982. In: Pauly D, Tsukayama I, editors. The Peruvian anchoveta and its upwelling ecosystem: three decades of changes. Manila: ICLARM. pp. 109–116.

75. Benavente-Villena AB, Pido MD (2004) Poaching in Philippine marine waters: intrusion of Chinese fishing vessels in Palawan waters. In: Silvestre G, Luna C, editors. In turbulent seas: the status of Philippine marine fisheries. Cebu City: Coastal Resource Management Project. pp. 265–268.

76. Kalentchenko M, Nagoda D, Esmark M (2005) Analysis of illegal fishery for cod in the Barents Sea. Moscow: WWF. WWF Russia and WWF Barents Sea Program. Available: http://assets.panda.org/downloads/wwf_russia_iuu_fishing_barents_2005_august_2005.pdf. Accessed 20 September 2008.

77. Hønneland G (2004) Russian Fishery Management: the precautionary approach in theory and practice. Boston: Martinus Nijhoff Publishers/Brill Academic Publishers.

78. Matishov G, Golubeva N, Titova G, Sydnes A, Voegele B, et al.. (2004) Barents Sea, GIWA Regional assessment 11. Kalmar: Univ. of Kalmar/UNEP.

79. Morato T, Guénette S, Pitcher TJ (2002) Fisheries of the Azores (Portugal), 1982–1999. In: Zeller D, Watson R, Pauly D, editors. Fisheries Impacts on North Atlantic Ecosystems: Catch, Effort and National/Regional Data Sets. vol. 9(3)Vancouver: Univ. British Columbia. pp. 214–220. Fisheries Centre Research Reports.

80. Morato T, Pauly D, editors. (2004) Seamounts: Biodiversity and fisheries. vol. 12(5)Vancouver: Univ. British Columbia. Fisheries Centre Research Reports.

81. Joseph L (1999) Management of shark fisheries in Sri Lanka. In: Shotton R, editor. Case studies of the management of elasmobranch fisheries. Rome: FAO, Rome. pp. 339–367.

82. Flewweling P, Hosch G (2006) Country review: Sri Lanka. In: De Young C, editor. Review of the State of the World Marine Capture Fisheries Management: Indian Ocean. Rome: FAO. pp. 163–174. FAO Tech Rep 488.

83. Kuo CL (2001) Development and Management of Coastal Fisheries in Taiwan. In: Liao IC, Baker J, editors. Aquaculture and Fisheries Resources Management. Keelung: Taiwan Fisheries Research Institute. pp. 241–248. Conference Proceedings Taiwan Fisheries Research Institute no. 4.

84. Tudela S (2004) Ecosystem effects of fishing in the Mediterranean: an analysis of the major threats of fishing gear and practices to biodiversity and marine habitats. Rome: FAO. GFCM Studies and Reviews 74.

85. Birkun A (2002) Cetaceans of the Mediterranean and Black Seas: State of Knowledge and Conservation Strategies. In: Notarbartolo di Sciara G, editor. Cetaceans of the Mediterranean and Black Seas: state of knowledge and conservation strategies. Monaco: ACCOBAMS Secretariat. Section 10.

86. Morgan G (2006) Country review: Yemen. In: De Young C, editor. Review of the State of the World Marine Capture Fisheries Management: Indian Ocean. Rome: FAO. pp. 337–348. FAO Tech Rep 488.

87. FAO (2007) Commodities 2005 FAO Yearbook of Fishery Statistics Vol. 101. Rome: FAO.

Chapter 3

Fish Communities in Coastal Freshwater Ecosystems: The Role of the Physical and Chemical Setting

Kristin K. Arend and Mark B. Bain

ABSTRACT

Background

We explored how embayment watershed inputs, morphometry, and hydrology influence fish community structure among eight embayments located along the southeastern shoreline of Lake Ontario, New York, USA. Embayments differed in surface area and depth, varied in their connections to Lake Ontario and their watersheds, and drained watersheds representing a gradient of agricultural to forested land use.

Results

We related various physicochemical factors, including total phosphorus load, embayment area, and submerged vegetation, to differences in fish species

Originally published as Arend, K.K., Bain, M.B. Fish communities in coastal freshwater ecosystems: the role of the physical and chemical setting. BMC Ecol 8, 23 (2008). https://doi.org/10.1186/1472-6785-8-23. © BioMed Central Ltd. http://creativecommons.org/licenses/by/2.0

diversity and community relative abundance, biomass, and size structure both among and within embayments. Yellow perch (Perca flavescens) and centrarchids numerically dominated most embayment fish communities. Biomass was dominated by piscivorous fishes including brown bullhead (Ameiurus nebulosus), bowfin (Amia calva), and northern pike (Esox lucius). Phosphorus loading influenced relative biomass, but not species diversity or relative abundance. Fish relative abundance differed among embayments; within embayments, fish abundance at individual sampling stations increased significantly with submerged vegetative cover. Relative biomass differed among embayments and was positively related to total phophorus loading and embayment area. Fish community size structure, based on size spectra analysis, differed among embayments, with the frequency of smaller-bodied fishes positively related to percent vegetation.

Conclusion

The importance of total phosphorus loading and vegetation in structuring fish communities has implications for anthropogenic impacts to embayment fish communities through activities such as farming and residential development, reduction of cultural eutrophication, and shoreline development and maintenance.

Background

Physicochemical features at multiple spatial scales (e.g., watershed, embayment, and habitat) can be important for fish community structure [1-3]. Variability in nutrient inputs, hydrology, and morphometry among and within aquatic ecosystems can shape fish communities [1,4-6]. In turn, fish community structure influences ecosystem function, such as energy transfer and nutrient cycling [7-9] via trophic interactions and, in some cases, habitat modification [10,11]. Consequently, fish communities are important indicators of and interactors in aquatic ecosystems.

We explored how physicochemical features shaped fish communities in Lake Ontario embayments. Great Lakes embayments are relatively shallow, inshore ecosystems located between the shorelines of the lakes and their watersheds. Embayments vary considerably in nutrient loading, hydrology, and morphometry. Additionally, embayments serve as conduits of nutrients and other materials from their watersheds [12,13], support high fish species diversity [12], provide spawning and nursery habitats for both nearshore and offshore Great Lakes fishes [14,3,15], and are concentrated areas of human activities [16]. These characteristics make embayments ideal systems with which to address physicochemical effects on fish community structure in the context of ecosystem function.

Great Lakes embayments range in hydrogeomorphic type, including flooded river mouths, coastal wetlands, and large, deep enclosed bays [17,18,15]. Embayments are connected to their watersheds by tributary inflow, surface runoff, and/or groundwater flow. While some embayments lack direct, surface water connections to the main lake, most embayments have either man-made or natural connections that can be permanent, seasonal, or ephemeral [19]. This combination of morphometric and hydrologic variability results in physicochemical habitat conditions that differ both among and within embayments [16,20]. For example, morphometry and water inflow from tributaries and the lake (via seiches) interact to influence water chemistry, submerged aquatic vegetation, and dissolved oxygen and temperature profiles [20].

We posed the question: how do watershed inputs, hydrology, and embayment morphometry affect fish community structure in eight embayments located along the southeastern coast of Lake Ontario? We expected that variation in these physicochemical factors across spatial scales would influence multiple metrics of fish community structure, including diversity, relative abundance, biomass, and size structure. At the watershed scale, watershed size, discharge and land use affect productivity, which in turn, can influence fish community structure and dynamics. We hypothesized that high nutrient inputs to embayments, from either high watershed flows (i.e., short water residence time) or high nutrient concentrations due to land use, would positively affect fish abundance and biomass [21,22,1] and negatively affect species diversity through loss of intolerant species [4,23]. At the system (i.e., embayment) scale, greater surface area with a more complex depth profile can increase habitat and resource heterogeneity, which positively impact fish abundance, biomass, and diversity [24,6].

Within systems, availability of vegetated, littoral habitat also affects fish communities by increasing habitat and resource heterogeneity [14,25,26]. As such, we predicted that embayments with higher habitat heterogeneity (e.g., large surface area and/or abundant, vegetated littoral habitat) would support more diverse and abundant fish communities than small or more homogeneous embayments. Morphometry also impacts fish community size structure [1,6]. We hypothesized that a higher proportion of small-bodied than large-bodied fishes would occur in shallow embayments dominated by vegetated habitat [1]. In contrast, large embayments having deep, open habitat would provide support for large-bodied fishes [6], resulting in a low proportion of small-bodied fishes due to predation [10,27].

In this article, we used hierarchical mixed modelling to relate differences among embayment fish communities to abiotic and biotic factors at the watershed through sampling station scales. The response variables we considered were fish community species diversity, relative biomass and abundance, and size structure.

Predictor variables included total phosphorus load, embayment area, sampling station depth, percent aquatic vegetation, and percent littoral habitat and, for size structure only, piscivore relative biomass.

Methods

Study Site Hydrogeomorphic Classification

Study embayments were located in two clusters along the southeastern shoreline of Lake Ontario, New York, USA (Figure 1) and varied in several watershed and embayment characteristics. For purposes of this research, we classified embayments into general hydrogeomorphic types, based on Keough et al.. [17]: (1) drowned-river mouth embayments (Sterling, Floodwood); (2) pelagic-protected embayments (Blind Sodus, Little Sodus, South Sandy); and (3) littoral-protected embayments (Juniper, North Sandy, and South Colwell). Drowned-river mouth embayments receive high watershed inputs, have short water residence times, and have a surface water connection with Lake Ontario [17]. Protected embayments have longer water residence times than drowned-river mouths, are separated from Lake Ontario by a sand barrier, and vary in their hydrologic connections to their watershed and Lake Ontario [17]. We defined pelagic-protected embayments as having depths that exceed euphotic zone depth estimates for at least 10% of their area. Littoral-protected embayment depths do not exceed euphotic zone depth estimates.

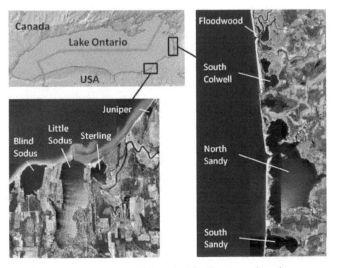

Figure 1. Embayment locations. Location map of the eight Lake Ontario study embayments.

Embayment-Scale Characteristics

Morphometry

Morphometric measurements included watershed area, embayment area, maximum depth, and percent littoral habitat. Watershed areas were calculated from digital elevation maps using ESRI ArcHydro tools [28]. Annual embayment area and maximum depth were calculated from bathymetric maps and annual averages of Lake Ontario water level for each year of the study (NOAA, Oswego, NY). Bathymetric maps were generated from elevations that were calculated using depth measurements taken in the field and, for reference over time, the 1985 Lake Ontario water level of 74.67 meters (m; NOAA, Oswego, NY).

Percent littoral habitat was calculated from bathymetric maps based on embayment-specific estimates of euphotic zone depth, i.e., the depth at which 1% incident light intensity occurs. Mean Secchi disk depth (m) for each embayment was calculated using data collected weekly from May through mid-October in 2001 and 2002, and biweekly from June through mid-October 2003 at centrally located stations in each embayment. Euphotic zone depth was estimated as 2.7 * mean Secchi disk depth [29].

Depths less than or equal to the euphotic zone depth in each embayment were defined as littoral; depths greater than the euphotic zone depth were defined as pelagic. Littoral and pelagic areas (km2) in each embayment were estimated from bathymetric maps using Arcview GIS 3.x [30]. Euphotic zone depth estimates for Juniper and South Colwell exceeded maximum depths in these embayments; therefore, 100% of the habitat was considered littoral, which matched field observations.

Water Residence Time and Water Chemistry

Annual water residence time and water chemistry were estimated from monthly sampling data provided by X. Chen (Syracuse University, unpublished data). Embayment water residence time was estimated from of the relative contributions of the watershed, Lake Ontario, and direct precipitation to each embayment, as determined by fluoride mass balance calculations (X. Chen, Syracuse University, personal communication). Total phosphorus loading to the embayments was calculated by multiplying stream discharge into the embayments by the input phosphorus concentration (X. Chen, Syracuse University, personal communication).

Station-Scale Characteristics

On one or two consecutive dates in July 2001, 2002, and 2003, each embayment was sampled at between three and eight stations, based on embayment size.

Embayment bathymetric maps superimposed with a 30 × 30 m grid were divided into three to eight strata. Grid intersection points at which the water column depth was less than or equal to four m were assigned numbers. Each year, sample stations were determined by randomly selecting one grid intersection point from each stratum. This design ensured that stations sampled were distributed throughout the embayments. Embayment sample stations were located using a Global Positioning System (GPS) set to the Universal Transverse Mercator coordinate system, and marked with a buoy (henceforth center).

Habitat data were collected at all stations within each embayment between 0800 and 1800 hours. Mean bottom depth was calculated from bottom depth measurements taken to the nearest 0.1 m at four locations 30 m out in each direction from the center. Visual estimates of the percent of sediment surface supporting submerged aquatic vegetation growth were made at one second time intervals while driving along a circular path approximately 30 m radius from the center. Mean cover for each station was calculated from the estimates. At the center, surface temperature was measured to the nearest 0.1°C with a standard thermometer and Secchi depth was measured to the nearest 0.1 m following standard methods.

Fish Sampling

Fish sampling coincided with sampling for station-level characteristics. At each station, fish were collected using a 4.6 m boat equipped with a Smith-Root Type VI-A electrofishing unit and a 5000 watt generator. The transformer was set at 120 pulses per second direct current electricity, with either 125 or 250 volts and pulse width varying between 7–9 milliseconds. Several factors that can influence fish captured using electrofishing include fish size, water clarity, water depth, and macrophyte density. We tried to minimize unequal bias in catch among embayments by focusing our sampling to concentrated areas at depths less than or equal to four m. Fish were collected along 15 minute (min) inward spirals starting at approximately a 30 m radius from the center of each station. In 2001 and 2002, fish were sampled between 0800 and 1800 hours; in 2003, fish were sampled between 1300 and 2300 hours. Fish were identified to species and total length (TL) was measured to the nearest 1.0 millimeter (mm). In 2002 and 2003, the wet weight in grams (g) of all fish was measured to the nearest 0.1 or 0.5 g. Fish were not weighed in 2001.

As all of our sampling stations were in less than four m deep water, discussions in this paper focus on littoral fish assemblages. However, gill net sampling conducted in littoral and pelagic habitats in Blind Sodus, Little Sodus, South Sandy, and North Sandy during late June – early July, 2002, yielded relatively few fish in pelagic habitat (89 fish, 3293 min total effort) compared with littoral habitat (215 fish, 3223 min total effort). Of fish captured in pelagic habitat, 18% were alewife (Alosa pseudoharengus) and 76% were yellow perch (Perca flavescens). Therefore, with the exception of alewife, the majority of fishes occurred at depths less than four m, and a unique offshore community was not detected. These data suggest that fishes captured at depths less than four m represented the majority of fishes that occupied the embayments during May through August.

Data Analysis

Fish data were analyzed either at the community or taxonomic levels. Taxonomic analyses concentrated on nine focal species. Eight of these occurred in relatively high numbers across all embayments and represented a range of trophic positions and feeding habits (e.g., planktivore, invertivore, piscivore): brown bullhead (Ameiurus nebulosus), bowfin (Amia calva), bluegill (Lepomis macrochirus), golden shiner (Notemigonus crysoleucas), largemouth bass (Micropterus salmoides), northern pike (Esox lucius), pumpkinseed (Lepomis gibbosus), and yellow perch. Walleye (Sander vitreus) was identified as a ninth focal species due to high densities in South Sandy.

Fish Community Structure

Fish species diversity was calculated by aggregating the number of fishes in each species collected across years. Diversity was estimated using Simpson's index (D^{-1}), because of its low sensitivity to sample size [31], which varied across embayments. We used ordinary least squares (OLS) linear regression to determine if species diversity was related to total phosphorus loading or to embayment area, which were log-transformed (ln) to reduce heterogeneity of variances.

Weights for all fish captured in 2001 were estimated using species-specific length-weight regressions either generated from data we collected in 2002 and 2003 or reported in the literature. We generated length-weight regressions either for individual embayments or for all embayments pooled, depending on the number of individuals per species captured within and across embayments. Annual catch per unit effort (number·min-1; CPUE) and biomass per unit effort (g·min-1; BPUE) were estimated for each species in each embayment. Within an embayment-year combination, CPUE and BPUE for each taxonomic group and all fish combined were calculated as the total number and biomass, respectively, of individuals caught in that group divided by total embayment sampling effort that year (min; summed across all stations).

Normalized size spectra (NSS) provide a quantitative way to evaluate the distribution of biomass within each embayment's fish community. The method identifies the size class that supports maximum biomass by sorting organisms (i.e., fish) into size classes and plotting total biomass in each size class versus size class. NSS were created for each embayment-year by transforming all fish weights by log base 2 [32]. The sum of transformed weights in each size class was plotted against the transformed weight of the heaviest fish actually recorded in that size class (sensu [33]). The maximum possible weight in a size class (e.g., 1-0.01 = 0.99) was used if no data existed for that size class. We then solved for the points h and k describing the location (x axis) and height (y axis), respectively, of the parabola vertex as $y = c + b \cdot x + a \cdot x^2$, using ordinary least squares regression. Values for the coefficients h and k were calculated as

$$h = \frac{-b}{2 \cdot a}$$

and

$$k = c + b \cdot h + a \cdot h^2.$$

Coefficient h approximates the weight class at which the majority of the fish community's biomass is concentrated, and henceforth will be referred to as maximum biomass weight class. Coefficient k estimates total biomass at maximum biomass weight class and is correlated with total fish community biomass [32]. Correlations between the coefficients and between each coefficient and fish biomass were calculated. Maximum biomass weight class also was used as a response variable in the community response analyses (see below).

Fish Community Response to Embayment and Habitat Characteristics

We conducted mixed model analyses using PROC MIXED in SAS [34] both to identify differences in fish community descriptors (e.g., CPUE) among embayments and to relate descriptors to embayment physicochemical features. We used a mixed model to account for the hierarchical structure of the data (stations within years within embayments) and for the use of both continuous and categorical variables. Sample sizes for each level were three years, eight embayments, and between three to eight stations per embayment-year. Community descriptors included: (1) CPUE of all fish combined; (2) BPUE of all fish combined; and (3) the maximum biomass weight class (coefficient h of the NSS). CPUE and BPUE data for all fish species combined were square-root transformed to meet the assumption of normality; maximum biomass weight classes were normally distributed. Similar analyses at the taxonomic level were not possible due to highly variable catch among stations, with many zero catches.

Embayment was specified as a random effect, because we assumed the study embayments represent Lake Ontario embayments in general [35]. Year was categorized as a fixed effect, because of the unlikelihood that three consecutive years of data represent a random sample of years [35]. We considered physicochemical variables at both the embayment and station scales to include as predictors in our analyses. Predictor variables were selected based on Pearson correlation coefficients. Selected variables included: year, water residence time, total phosphorus load (natural log transformed), embayment area (natural log transformed), station depth (natural log transformed), percent submerged aquatic vegetation (arcsine transformation), and percent littoral habitat. Percent littoral habitat, was converted to a binomial variable as either pelagic-dominated (less than 60% of bottom depth area falling within the euphotic zone) or littoral-dominated (for this study, all had greater than 95% of bottom depth area falling within the euphotic zone). The following factors were not considered due to correlation: embayment: watershed area, annual nitrogen loading (kg·y-1), embayment volume, and embayment mean depth. We did not include Secchi depth, because depth readings frequently were limited by bottom depth or dense macrophyte beds. We did not include temperature or dissolved oxygen in the models, because data collection was limited to sampling times and thus the data do not characterize the thermal and oxygen regimes of the embayments. From May through October, 2001 and 2002, mean temperature at depths less than or equal to four m ranged from 19–21°C across embayments. Dissolved oxygen differed by less than four mg·Liter (L)-1 across embayments, and was always greater than five mg·L-1 at zero to five m depths.

Piscivorous fish BPUE (untransformed) was included as a fixed effect in the model of the maximum biomass weight classes. Piscivore BPUE included American eel (Anguilla rostrata), bowfin, chain pickerel (Esox niger), grass pickerel (Esox americanus vermiculatus), largemouth bass, longnose gar (Lepisosteus osseus), northern pike, smallmouth bass (Micropterus dolomieu), walleye, and white perch (Morone americana). We included all largemouth bass, despite high catches of young-of-year bass, based on prey fish in diets of largemouth bass as small as 37 mm TL and on findings by Olive et al.. [27] that piscivory by high densities of small-bodied largemouth bass can structure fish communities.

For the mixed model analyses of CPUE and BPUE, data were classified according to embayment, year, and station. Mixed model analysis of maximum biomass weight classes was conducted at the embayment scale only, with data classified according to embayment and year. Degrees of freedom were adjusted using the Kenward-Rogers method. To ensure that only uncorrelated variables were included in each model, (1) one of two or more correlated variables was selected or (2) correlated variables were tested in separate model runs (e.g., percent littoral habitat and percent submerged aquatic vegetation). We selected those models that provided the best, most parsimonious fit to the data, based on model covariance estimates and number of parameters.

Two models were run for each response variable to test for variance in the responses among embayments. In the first model, embayment was specified as a random effect; in the second, no random effects were specified, to identify the variation explained by embayment. The test statistic was calculated as the difference between the two models' log-likelihood values. It follows a $\chi 2$ distribution, and its p-value is determined by dividing the probability of a greater $\chi 2$ for one degree of freedom by two [34]. We used analyses of the full mixed models to identify significant fixed effects for each response variable. The percent of the variation between embayments explained by each full mixed model was calculated as the difference in variance due to embayment between models with and without the physicochemical factors as predictor variables, expressed as a fraction of the variance due to embayment in the model without the physicochemical factors as predictor variables. Within embayment variation was calculated similarly, using the unexplained (i.e., residual) variance estimates for each model in place of variance due to embayment.

Results

Fish Community Structure

Across all embayments, we collected a total of 3475 fishes representing 42 different species and 16 families. Species diversity (Simpson's index) ranged from 2.7 – 6.7 among embayments, but was not related to total phosphorus loading (r^2 = 0.41, p = 0.09), embayment area (r^2 = 0.01; p = 0.81), or percent littoral area (r^2 = 0.01; p = 0.83).

Relative abundance (CPUE) and biomass (BPUE) of all fish combined and of individual species varied among embayments and years (Figure 2). Fish communities were numerically dominated by yellow perch, pumpkinseed, bluegill, and largemouth bass (Figure 2). With the exception of Floodwood and Juniper, yellow perch constituted between 20–60% (by number) of the fish community. In Floodwood, abundance was more evenly distributed across yellow perch and the centrarchid populations; in Juniper, golden shiner was the numerically dominant species (Figure 2). Large piscivores accounted for the majority of the biomass in all embayments except Juniper, an unconnected embayment where large piscivores were not captured (Figure 2). The most common non-focal species included alewife, common carp (Cyprinus carpio carpio), blacknose shiner (Notropis heterolepis), common shiner (Luxilus cornutus), banded killifish (Fundulus diaphanus diaphanus http://www.fishbase.org/ComNames/CommonNameSummary.cfm?autoctr=241393 webcite), black crappie (Pomoxis nigromaculatus), and smallmouth bass. Large common carp accounted for the high biomass of non-focal species in Blind Sodus, Little Sodus, South Sandy, and Floodwood.

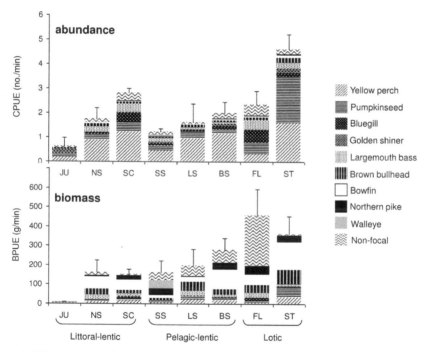

Figure 2. Fish community relative abundance and biomass. Relative abundance (total fish catch per unit effort) and biomass (total fish biomass per unit effort) of embayment fish communities. Embayments are shown from left to right in order of increasing phosphorus loading, within each hydrogeomorphic type. Embayment codes are: SC, South Colwell; JU, Juniper; NS, North Sandy; LS, Little Sodus; SS, South Sandy; BS, Blind Sodus; ST, Sterling; and FL, Floodwood. High non-focal species biomass typically is due to the presence of common carp.

Fish community size structure also varied among embayments and years. Normalized size spectra models captured between 17–71% of the variation in total biomass per weight class for each embayment-year combination. Juniper supported a small-bodied fish community (< 200 mm TL), indicated by a much lower maximum biomass weight class than for the other embayments (Figure 3). Maximum biomass weight class estimates for all other embayments varied among embayment-year combinations. Maximum biomass weight class in Blind Sodus, South Sandy, and Floodwood occurred at larger weight classes, indicating these fish communities contained a greater proportion of large-bodied fishes than in the other embayments (200–500 mm TL; Figure 3). Fish biomass was concentrated in medium-sized fish in Little Sodus, Sterling, North Sandy, and South Colwell (Figure 3). Juniper supported a small-bodied fish community (Figure 3). Excluding Juniper, total biomass at the maximum biomass weight class was negatively correlated with maximum biomass weight class (p = 0.002; Figure 3). For all embayments, neither maximum biomass weight class nor total biomass at the maximum biomass weight class was correlated with total fish biomass.

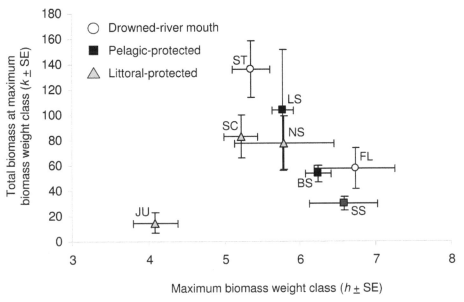

Figure 3. Embayment fish community biomass at maximum biomass weight class versus maximum biomass weight class. Total fish community biomass at maximum biomass weight class (k) versus maximum fish community biomass weight class (h) for drowned-river mouth (open circles), pelagic-protected (black squares), and littoral-protected (grey triangles) embayments. Points represent 3-year means ± standard error (SE). Embayment codes follow Figure 2.

Fish Community Response to Embayment and Habitat Characteristics

The models that provided the best, most parsimonious fit to the data included total phosphorus load, area, the interaction between total phosphorus load and area, and percent vegetation. Both CPUE and BPUE differed significantly among embayments, as indicated by the likelihood ratio test statistics (p < 0.0025 for both; Table 1). CPUE within embayments was positively related to percent vegetation (p = 0.02; Table 1; Figure 4). BPUE among embayments was positively related to embayment area (p = 0.04) and total phosphorus load (p = 0.02; Table 1; Figure 5) and negatively related to the interaction between area and total phosphorus load (p = 0.03; Table 1). Maximum biomass weight class (the NSS coefficient h) also differed among embayments (p = 0.004), and was negatively related to percent vegetation (p = 0.005; Table 1; Figure 6) between embayments.

Table 1. Fish community response to physicochemical factors

	Species diversity	CPUE (#·m²)	BPUE (g·m²)	Size-structure (h)
Variation by embayment	ns	p < 0.0025	p < 0.0025	p = 0.04
Physicochemical Factor				
TP load (0.88 – 4.34)	ns	ns	28 (0.02)	ns
Area (4.79 – 7.05)	ns	ns	9.6 (0.04)	ns
TPload*Area	ns	ns	-4.3 (0.03)	ns
% vegetation (0 – 90)	ns	0.01 (0.02)	ns	-0.05 (0.005)
Variation model	n/a			
Between		-0.146	0.86	0.83
Within		0.132	0.005	-0.004

"Variation by embayment" indicates if variability for each response among embayments significantly differed from zero. Effects of total phosphorus load (TPload; kg·y⁻¹), embayment area (m²), and percent vegetation on each characteristic are presented as slope estimates; ns indicates no significant effect of that factor; n/a indicates not applicable. P values for effects are indicated in parentheses under each response column; the range of transformed values for each physicochemical effect are indicated in parentheses to the right of each effect's name. The variation explained by each model ("Variation model") is divided into variation between embayments ("Between") and within embayments ("Within").

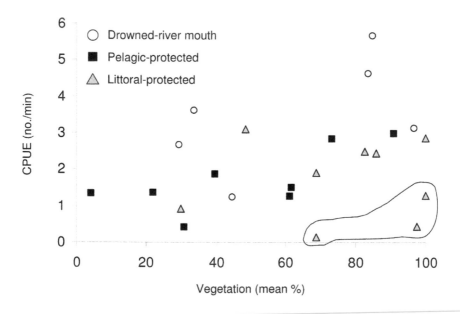

Figure 4. Fish relative abundance versus mean percent vegetation. Annual fish relative abundance (total fish catch per unit effort; #·min-1) in July, 2001–2003, versus mean percent vegetation for drowned-river mouth (open circles), pelagic-protected (black squares), and littoral-protected (grey triangles) embayments. Circled data outliers are from Juniper, from which relatively few fish were collected.

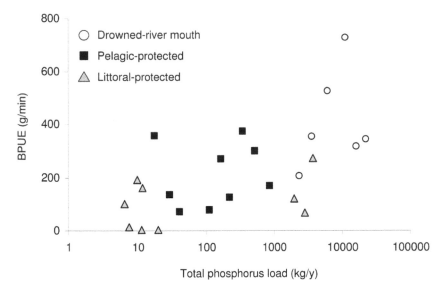

Figure 5. Fish relative biomass versus total phosphorus loading. Annual total fish biomass per unit effort (g·min-1) in July, 2001–2003, versus total phosphorus loading (kg/y) for drowned-river mouth (open circles), pelagic-protected (black squares), and littoral-protected (grey triangles) embayments.

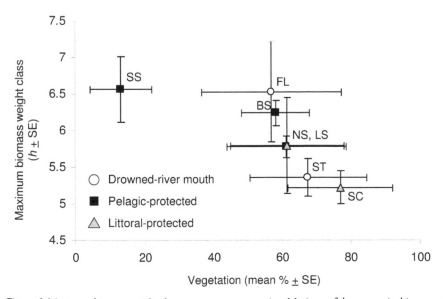

Figure 6. Maximum biomass weight class versus percent vegetation. Maximum fish community biomass weight class (h) versus percent vegetation for drowned-river mouth (open circles), pelagic-protected (black squares), and littoral-protected (grey triangles) embayments. Points are 3-year means ± SE. Embayment codes follow Figure 2.

Discussion

Despite similar fish species composition and diversity among embayments, community relative abundance, biomass, and size structure differed among embayments. These differences were correlated with physicochemical attributes at the watershed and embayment scales. Phosphorus loading influenced fish community relative biomass, but not species composition or community relative abundance. Greater fish biomass was supported in large, deep embayments and those receiving high phosphorus loading. Vegetated embayments supported more fish, with biomass concentrated in small-bodied fishes than did less vegetated embayments. Within embayments, stations with greater submerged vegetative cover supported more smaller-bodied fishes. Water residence time did not influence fish community characteristics directly, but could inversely affect phosphorus loading by phosphorus dilution or reduced phosphorus retention at high flows.

Species diversity was not significantly affected by total phosphorus loading, despite a large range in loading. Our sites are located in eastern Lake Ontario, which is less impacted by urban and agricultural activity and receives lower nutrient and sediment inputs from the watershed than western Lake Ontario [18,36]. Among our sites, embayments in the eastern cluster (South Sandy, North Sandy, South Colwell, and Floodwood) are less impacted by land use than those in the western cluster [37], which could explain slightly higher values for species diversity in those embayments. Anthropogenic eutrophication may not be great enough in these systems to alter species composition significantly. Phosphorus loading did affect biomass, however, suggesting that impacts of nutrient enrichment on fish communities can be detected before negative effects such as changes in fish community composition are evident. A slight positive trend in species diversity with phosphorus loading indicates even those embayments with higher loadings were not sufficiently eutrophied to experience loss of intolerant species. We would expect even greater variation among embayments in fish biomass and possibly loss of species diversity at the highest levels of loading, if the entire range of phosphorus loading to Lake Ontario embayments had been included in our study.

As shown in other studies, fish biomass increased with total phosphorus loading (e.g., [21,22,1]) and embayment area (e.g., [24,6]). In contrast, however, phosphorus inputs and area did not influence fish relative abundance. Relative fish biomass may have been more sensitive to phosphorus loading than abundance because biomass more accurately represents the amount of fish tissue that must be supported. The positive effect of embayment area on biomass was reduced as total phosphorus load increased, and vice versa. Differences among hydrogeomorphic types in the relative importance of area and phosphorus may explain this relationship. For example, fish biomass was greatest in the drowned-river mouth embayments (Sterling and Floodwood), which also received the highest nutrient loading but are two of the smaller embayments. Both area and productivity appeared to influence fish biomass in the pelagic-protected embayments (Little Sodus, South Sandy, and Blind Sodus). For example, high phosphorus loading to Blind Sodus resulted in high biomass despite it being the smallest of the three embayments, whereas the large size of Little Sodus resulted in it supporting an intermediate amount of biomass, despite very low phosphorus loading. Neither size nor productivity seems to explain fish biomass in the littoral-protected embayments (Juniper, South Colwell, and North Sandy). Although North Sandy is the largest embayment and receives high phosphorus loading, it supports similar fish biomass to South Colwell, a small embayment with little loading. Therefore, other factors, such as habitat availability within embayments, may be more important in structuring littoral-protected fish communities.

Indeed, aquatic vegetation has been identified as an important factor in structuring fish communities in shallow, littoral-dominated systems [1,26]. Randall et al.. [1] found that fish were more numerous and smaller sized in vegetated versus unvegetated littoral habitat in Lake Ontario and Lake Huron bays, but that fish biomass did not differ. Our results complement those findings, even when considering more pelagic-dominated systems. For example, the two drowned-river mouth systems, Sterling and Floodwood, supported similarly high fish biomass; however, numerous, small-bodied fishes dominated the fish community in Sterling (with dense macrophyte beds), whereas fewer but larger-bodied fishes occupied Floodwood (with less vegetation, mostly concentrated at channel edges). Furthermore, fish abundance and size structure appear to be related to vegetation itself, and not simply shallow habitat (e.g., 100% of Sterling and 97% of Floodwood bottom depths are within the euphotic zone). Vegetation may be of greater benefit to small-bodied than large-bodied fishes, because it provides zoobenthivores, such as the numerically dominant yellow perch and pumpkinseed, with diverse, abundant prey and protection from predation [1,26]. In embayments supporting a greater proportion of large-bodied fishes (e.g., Floodwood, Blind Sodus, and South Sandy), peak biomass was concentrated in fewer, but larger individuals. Neither embayment area nor piscivore biomass explained maximum biomass weight class, suggesting that both medium- and large-bodied fishes benefited from any deeper habitat associated with larger surface area.

Different distributions of biomass across fish size classes among embayments certainly could have implications for trophic cascades [38,10] and the susceptibility of some of these systems to shift from a macrophyte-to phytoplankton-dominated stable state [39]. For example, an ecosystem in which peak biomass occurs at larger size classes may be primarily structured by top-down effects. Such systems, such as Floodwood, may be less prone to undesirable eutrophication effects (e.g., algal blooms) due to piscivory of planktivorous and benthivorous fishes [10]. In contrast, a system such as Sterling, in which fish biomass is concentrated in smaller-bodied fishes may be more susceptible to eutrophication effects. In fact, zooplankton biovolume is low and phytoplankton biovolume is high in this embayment compared to the others (R. Doyle-Morin, Cornell University, personal communication). Certain fish communities may be an indication of top-down effects, while bottom-up (e.g., nutrient loading) control may play a greater role in other fish communities. In a study of yellow perch growth and size structure in four Lake Ontario embayments greater bottom-up control was observed in shallow embayments, whereas predation may play a more important role in deep, less vegetated embayments [40].

Conclusion

Our study contributes to general understanding of how fish communities respond to physicochemical features both at the watershed and lake levels. Our findings suggest that fish communities are structured by factors operating at multiple spatial scales and on multiple community characteristics. Additionally, the importance of these factors appears to differ with hydrogeomorphology. Therefore, the relative impacts of natural variability and anthropogenic activity on fish communities in shallow, vegetated aquatic ecosystems are likely to differ somewhat from those in large, deep lakes. Influential factors of particular importance are those subject to human modification, such as percent vegetation and total phosphorus loading. For example, as water clarity has improved in the Great Lakes, macrophyte densities have increased to the extent that they are now considered a nuisance to nearshore activities and are being controlled through mechanical harvesting. Shoreline development and modification of connections between embayments and the main lake impact the quality of littoral habitat, integrity of adjacent wetland habitat, and water residence time. Additionally, these changes could alter fish movement into and out of the embayments as well as the quality of spawning habitat, two factors that were not considered in our study. Changing land use, such as the transformation of farmland to forested or urban land will continue to alter water discharge and nutrient and sediment loading. Identifying the actual mechanisms by which morphological and hydrological variables operate is challenging due to the degree to which many of these factors are correlated. However, developing a more explicit understanding of how these factors structure fish communities is important not only for coastal reclamation or restoration efforts along the Great Lakes coastline, but also for anticipating effects of future changes to inland, coastal, and offshore freshwater habitats and fish communities.

Author Contributions

KA contributed to the study's conception and design, conducted the research and data analysis, and wrote the manuscript. MB contributed to the study's conception and design, provided funding and resources for the research, and assisted with manuscript preparation.

Acknowledgements

This research would not have been possible without the assistance of a large number of undergraduate interns, graduate students, and research staff associated with the Lake Ontario Biocomplexity project. In particular, we thank Marci Meixler, Andrea Parmenter, and Gail Steinhart for their work on watershed scale characteristics and embayment morphometry metrics and Xiaoxia Chen and Charley Driscoll for their generosity with their water chemistry data. Françoise Vermeylen provided valuable assistance with the statistical analyses. This research was supported by the National Science Foundation (NSF) under Grant No. 0083625 to MB.

References

1. Randall RG, Minns CK, Cairns VW, Moore JE: The relationship between an index of fish production and submerged aquatic macrophytes and other habitat features at three littoral areas. Canadian Journal of Fisheries and Aquatic Sciences 1996, 53(Suppl. 1): 35–44.

2. Breneman D, Richards C, Lozano S: Environmental influences on benthic community structure in a Great Lakes embayment. J Gt Lakes Res 2000, 26(3):287–304.

3. Höök TO, Eagan NM, Webb PW: Habitat and human influences on larval fish assemblages in northern Lake Huron coastal marsh bays. Wetlands 2001, 21(2):281–291.

4. Rosenzweig ML, Abramsky Z: How are diversity and productivity related? In Species Diversity in Ecological Communities: Historical and Geographical Perspectives. Edited by: Ricklefs RE, Schluter D. Chicago (IL): The University of Chicago Press; 1993:52–65.

5. Bachmann RW, Jones BL, Fox DD, Hoyer M, Bull LA, Canfield DE: Relations between trophic state indicators and fish in Florida (USA) lakes. Canadian Journal of Fisheries and Aquatic Sciences 1996, 53(4):842–855.

6. Holmgren K, Appelberg M: Size structure of benthic freshwater fish communities in relation to environmental gradients. Journal of Fish Biology 2000, 57(5):1312–1330.

7. Polis GA, Anderson WB, Holt RD: Toward an integration of landscape and food web ecology: The dynamics of spatially subsidized food webs. Annu Rev Ecol Syst 1997, 28:289–316.

8. Vanni MJ: Nutrient cycling by animals in freshwater ecosystems. Annu Rev Ecol Syst 2002, 33:341–370.

9. Vanni MJ, Arend KK, Bremigan MT, Bunnell DB, Garvey JE, Gonzalez MJ, Renwick WH, Soranno PA, Stein RA: Linking landscapes and food webs: Effects of omnivorous fish and watersheds on reservoir ecosystems. Bioscience 2005, 55(2):155–167.

10. Carpenter SR, Cole JJ, Hodgson JR, Kitchell JF, Pace ML, Bade D, Cottingham KL, Essington TE, Houser JN, Schindler DE: Trophic cascades, nutrients, and lake productivity: whole-lake experiments. Ecological Monographs 2001, 71(2):163–186.

11. Lougheed VL, Theysmeyer TS, Smith T, Chow-Fraser P: Carp exclusion, food-web interactions, and the restoration of Cootes Paradise Marsh. J Gt Lakes Res 2004, 30(1):44–57.

12. Jude DJ, Pappas J: Fish utilization of Great Lakes coastal wetlands. J Gt Lakes Res 1992, 18:651–672.

13. Uzarski DG, Burton TM, Cooper MJ, Ingram JW, Timmermans STA: Fish habitat use within and across wetland classes in coastal wetlands of the five Great Lakes: Development of a fish-based index of biotic integrity. J Gt Lakes Res 2005, 31(Suppl. 1): 171–187.

14. Brazner JC, Beals EW: Patterns in fish assemblages from coastal wetland and beach habitats in Green Bay, Lake Michigan: a multivariate analysis of abiotic and biotic forcing factors. Canadian Journal of Fisheries and Aquatic Sciences 1997, 54(8):1743–1761.

15. Klumb RA, Rudstam LG, Mills EL, Schneider CP, Sawyko PM: Importance of Lake Ontario embayments and nearshore habitats as nurseries for larval fishes with emphasis on alewife (Alosa pseudoharengus). J Gt Lakes Res 2003, 29(1): 181–198.

16. Mackey SD, Goforth RR: Great Lakes nearshore habitat science – Foreword. J Gt Lakes Res 2005, 31(Suppl. 1): 1–5.

17. Keough JR, Thompson TA, Guntenspergen GR, Wilcox DA: Hydrogeomorphic factors and ecosystem responses in coastal wetlands of the Great Lakes. Wetlands 1999, 19(4):821–834.

18. Hall SR, Pauliukonis NK, Mills EL, Rudstam LG, Schneider CP, Lary SJ, Arrhenius F: A comparison of total phosphorus, chlorophyll a, and zooplankton in embayment, nearshore, and offshore habitats of Lake Ontario. J Gt Lakes Res 2003, 29(1):54–69.

19. Trebitz AS, Morrice JA, Cotter AM: Relative role of lake and tributary in hydrology of Lake Superior coastal wetlands. J Gt Lakes Res 2002, 28(2):212–227.

20. Trebitz AS, Morrice JA, Taylor DL, Anderson RL, West CW, Kelly JR: Hydromorphic determinants of aquatic habitat variability in Lake Superior coastal wetlands. Wetlands 2005, 25(3):505–519.

21. Oglesby RT: Relationships of fish yield to lake phytoplankton standing crop, production, and morphoedaphic factors. Journal of the Fisheries Research Board of Canada 1977, 34:2255–2270.

22. Ney JJ: Oligotrophication and its discontents: effects of reduced nutrient loading on reservoir fisheries. American Fisheries Society Symposium 1996, 16:285–295.

23. Ludsin SA, Kershner MW, Blocksom KA, Knight RL, Stein RA: Life after death in Lake Erie: Nutrient controls drive fish species richness, rehabilitation. Ecological Applications 2001, 11(3):731–746.

24. Eadie JM, Keast A: Resource heterogeneity and fish species-diversity in lakes. Canadian Journal of Zoology-Revue Canadienne De Zoologie 1984, 62(9):1689–1695.

25. Pierce RB, Tomcko CM: Density and biomass of native northern pike populations in relation to basin-scale characteristics of north-central Minnesota lakes. Transactions of the American Fisheries Society 2005, 134(1):231–241.

26. Zambrano L, Perrow MR, Sayer CD, Tomlinson ML, Davidson TA: Relationships between fish feeding guild and trophic structure in English lowland shallow lakes subject to anthropogenic influence: implications for lake restoration. Aquatic Ecology 2006, 40(3):391–405.

27. Olive JA, Miranda LE, Hubbard WD: Centrarchid assemblages in Mississippi state-operated fishing lakes. North American Journal of Fisheries Management 2005, 25(1):7–15.

28. Maidment DR: ArcHydro: GIS for Water Resources. Redlands, CA: ESRI Press; 2002.

29. Cole GA: Textbook of Limnology. 4th edition. Long Grove (IL): Waveland Press, Inc; 1994.

30. ESRI: Arcview. 3.x edition. Redlands, CA: Environmental Systems Research Institute; 2001.

31. Stiling PD: Ecology: Theories and Applications. 3rd edition. Upper Saddle River (NJ): Prentice Hall; 1999.

32. Duplisea DE, Castonguay M: Comparison and utility of different size-based metrics of fish communities for detecting fishery impacts. [http://rparticle. web-p.cisti.nrc.ca/rparticle/AbstractTemplateServlet?calyLang=eng&journal=cj fas&volume=63&year=0&issue=4&msno=f05-261] Canadian Journal of Fisheries and Aquatic Sciences 2006, 63(4):810–820.

33. Kimmel DG, Roman MR, Zhang XS: Spatial and temporal variability in factors affecting mesozooplankton dynamics in Chesapeake Bay: Evidence from biomass size spectra. Limnology and Oceanography 2006, 51(1):131–141.

34. Littell RC, Milliken GA, Stroup WW, Wolfinger RD: SAS System for Mixed Models. Cary (NC): SAS Institute, Inc; 1996.

35. Wagner T, Hayes DB, Bremigan MT: Accounting for multilevel data structures in fisheries data using mixed models. Fisheries 2006, 31(4):180–187.

36. Minns CK, Wichert GA: A framework for defining fish habitat domains in Lake Ontario and its drainage. J Gt Lakes Res 2005, 31 (Suppl. 1): 6–27.

37. Niemi GJ, Kelly JR, Danz NP: Environmental Indicators for the Coastal Region of the North American Great Lakes: Introduction and Prospectus. J Gt Lakes Res 2007, 33(Suppl. 3): 1–12.

38. McQueen DJ, Post JR, Mills EL: Trophic relationships in freshwater pelagic ecosystems. Canadian Journal of Fisheries and Aquatic Sciences 1986, 43:1572–1581.

39. Scheffer M: Multiplicity of stable states in fresh-water systems. Hydrobiologia 1990, 200:475–486.

40. Arend KK: The role of environmental characteristics on fish community structure and food web interactions in Lake Ontario embayments. In Dissertation. Ithaca: Cornell University; 2008.

Chapter 4

Recovery of a U.S. Endangered Fish

Mark B. Bain, Nancy Haley, Douglas L. Peterson,
Kristin K. Arend, Kathy E. Mills and Patrick J. Sullivan

ABSTRACT

Background

More fish have been afforded US Endangered Species Act protection than any other vertebrate taxonomic group, and none has been designated as recovered. Shortnose sturgeon (Acipenser brevirostrum) occupy large rivers and estuaries along the Atlantic coast of North America, and the species has been protected by the US Endangered Species Act since its enactment.

Methodology/Principal Findings

Data on the shortnose sturgeon in the Hudson River (New York to Albany, NY, USA) were obtained from a 1970s population study, a population and fish distribution study we conducted in the late 1990s, and a fish monitoring program during the 1980s and 1990s. Population estimates indicate a late 1990s abundance of about 60,000 fish, dominated by adults. The Hudson

Originally published as Bain MB, Haley N, Peterson DL, Arend KK, Mills KE, Sullivan PJ (2007) Recovery of a US Endangered Fish. PLoS ONE 2(1): e168. https://doi.org/10.1371/journal. pone.0000168. © 2007 Bain et al. https://creativecommons.org/licenses/by/4.0/

River population has increased by more than 400% since the 1970s, appears healthy, and has attributes typical for a long-lived species. Our population estimates exceed the government and scientific population recovery criteria by more than 500%, we found a positive trend in population abundance, and key habitats have remained intact despite heavy human river use.

Conclusions/Significance

Scientists and legislators have called for changes in the US Endangered Species Act, the Act is being debated in the US Congress, and the Act has been characterized as failing to recover species. Recovery of the Hudson River population of shortnose sturgeon suggests the combination of species and habitat protection with patience can yield successful species recovery, even near one of the world's largest human population centers.

Introduction

In the last 100 years, three genera, 27 species, and 13 subspecies of fish have been extirpated from North America [1]. The US government currently lists more fish (101 [2]) as threatened and endangered species than any other vertebrate taxonomic group. A total of 149 [3] species and distinct populations are currently under federal government protection provided by the US Endangered Species Act, and many have been listed for decades. However, none of these fish species or populations have been designated as recovered and delisted in the three decades since passage of the US Endangered Species Act. Five fish species have been removed from the endangered species list: four by extinction and one by taxonomic revision [3]. Independent review of imperiled fishes [4] in North America also concluded that species recovery is lacking. However, data and research findings reported here on the endangered shortnose sturgeon (Acipenser brevirostrum) in the Hudson River of New York indicates this population meets government and scientific criteria for recovery.

The shortnose sturgeon was formally protected with the passage of the 1966 US Endangered Species Preservation Act and later designated as endangered under the current 1973 US Endangered Species Act [5]. The species was considered to be in peril or absent in coastal rivers throughout its range due to overfishing, pollution, and habitat losses from river damming. It is also on the IUCN (International Union for Conservation of Nature and Natural Resources) Red List of Threatened Species [6] because of reduced population size, decline in range and number of locations, and continued decline. Evidence reported here suggests this charter member of the US Endangered Species Act is the first fish to clearly merit designation as a recovered distinct population. The nature of the species, its habitat, and the evidence for a large and secure population is reported as an example of successful protected species management.

The shortnose sturgeon inhabits rivers along the North American Atlantic coast, from the Saint John River, New Brunswick to the St. John's River, Florida. The shortnose sturgeon is best described as an amphidromous [7] species because its use of marine waters is limited to the estuaries of natal rivers [8]. Captures in coastal marine waters and non-natal rivers have occurred but are rare. A long-lived species, shortnose sturgeon maturity is attained in 8 to 10 years and adults may live for 60 years or more [9]. Shortnose sturgeon occupy the lower Hudson River: 246 kilometers of tidal freshwater river and brackish estuary habitats. From late spring through early fall, shortnose sturgeon are dispersed throughout the deep, channel habitats of the freshwater and brackish reaches of this river-estuary [9]. Diet includes insects and crustaceans with mollusks being a major component (25 to 50% of the diet; [10], [11]). In the late fall, most or all adult shortnose sturgeon congregate at a single wintering site near Sturgeon Point (river kilometer, rkm, 139). These fish migrate upstream to spawn in the spring and later disperse throughout much of the estuary.

Hudson River shortnose sturgeon spawn in the spring (late-April to early Mary) downstream of the Troy Dam [9] where the river turbulent and relatively shallow. Eggs adhere to the river bottom, as do the newly hatched larvae [12], [13]. Hatching size ranges from 7 to 11 mm total length (TL; [12], [13]), with Hudson River larvae ranging in size from 15 to 18 mm TL at 10 to 15 days of age [14]. After hatching, larvae gradually disperse downstream over much of the Hudson River Estuary [15]. Larval shortnose sturgeon captured in the Hudson River were associated with deep waters and strong currents [14], [15].

Juvenile shortnose sturgeon (2–55 cm TL), use a large portion of the tidal reach of the Hudson River. Yearling juvenile sturgeon grow rapidly (to 30 cm TL in first year) and disperse downriver to about rkm 55 by fall [16]. Juvenile distribution during the summer centers on the mid-river region [17] and shifts downriver (Haverstraw Bay, rkm 55–63 [16], [17]) for the late fall and winter seasons.

Methods

From the Battery in New York (rkm 0) to the Troy Dam above Albany (rkm 246), the Hudson River (Figure 1) spans a river-estuary gradient providing tidal habitats that include freshwater river channels, a brackish fjord, and a rock confined estuary [18]–[20]. Although largely a glacially scoured channel, the Hudson River estuary varies inversely in width relative to depth; maximum width is 4.8 km (rkm 50) while the maximum depth is 66 m (rkm 81). The U.S. Army Corps of

Engineers maintains a navigation channel depth of 9 to 11 m although much of the channel in much deeper [20]. Mean ebb and flood current velocities are 0.4 m/s and 0.36 m/s, respectively. The normal tidal amplitude ranges from 0.82 to 1.43 m causing a tidal volume (mean 5,670–8,500 m3/s depending on location) from 10 to 100 times river discharge (mean 623 m³/s; [20]). Saltwater intrusion extends from rkm 80 to 100 during the summer months (Figure 1) and varies with river discharge. Generally, the limnetic zone (<0.3l ppt) occurs upriver of rkm 80 (Croton Point). An oligohaline zone (0.3–5 ppt) ranges from rkm 40 to 80 with higher salinity (5–18 ppt, mesohaline) further downstream. Sediment characteristics of the Hudson River channel vary along the estuary from sand (dominant above rkm 164) to silty sand (rkm 164 to 148) to clayey silt (below rkm 148 to 64). Larger shell fragments and sandier sediments comprise a larger percentage of channel sediments below rkm 64. Isolated patches of coarser material (sand, gravel) occur near tributary mouths, within the Hudson Highlands, and near Peekskill.

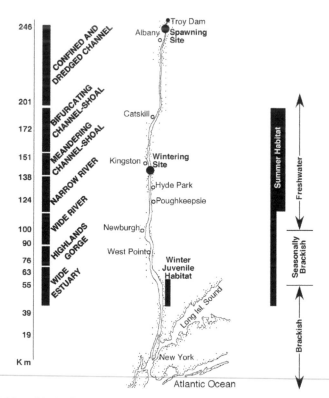

Figure 1. Map of the Hudson River estuary with key habitats used by shortnose sturgeon and the salinity zones in the system. Summer habitat, winter juvenile habitat, and salinity zones match horizontally on the figure with locations in the river. The width of the summer habitat designation corresponds with most and least heavily used sections of the river.

Data on the shortnose sturgeon population in the Hudson River estuary were obtained from a field study we conducted from 1994 to 1997, a shortnose sturgeon population study conducted by William Dovel and others during the 1970s [16], and a standardized fish monitoring program [21], [22] by the Hudson River electric utilities (Central Hudson Gas and Electric Corporation, Consolidated Edison Corporation of New York, New York Power Authority, Niagara Mohawk Power Corporation, and Southern Energy New York). These studies provide a record of the shortnose sturgeon population spanning almost two decades with thorough population estimates made at the beginning and end of the period, and relative abundance data covering many of the intervening years.

Our shortnose sturgeon sampling was completed in two ways: (1) randomly dispersed sampling from June to mid-September (1995 and 1996) throughout the river when the sturgeon were feeding and widely distributed; and (2) targeted sampling of adult sturgeon at their wintering site in December, March, and early April, and their spawning grounds near Albany from mid-April through May (1994 to 1997, Table 1). For both types of sampling we used gill nets (3 m high by 91 m long) with mesh sizes measuring 5-, 10-, and 15-centimeters (stretch mesh). For random sampling, one gill net of each mesh size was anchored and set perpendicular to shore, positioned between mid-channel and the shoreline, parallel to one another and approximately 30 m apart, and deployed in daylight during slack tides (30 to 90 minutes). Targeted gill netting was done in a similar manner but on some occasions a single net was used because catch often exceeded the time available to safely process the fish.

Table 1. Numbers of shortnose sturgeon marked and recaptured in targeted and random sampling during the study.

Year	Season	Location	Type of sampling	Number caught	Recaptures	New marks	Total marks
1994	Spring	Spawning site	Targeted	240	0	240	0
1994	Summer	Estuary-wide	Random	118	0	82	240
1994	Fall	Wintering site	Targeted	424	0	424	322
1995	Spring	Wintering site	Targeted	1024	13	1025	746
1995	Spring	Spawning site	Targeted	783	29	754	1771
1995	Summer	Estuary-wide	Random	180	1	164	2525
1995	Fall	Wintering site	Targeted	664	27	637	2689
1996	Spring	Wintering site	Targeted	808	33	775	3326
1996	Spring	Spawning site	Targeted	294	24	270	4101
1996	Summer	Estuary-wide	Random	194	10	184	4371
1996	Fall	Wintering site	Targeted	916	68	848	4555
1997	Spring	Spawning site	Targeted	620	64	556	5403
Totals				6265	269	5959	

Fish were removed from gill nets and were either processed immediately on the boat or placed in floating mesh pens along side the boat until being processed. Fish were checked for the presence of PIT (passive integrated transponder) tags,

Carlin-Ritchie dangler tags, and Floy tags; PIT tags were applied if one was not present. Fork length (FL) and sometimes total length (TL) were measured to the nearest millimeter and weight measured to the nearest gram. All fish were measured and tagged unless the number of fish caught was so large that processing all of them would take many hours and delay release. At such times, only a subset of the catch was processed, but all were checked for existing tags.

Randomly dispersed sampling occurred between rkm 43 (Tappan Zee Bridge, Nyack, NY) and rkm 246 (Troy Dam) using seven strata based on geomorphological characteristics [18] of the Hudson system. The stratified random sampling design apportioned effort throughout the river. Individual sampling stations (located at river kilometers) were selected using a random numbers table and alternated in orientation to each shore when possible. An equal number of samples were taken in each stratum per month (i.e., June, July, August/September) to ensure equivalent effort throughout the study period.

William Dovel and his associated investigators [16] collected shortnose sturgeon in the Hudson River from 1975 through 1980. Sturgeon were sampled using 6.4 m and 10.7 m otter trawls and drifted, anchored, or staked gill nets of 5.1, 6.4, 7.6, 8.9 cm bar monofilament meshes. Sampling varied among years with trawling occurring between rkm 19 and 246, gill nets between rkm 208 and 246, and some gill net sets below rkm 64. Total or fork lengths were measured to the nearest millimeter and weight was measured to the nearest gram or ounce. Adult and juvenile fish greater than 228 mm TL were marked with Carlin-Ritchie dangler tags attached at the base of the anterior portion of the dorsal fin. Any recaptures were recorded. Sampling in 1979 was conducted from late April through June at the spawning site (rkm 246). For four days each week, two to four drift gill nets were set during slack tide and allowed to drift along the channel bottom for at least 15 minutes [14]. Anchored gill nets were set parallel to shore on both sides of the river in at least six locations each day and allowed to fish overnight. Extensive sampling was conducted between 24 October 1979 and 13 May 1980 at the wintering site near Esopus Meadows (rkm 140; [23]) to capture large numbers of adults.

The standardized fish monitoring program of the Hudson River electric utilities provided annual shortnose sturgeon catch data for years 1985 through 1996. Samples were collected biweekly for 15 weeks from midsummer through fall using a 3.0-m beam trawl. At least three samples were collected in the channel of each of 12 river sampling strata ranging from river rkm 1 through 245 for an annual total of about 1,240 samples. All shortnose sturgeon were recorded with total length in millimeters and weight in grams.

Data analyses were conducted to make comparisons across time and studies, and to provide the best possible population estimates with different data sets.

Total length measurements were converted to fork length using the conversion formulae, FL = 0.90(TL) [24], as this relationship corresponds well with TL and FL measurements from double-measured sturgeon in our data sets. Sturgeon less than 500 mm FL were considered juveniles [9]. Fish body condition was calculated using Fulton's Condition Factor K [25], where K(FL) = (weight • 105)/FL3.

The shortnose sturgeon population was estimated from mark and recapture data using the Schnabel method that assumes a closed population [26]. This closed population method allowed direct comparison of population estimates from our data and those from the study by Dovel et al.. [16], [23]. They also provide precise estimates when assumptions are largely satisfied. Mark and recapture periods were defined by season and location: wintering site in late fall, wintering site in early spring, spawning site in mid to late spring, and summer and early fall dispersed sampling. All marked fish captured in the same sampling period as the period of marking were deleted from the record as recaptures. Multiple recaptures of the same fish were counted as separate recaptures so long as each recapture occurred in separate sampling periods. Comparisons of our estimated population sizes to a population size of 10,000 fish (considered adequate and safe under Endangered Species Act actions for shortnose sturgeon) were made by computing the probability of this observation under our estimated population parameters. A mean and confidence interval for the estimated change in population size between studies was calculated using the distribution of a 1000 randomly selected values from 95% confidence intervals of the population estimates [27].

Closed population estimates assume no significant change in population size occurs during the estimation period due to recruitment, mortality, and movements in or out of the study area. Our study population would not strictly be closed, but shortnose sturgeon are known to be very long-lived fish with low rates of annual mortality and recruitment. Nonetheless, we investigated the potential for bias in our closed population estimates using a series of open population estimates (Jolly-Seber method [26]) and by analyzing the ratio of marked fish in the catch and the known number of marked fish in the estuary through the study period [26]. Finally, we assessed population trend over most of the study period using annual catch rates in the fish monitoring survey of the Hudson River electric utilities.

T-tests were used to test for differences in fish lengths and body condition of sturgeon from our samples and those of Dovel et al.. [16], [23]. Paired t-tests were used to determine if there was a significant increase in mean fish length between a series of individual fish marked in the 1970s and recaptured in the 1990s. Differences in fish condition were calculated only from summer catches to avoid potential biases associated with measures of body weights collected immediately prior to or after the spawning season. The dispersed summer distribution of sturgeon was analyzed with a chi-square frequency analysis (samples with and without sturgeon) against a uniform distribution. The presence versus absence data format was used in this analysis so that sites with multiple captures would not bias results.

Results

We captured 6,265 different shortnose sturgeon and marked 5,959 of these fish. Most (3,836) shortnose sturgeon were captured and marked at the wintering site, high numbers (1,937) were captured and marked at the spawning site, and relatively few (492) sturgeon were handled in the summer random sampling that covered the estuary (Table 1). Recaptures started appearing in the second year of the study (1995) and increased to a total of 269 by the end of our study. Shortnose sturgeon captured during the targeted sampling were adults (Fig. 2), while the summer random sampling captured a broader size range of sturgeon including some juveniles (≤50 cm FL, 4% of total catch).

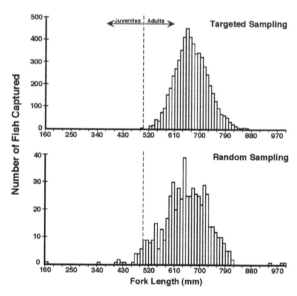

Figure 2. Size distribution of adult shortnose sturgeon captured in targeted sampling in spawning and adult wintering habitats, and the size distribution of shortnose sturgeon captured in random sampling during summer. Shortnose sturgeon greater than 50 cm fork length (FL) were classified as adults. During summer sampling, all life stages of shortnose sturgeon are well distributed in the river system.

A closed population estimate (Schnabel method, [26]) based on nine targeted sampling periods yielded 56,708 adult fish with a narrow 95% confidence interval: 50,862–64,072 (Fig. 3). Using the same methods and algorithm, Dovel et al..

[16], [23] estimated the number of adult shortnose sturgeon at 12,669 and 13,844 (Fig. 3, 95% confidence intervals of 9,080–17,735 and 10,014–19,224 respectively) in 1979 and 1980. The probability of our sturgeon population was within the range (95% interval) of the Dovel et al.. estimates was remote (P<0.001). The population estimates yielded a mean adult sturgeon abundance increase of 407% (95% confidence interval of 290 to 580%) from the late 1970s to the 1990s. Also, the probability that the Hudson River shortnose sturgeon population was 10,000 or fewer fish is highly unlikely (P<0.001) indicating the population was clearly larger than the size considered adequate in Endangered Species Act rulings.

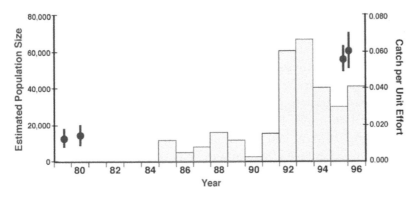

Figure 3. Population estimates and abundance trend for Hudson River shortnose sturgeon in the 1980s and 1990s. The paired symbols of circles (means) and heavy lines (95% confidence intervals) show the results of population estimates in the late 1970s and late 1990s. The catch per unit effort histogram bars are the average catch of shortnose sturgeon per trawl haul in a riverwide fish survey conducted annually by the Hudson River electric utilities.

A second closed population estimate was computed using all 12 sampling periods resulting in an estimate of 61,057 shortnose sturgeon with a narrow 95% confidence interval: 52,898–72,191 (Fig. 3). This estimate is larger than the corresponding 9-period estimate, includes juveniles and possibly adults not using the wintering and spawning sites, and is our best estimate of the whole shortnose sturgeon population of the Hudson River. The addition of juvenile and possible non-spawning adult sturgeon in the population was minor (ca. 7% of the overall estimate) indicating that all or nearly all adult shortnose sturgeon are present annually at the overwintering and spawning sites. Also, the summer sampling included some juveniles (4% of total catch) which could account for much of the difference in the 9 and 12 period closed population estimates (Fig. 2).

Analyses addressing the closed population assumption support our population estimates. A regression of the number of marked fish in our targeted sampling catches and the known number of marked fish in the river was linear (R2 = 0.96) indicating minimal effect of changing population size during the study. The relation was also linear (R2 = 0.84) but less precise when all sampling periods were included. A series of six open population estimates (Jolly-Seber method) varied in results as expected for this method [26] with initial and ending estimates in the series showing high variance. A mid-series set of three estimates had consistent results: population sizes centered on 59,545 with modest variation (coefficient of variation 27 to 30%). Findings using the open population estimates were not different than those using the closed estimates: probabilities of the sturgeon population being 10,000 or fewer fish was remote (P<0.003) and unlikely (P<0.05) to be within the range of the Dovel et al.. estimates.

Shortnose sturgeon captured in the 1970s and in our 1990s sampling were very similar in size composition with a slight (mean FL 655 and 665 mm, respectively) but significant (t-test, P≤0.0001) increase in average size. A more equivalent comparison of shortnose sturgeon was made by comparing only those fish captured and measured at the wintering site. In the 1970s, 1,220 captured shortnose sturgeon had a mean fork length of 645 mm while 4,310 sturgeon recorded at the same location in the 1990s had a mean fork length of 663 mm. Again, there was a slight (18 mm) and significant (t-test, P≤0.0001) difference among these large groups of sturgeon. Measures of body condition (Fulton K, [25]) for shortnose sturgeon captured during summers in the 1970s (Mean = 0.845, 95% CI = 0.813–0.877, 13) and 1990s (Mean = 0.835, 95% CI = 0.826–0.845) were similar and are comparable with other populations [24].

Some (37) shortnose sturgeon marked in 1979 and 1980 during the study by Dovel et al.. [16] were recovered in our sampling in 1996 and 1997. The fork lengths of these 37 fish after 17 or 18 years in the river indicated very little growth on average (mean increase in FL = 28 mm, P = 0.038). Of these 37 fish, four were juveniles at the time of capture and all of these fish grew (mean increase of 178 mm). There was no increase in length (P = 0.8243) for the 33 sturgeon that were adults when initially measured and marked in the 1970s. Overall, there was very little growth found in fish recovered after 17 to 18 years except for some individuals that were small when initially caught.

From 1985 through 1996, the Hudson River electric utilities conducted an annual trawl survey typically composed of about 1,240 (range 1185–1549) highly standardized samples per year. These data show (Fig. 3) a clear increase in abundance of shortnose sturgeon during the period. Catch ranged from a low of 2 shortnose sturgeon in 1990 to a maximum catch of 82 sturgeon in 1993. The increase in average catch rate was more than four fold higher in the second half of the survey period. The trawl samples captured almost exclusively adult sturgeon with an average total length about 670 mm across years.

Shortnose sturgeon captured during randomly dispersed summer sampling (166 stations, 498 net sets) were distributed non-randomly ($X2 = 16.87$, $P<0.01$) among seven distinct river strata (Fig. 1). Shortnose sturgeon were most frequently captured (63% of catch, present in 71% of samples) in the middle section of the estuary (Fig. 1, 3 strata from rkm 108 to 189) and were well represented (35% of catch, 51% of samples) in habitats downstream to persistently brackish waters (3 strata from rkm 43 to 107). The primary summer habitat for Hudson River shortnose sturgeon is a deep (regularly 13 to 42 m) tidal freshwater river channel. Downstream the estuary becomes brackish, deeper (regularly 18 to 48 m), and variable in width. The summer distribution of shortnose sturgeon in the Hudson River estuary combined with the wintering and spawning location forms a complete record of major habitats supporting almost all of the population.

Discussion

Our different population estimates made under varying assumptions indicate a late 1990's shortnose sturgeon population in the Hudson River estuary of about 60,000 fish with adults comprising a very large portion (>90%) of the population. Compared to population estimates in the late 1970s, we conclude the Hudson River population has increased by more than 400%. Independent data from the Hudson River electric utilities annual trawl survey also indicate more than a four fold increase in abundance and again mainly in the adult segment of the population. For the species overall, the Hudson River population is very large and dominant to all others. The number of sturgeon marked during this study exceeds the estimated size of most other populations of shortnose sturgeon [5], and our population estimates are larger than the sum of all other estimated populations. Therefore, it is safe to conclude that Hudson River supports by far the largest population of shortnose sturgeon, and the system may harbor most individuals of the species.

While we assembled multiple lines of evidence supporting a large population increase over two decades, other findings suggest the population of shortnose sturgeon in the estuary is healthy. Shortnose sturgeon captured in the 1970s and in our 1990s sampling were very similar in size composition with a slight increase in average size. Measures of body condition for shortnose sturgeon captured during summers in the 1970s and 1990s were similar and are comparable with other populations [24]. A surprising number of adult sturgeon tagged in the 1970s were recaptured in our 1990s sampling, suggesting that many individual fish have lived for decades in the estuary without growing a measurable amount. These findings depict a population of long-lived fish that has increased in number over decades reaching a high abundance for the species.

Most shortnose sturgeon captured in the Hudson River estuary in research and monitoring programs have been adults ([17], [24], Utilities data set, and this study) regardless of sampling gear and time period. Shortnose sturgeon reach maturity at age-6 or younger with an adult lifespan of several decades [9]. Few unexploited populations of long-lived and large fish have been studied. Some fish populations like this were found to be composed overwhelmingly of slow growing, long-lived adults displaying a normal-shaped size distribution as in Figure 2 [28]. Few young are found in such populations and juveniles slowly add to the adult group, maintaining a very consistent population size structure. Hence, the Hudson River population of shortnose sturgeon displays the characteristics of an unexploited, long-lived fish population.

The availability and security of habitat is an important consideration in US Endangered Species Act decisions. The spawning and wintering habitats of shortnose sturgeon have been well known since the late 1800s when an intense sturgeon fishery operated in the estuary. The juvenile wintering habitat has been described [16], but the spatial extent of summer sturgeon habitat had not been documented. The sections of the Hudson River primarily used by shortnose sturgeon have remained physically intact with shoreline land use established early in the last century. Many historic residential structures and estates are located along the Hudson River, and very limited portions of the waterfront have been used for industrial uses. The spawning site for shortnose sturgeon is removed from the other habitats, because it is centered on turbulent river habitat between the head of tide and the Troy Dam. This section of the Hudson River is surrounded by urban areas and it is immediately upstream of a river section modified to accommodate a port facility. Nevertheless, the spawning site appears to be supporting adequate spawning in its current modified condition.

Section 7(a)(2) of the Endangered Species Act requires Federal agencies to ensure that actions they authorize, fund, or carry out do not jeopardize the continued existence of an endangered species or result in the destruction or adverse modification of critical habitat. The National Oceanic and Atmospheric Administration (NOAA), National Marine Fisheries Service is the responsible federal agency for planning recovery and implementing protection measures for shortnose sturgeon. Since 2000, the NOAA Fisheries Service has reviewed more than 50 proposed actions (e. g., dredging, shoreline stabilization and docks, pollution discharge permits, [29]) potentially affecting shortnose sturgeon in the Hudson River, often specifying protection measures (e.g., construction timing, design changes, local water quality standards). Shortnose sturgeon have also benefited from a cessation of fishing and other harm to individuals by capture, handling, and disturbance. Overall, the approach to recovery of shortnose sturgeon in the Hudson River has been to minimize interference with natural population processes and maintain habitat conditions able to support the species. This protect-and-wait approach to population recovery is in contrast to strategies employed for other species using hatchery-reared fish to actively promote population increases.

The US Endangered Species Act recognizes for listing and delisting populations that are discrete from other populations, and significant in relation to the entire species [distinct populations, 5]. Endangered species recovery plans specify the criteria to remove a species or a distinct population from the list of threatened and endangered species [30] making them key documents defining recovery [31]. The shortnose sturgeon plan [5] names 19 distinct populations and specifies three recovery criteria: adequate size with a favorable trend in abundance; habitat sufficient to support a recovered population; and potential causes of mortality insufficient to reduce the population. A shortnose sturgeon population composed of 10,000 spawning adults has been considered large enough to be at a low risk of extinction by the NOAA [32] and adequate for delisting under the US Endangered Species Act [32], [33]. This population threshold was based on analyses of minimum viable adult population sizes of vertebrates [34] applied to fish [35]. Population viability analysis was found to be an effective and realistic tool for endangered species protection in an analysis of 21 long-term population studies [36]. Other minimum population analyses have identified abundances less than the NOAA criteria for shortnose sturgeon [30], [37]–[42]. Following the criteria used by the NOAA for shortnose sturgeon, our total and spawning population estimates exceed the safe level by a wide margin (≥500%), clearly indicating recovery of this shortnose sturgeon population.

Aside from population size, estuary fish monitoring and the population estimates we report over two decades indicate a positive trend in population abundance. Shortnose sturgeon habitats in the Hudson River have supported the growing and now large population, and both the specific spawning and wintering areas and the widely dispersed growing season habitats have remained intact. No major changes are expected in the tidal portion of the Hudson River that would greatly alter or eliminate deep channel waters or the turbulent spawning reach. Finally, likely future causes of high mortality such as unregulated harvest, bycatch in active fisheries, and pollution stress have been and can be controlled through established fishery management and water quality regulations. By all three criteria specified in the shortnose sturgeon recovery plan, we believe the Hudson River estuary population merits designation as 'recovered' and qualifies for delisting from the US Endangered Species Act protection.

The NOAA Fisheries Service periodically reports on the status of shortnose sturgeon throughout their range [5], [43]–[45] using the latest information from field studies. A complex three-river estuary in Maine (Sheepscot, Kennebec, and Androscoggin Rivers) has had increasing numbers (7,222 fish in 1981 to 9,488 in 2000) of shortnose sturgeon recently approaching the safe population size, although there appears to be two distinct spawning populations contributing to the total numbers [46]. Substantial and stable populations occur in the Delaware River (6,408–14,080 in 1981–1984, near 10,000 in 2002, and 8,445 in 2004) and the Saint John River, New Brunswick (18,000 in 1970s). The Connecticut River appears to have a small (<150 fish) stable population isolated above the Holyoke Dam, and an increasing (895 in 1993, 1,800 in 2003[47]) population in the lower river. The Savannah River (South Carolina and Georgia) was stocked with 97,000 shortnose sturgeon between 1984 and 1992 but the most recent population estimate is modest (3,000 in 1999). The large Altamaha River of Georgia supports a modest population (798 in 1990, 468 in 1993, as many as 2,000 in 2004) of shortnose sturgeon. Another 12 mostly small Atlantic coast rivers have some evidence of shortnose sturgeon presence in low numbers (ca.<100) with increasingly frequent captures after decades of no records. Notable is the near lack (18 fish captured since 1996) of shortnose sturgeon in the largest Chesapeake Bay rivers (James, Potomac, and Susquehanna Rivers) although these rivers have dams and obstructions on or close to the tidal zone. What may make the Hudson River unique for shortnose sturgeon is the large area of tidal freshwater habitat used as the summer foraging range: the most commonly occupied 81 km of the tidal freshwater Hudson River. Other rivers with large summer habitat have sizable and near safe level populations (Maine rivers, Delaware River, Saint John River) except in the large southern rivers (Savannah, Altamaha Rivers) where mortality in river gillnet fisheries for shad (Alosa spp) is believed a critical impediment [5], [8], [45]. Overall, shortnose sturgeon in the Hudson River and across the species range suggest that slowly increasing populations could reach recovered status where they are managed under full protection in substantial foraging habitat.

Calls to change the US Endangered Species Act have come from scientists and legislators for more than a decade [48]–[50], and changes to this law are being debated in the US Congress [51], [52]. The Act has been characterized as failing to recover species [50], [52], [53], promote effective recovery programs [54]–[56], or properly assess species endangerment [57], [58]. One commonly reported flaw in government species recovery plans is that not enough is being done to increase population size and viability. Foin et al.. [58] predict that most (63%) endangered species will not reach recovery criteria through habitat protection alone, and that more active management such as habitat restoration and population augmentation will be needed. Despite the multitude of anthropogenic influences on the Hudson River ecosystem, the shortnose sturgeon population appears to have achieved recovery and may merit removal from the list of threatened and endangered species. Other rivers with shortnose sturgeon appear to be slowly developing larger populations or have impediments that can be addressed with more determined species protection measures. Extension of a protect-and-wait conservation strategy seems viable for recovering shortnose sturgeon populations in the largest un-dammed rivers scattered along the Atlantic Coast.

Another assessment [59] of the Endangered Species Act concludes it is working more often than recognized because of poor reporting on the status and trends of endangered species populations. Few data have been collected following recovery efforts [31], [60], [61] making recovery and species management success difficult to recognize. The population status and trend of shortnose sturgeon in the Hudson River estuary had not been well documented prior to this study. The status of other shortnose sturgeon populations has been widely scattered through time and lacking for about half of the rivers suspected of harboring shortnose sturgeon [5]. More thorough and encompassing assessments of species status and trends could reveal additional recovery successes over time. Such findings provided evidence and optimism that public efforts for endangered species conservation can work. Our analysis of the shortnose sturgeon population in the Hudson River provides the first well documented case that fish species and habitat protection, combined with patience, can result in endangered species recovery; even in a human dominated ecosystem associated with one of the World's largest and most prominent cities.

Acknowledgements

We thank William Dovel for providing data and study details from his 1970s studies, and the Consolidated Edison Corporation of New York for river fishery monitoring data. We also thank David Lusseau for computing the population growth values. Permits allowing our field research were granted by the Cornell University Institutional Animal Care and Use Committee, the NOAA National Marine Fisheries Service, and the New York Department of Environmental Conservation.

Author Contributions

Conceived and designed the experiments: MB NH DP. Performed the experiments: MB NH DP KA. Analyzed the data: MB KM KA. Contributed reagents/materials/analysis tools: KM PS. Wrote the paper: MB. Other: Arranged funding: MB.

References

1. Miller RR, Williams JD, Williams JE (1989) Extinctions of North American Fishes during the past century. Fisheries 14(6): 22–38.

2. U. S. Fish and Wildlife Service (2002) Report to Congress on the recovery and threatened and endangered species. Arlington (Virginia): U. S. Fish and Wildlife Service.

3. U. S. Fish and Wildlife Service (2006) Threatened and endangered species system (TESS). Arlington (Virginia): U. S. Fish and Wildlife Service. Available: http://ecos.fws.gov/tess_public/StartTESS.do. Accessed 2006 Sept 9.

4. Williams JE, Johnson JE, Hendrickson DA, Contreras-Balderas S, Williams JD, et al.. (1989) Fishes of North America – endangered, threatened, or of special concern. Fisheries 14(6): 2–20.

5. NOAA National Marine Fisheries Service (1998) Final recovery plan for the shortnose sturgeon Acipenser brevirostrum. Silver Springs (Maryland): National Oceanic and Atmospheric Administration.

6. Friedland KD, Kynard B (2004) Acipenser brevirostrum. 2006 IUCN red list of threatened species. Available: http://www.iucnredlist.org. Accessed 2006 Sept 3.

7. McDowall RM (1987) The occurrence and distribution of diadromy among fishes. Am Fish Soc Symp 1: 1–13.

8. Kynard B (1997) Life history, latitudinal patterns, and status of shortnose sturgeon, Acipenser brevirostrum. Environ Biol Fish 48: 319–334.

9. Bain MB (1997) Atlantic and shortnose sturgeons of the Hudson River: common and divergent life history attributes. Environ Biol Fish 48: 347–358.

10. Curran HW, Ries DT (1937) Fisheries investigations in the lower Hudson River. A biological survey of the lower Hudson watershed. Albany (New York): Supplement to the 26th Annual Report of the New York State Conservation Department. pp. 124–145.

11. Townes AK, Jr (1937) Fisheries investigations in the lower Hudson River. A biological survey of the lower Hudson watershed. Albany (New York): Supplement to the 26th Annual Report of the New York State Conservation Department. pp. 217–230.

12. Buckley J, Kynard B (1981) Spawning and rearing of shortnose sturgeon from the Connecticut River. Prog Fish-Cult 43: 74–76.

13. Taubert BD (1980) Reproduction of shortnose sturgeon, Acipenser Brevirostrum, in the Holyoke Pool, Connecticut River, Massachusetts. Copeia 1980: 114–117.

14. Pekovitch AW (1979) Distribution and some life history aspects of the shortnose sturgeon (Acipenser brevirostrum). Northbrook (Illinois): Hazleton Environmental Sciences Corp.

15. Hoff TB, Klauda RJ, Young JR (1988) Contribution to the biology of shortnose sturgeon in the Hudson River estuary. In: Smith CL, editor. Fisheries research in the Hudson River. Albany (New York): State University of New York Press. pp. 171–189.

16. Dovel WL, Pekovitch AW, Berggren TJ (1992) Biology of the shortnose sturgeon (Acipenser brevirostrum Leseur, 1818) in the Hudson River estuary, New York. In: Smith CL, editor. Estuarine research in the 1980s. Albany (New York): State University of New York Press. pp. 187–216.

17. Geoghegan P, Mattson MT, Keppel RG (1992) Distribution of the shortnose sturgeon in the Hudson River estuary, 1984–1988. In: Smith CL, editor. Estuarine research in the 1980s. Albany (New York): State University of New York Press. pp. 217–277.

18. Coch NK, Bokuniewicz HJ (1986) Oceanographic and geologic framework of the Hudson system. Northeastern Geol 8: 96–108.

19. Cooper JC, Cantelmo FR, Newton CE (1988) Overview of the Hudson River estuary. Amer Fish Soc Monogr 4: 11–24.

20. Limburg KE, Levin SA, Brandt RE (1989) Perspectives on management of the Hudson River ecosystem. In: Dodge DP, editor. Proceedings of the international large river symposium. Ottawa (Ontario): Can Special Publ Fish Aqu Sci 106. pp. 265–291.

21. Klauda RJ, Muessig PH, Matousek JA (1988) Fisheries data sets compiled by utility-sponsored research in the Hudson River Estuary. In: Smith CL, editor. Fisheries research in the Hudson River. Albany (New York): State University of New York Press. pp. 7–85.

22. Applied Science Associates (1999) 1996 Year Class Report of the Hudson River Estuary monitoring program. Poughkeepsie (New York): Annual report to the Central Hudson Gas and Electric Corporation.

23. Dovel WL (1981) The endangered shortnose sturgeon of the Hudson estuary: Its life history and vulnerability to the activities of man. San Francisco (California): Report by the Oceanic Society to the US Federal Energy Regulatory Commission.

24. Dadswell MJ, Taubert BD, Squiers TS, Marchette D, Buckley J (1984) Synopsis of biological data on shortnose sturgeon, Acipenser brevirostrum Le Sueur 1818. Silver Springs (Maryland): National Oceanic and Atmospheric Administration, Technical Report NMFS 14.

25. Ricker WE (1975) Computation and interpretation of biological statistics of fish populations. Ottawa (Ontario): Bull Fish Res Bd Can 191.

26. Krebs CJ (1989) Ecological methodology. New York (New York): Harper Collins Publishers.

27. Manly BFJ (1997) Randomization, bootstrap, and monte carlo methods in biology. London: Chapman & Hall.

28. Johnson L (1976) Ecology of arctic populations of lake trout, Salvelinus namaycush, lake whitefish, Coregonus clupeaformis, arctic char, s. alpinus, and associated species in unexploited lakes of the Canadian Northwest Territories. J Fish Res Bd Can 33: 2459–2488.

29. Personal communication (MBB). November 2006: Julie Crocker, NOAA National Marine Fisheries Service, Northeast Regional Office, Gloucester, MA.

30. Gerber LR, DeMaster DP (1999) A quantitative approach to Endangered Species Act classification of long-lived vertebrates: application to the North Pacific humpback whale. Cons Biol 13: 1203–1214.

31. Gerber LR, Hatch LT (2002) Are we recovering? An evaluation of recovery criteria under the U. S. Endangered Species Act. Ecol Appl 12: 668–673.

32. National Oceanic and Atmospheric Administration (1996) Listing endangered and threatened species; shortnose sturgeon in the Androscoggin and Kennebec Rivers, Maine. Federal Register 61(201): 53893–53896.

33. NOAA National Marine Fisheries Service (1996) Status review of shortnose sturgeon in the Androscoggin and Kennebec Rivers. Gloucester (Massachusetts): Northeast Regional Office, National Marine Fisheries Service.

34. Thomas CD (1990) What do real population dynamics tell us about minimum viable population sizes? Cons Bio 4: 324–327.

35. Thompson GG (1991) Determining minimum viable populations under the Endangered Species Act. Seattle (Washington): National Oceanic and Atmospheric Administration Technical Memorandum NMFS F/NWC-198.

36. Brook BW, O'Grady JJ, Chapman AP, Burgman MA, Akçakaya, et al.. (2000) Predictive accuracy of population viability analysis in conservation biology. Nature 404: 385–387.

37. Franklin IR (1980) Evolutionary change in small populations. In: Soulé MA, Wilcox BA, editors. Conservation biology: an evolutionary-ecological perspective. Sunderland, (Massachusetts): Sinauer Associates.

38. Mace GM, Lande R (1991) Assessing extinction threats: toward a reevaluation of IUCN threatened species categories. Cons Biol 5: 148–157.

39. Wilcove DS, McMillan M, Winston KC (1993) What exactly is an endangered species? An analysis of the Endangered Species List, 1985–1991. Cons Biol 7: 87–93.

40. Lande R (1994) Mutation and conservation. Cons Biol 9: 782–791.

41. Ralls K, DeMaster DP, Estes JA (1996) Developing a criterion for delisting the southern sea otter under the U. S. Endangered Species Act. Cons Biol 10: 1528–1537.

42. Shelden KEW, DeMaster DP, Rugh DJ, Olson AM (2001) Developing classification criteria under the U. S. Endangered Species Act: Bowhead whales as a case study. Cons Biol 15: 1300–1307.

43. National Marine Fisheries Service (1987) Status review of shortnose sturgeon (Acipenser brevirostrum LeSueur 1818). Silver Springs (MD): National Oceanic and Atmospheric Administration.

44. National Marine Fisheries Service (2002) Biennial report to Congress on the recovery program for threatened and endangered species. Silver Springs (MD): National Oceanic and Atmospheric Administration.

45. National Marine Fisheries Service (2004) Biennial report to Congress on the recovery program for threatened and endangered species. Silver Springs (MD): National Oceanic and Atmospheric Administration.

46. Walsh M, Bain M, Squiers T Jr, Waldman JR, Wirgin I (2001) Morphological and genetic variation among shortnose sturgeon Acipenser brevirostrum from adjacent and distant rivers. Estuaries 24: 41–48.

47. Connecticut Department of Environmental Protection (2003) Working with nature: shortnose sturgeon. Hartford: Connecticut Department of Environmental Protection. Available: http://dep.state.ct.us/whatshap/press/2003/mf0730.htm. Accessed 2006 December 15.

48. Gibbons A (1992) Mission impossible: saving all endangered species. Science 256: 1386.

49. National Research Council (1995) Science and the Endangered Species Act. Washington (DC): National Academy Press.

50. Scott JM, Goble DD, Weins JA, Wilcove DS, Bean M, et al.. (2005) Recovery of imperiled species under the Endangered Species Act: the need for a new approach. Frontiers Ecol Environ 3: 383–389.

51. Stokstad E (2005) What's wrong with the Endangered Species Act? Science 309: 2150–2152.

52. Bean MJ (2006) The Endangered Species Act under threat. BioScience 56: 98.

53. Tear TH, Scott JM, Hayward PH, Griffith B (1993) Status and prospects for success of the Endangered Species Act: a look at recovery plans. Science 262: 976–977.

54. Carroll R, Augspurger C, Dobson A, Franklin J, Orians G, et al.. (1996) Strengthening the use of science in achieving the goals of the Endangered Species Act: an assessment by the Ecological Society of America. Ecol Appl 6: 1–11.

55. Hoekstra JM, Clark JA, Fagan WF, Boersma PD (2002) A comprehensive review of endangered species act recovery plans. Ecol Appl 12: 630–640.

56. Doremus H, Pagel JE (2001) Why listing may be forever: perspectives on delisting under the U. S. Endangered Species Act. Cons Biol 15: 1258–1268.

57. Christy CA, Power AG, Hunter A (2002) Evaluating the internal consistency of recovery plans for federally endangered species. Ecol Appl 12: 648–654.

58. Foin TC, Seth PD, Pawley AL, Ayres DR, Carlsen TM, et al.. (1998) Improving recovery planning for threatened and endangered species. BioScience 48: 177–184.

59. Male TD, Bean MJ (2005) Measuring progress in US endangered species conservation. Ecol Let 8: 986–992.

60. Campbell SP, Clark JA, Crampton LH, Guerry AD, Parviez LT, et al.. (2002) An assessment of monitoring efforts in endangered species recovery plans. Ecol Appl 12: 674–681.

61. Tear TH, Scott JM, Hayward PH, Griffith B (1995) Recovery plans and the Endangered Species Act: are criticisms supported by data? Cons Biol 9: 182–195.

Chapter 5

Management Effectiveness of the World's Marine Fisheries

Camilo Mora, Ransom A. Myers, Marta Coll,
Simone Libralato, Tony J. Pitcher, Rashid U. Sumaila,
Dirk Zeller, Reg Watson, Kevin J. Gaston and Boris Worm

ABSTRACT

Ongoing declines in production of the world's fisheries may have serious ecological and socioeconomic consequences. As a result, a number of international efforts have sought to improve management and prevent overexploitation, while helping to maintain biodiversity and a sustainable food supply. Although these initiatives have received broad acceptance, the extent to which corrective measures have been implemented and are effective remains largely unknown. We used a survey approach, validated with empirical data, and enquiries to over 13,000 fisheries experts (of which 1,188 responded) to assess the current effectiveness of fisheries management regimes worldwide; for each of those regimes, we also calculated the probable sustainability of reported catches to determine how management affects fisheries sustainability. Our survey shows that 7% of all coastal states undergo rigorous scientific

Originally published as Mora C, Myers RA, Coll M, Libralato S, Pitcher TJ, Sumaila RU, et al. (2009) Management Effectiveness of the World's Marine Fisheries. PLoS Biol 7(6): e1000131. https://doi.org/10.1371/journal.pbio.1000131. © 2009 Mora et al. https://creativecommons.org/licenses/by/4.0/

assessment for the generation of management policies, 1.4% also have a participatory and transparent processes to convert scientific recommendations into policy, and 0.95% also provide for robust mechanisms to ensure the compliance with regulations; none is also free of the effects of excess fishing capacity, subsidies, or access to foreign fishing. A comparison of fisheries management attributes with the sustainability of reported fisheries catches indicated that the conversion of scientific advice into policy, through a participatory and transparent process, is at the core of achieving fisheries sustainability, regardless of other attributes of the fisheries. Our results illustrate the great vulnerability of the world's fisheries and the urgent need to meet well-identified guidelines for sustainable management; they also provide a baseline against which future changes can be quantified.

Introduction

Fisheries play an important role in the global provision of food, directly accounting for at least 15% of the animal protein consumed by humans and indirectly supporting food production by aquaculture and livestock industries [1],[2]. Demand for fish is expected to grow given escalating animal protein demands in developing countries and the rapidly increasing human population [1]–[4]. However, reported global marine fisheries landings have declined by about 0.7 million tonnes per year since the late 1980s [5], with at least 28% of the world's fish stocks overexploited or depleted, and 52% fully exploited by 2008 [1]. Severe reductions in abundance can change population genetic structure [6], harm the recovery potential of stocks [7], trigger broader ecosystem changes (e.g., [8]–[10]), threaten livelihoods [1], and endanger food security [11] and efforts towards the reduction of hunger [11],[12]. Given the different ecological and socioeconomic consequences of a global fisheries crisis, a number of international efforts have sought to improve management in the hope of moving towards sustainable marine fisheries (sensu Pauly et al.. [13]). Some of these initiatives, which incorporated to varying degrees the improvement of marine fisheries management, include the United Nations Code of Conduct for Responsible Fisheries from the Food and Agriculture Organization [14], the Convention on Biological Diversity (http://www.cbd.int/), and the Millennium Ecosystem Assessment (http://www.millenniumassessment.org). Although these initiatives have received broad acceptance, the extent to which corrective measures are implemented and effective remains poorly known [15]–[17]. Using a survey approach, validated with empirical data and enquiries to fisheries experts, we quantified the status of fisheries management in each nation worldwide that has an exclusive economic zone (EEZ). We also related our measurements of management effectiveness to a recently developed index of fisheries sustainability. To our knowledge, these results represent the first global assessment of how fisheries management attributes influence sustainability, while providing a baseline against which future changes can be quantified.

Results and Discussion

Approach and Validation

We evaluated the effectiveness of national fisheries management regimes by quantifying their degree of compliance with a well-recognized set of conditions necessary for sustainable fisheries: (1) robust scientific basis for management recommendations, (2) transparency in turning recommendations into policy, (3) capacity to enforce and ensure compliance with regulations, and minimizing the extent of (4) subsidies, (5) fishing overcapacity, and (6) foreign fishing in the form of fisheries agreements [8],[14]. The extent to which individual countries met or were affected by these conditions was quantified using a set of normative questions assembled in an Internet survey, which was systematically distributed to fisheries experts worldwide. Over 13,000 experts were contacted as part of this survey, of which 1,188 responded from each country bordering the ocean (i.e., EEZ; see Materials and Methods for additional details on areas surveyed). Experts were mostly fisheries managers, university professors, and governmental and nongovernmental researchers. Despite these diverse backgrounds, responses were highly consistent within each country (i.e., where multiple responses were given, 67% of experts chose the same answer to any given question and 27% chose the next closest response; Figure 1A and 1B) and in accordance with independent empirical data (we found a strong correlation between experts' opinions and empirical data [$r = 0.74$, $p<0.00001$, $n = 28$ countries; Figure 1C]). Justification, extended results, and discussion on the reliability and validity of the experts' data are presented in Materials and Methods. We also used a Monte Carlo simulation approach to include score uncertainty estimates in the results. We provide the main results and general conclusions in the text.

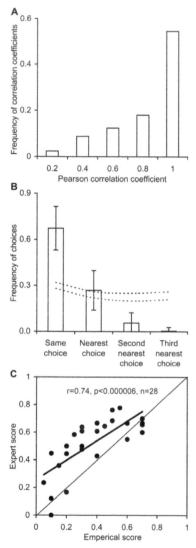

Figure 1. Reliability and validity of the expert's answers. Validity refers to the degree to which the responders' answers approach the truth. Reliability refers to the extent to which different experts agreed in their answers. (A) Using countries for which duplicated responses were obtained, we show the frequency distribution of the Pearson correlation coefficients contrasting each responder to other responders in the same country. (B) depicts the frequency with which responders chose the same score or the next closest choice. Dotted lines in the plot indicate the confidence limits of a null model in which the levels of agreement were measured when choices were made randomly. The confidence limits are based on 1,000 repetitions of this null model. The error bars indicate standard deviation. (C) Using empirical data collected by another study [15], we show the similarities between our expert-based score and an empirically based score for a particular question (see Materials and Methods). The diagonal line indicates the 1:1 ratio.

Scientific Robustness

Critical to the success of fisheries management is the scientific basis on which management recommendations are made [18],[19]. Preventing the collapse of fisheries and ecosystem-wide impacts requires scientific advice in which uncertainty is minimized by using skilled personnel, models that include, not only the dynamics of fished stocks, but also their embedded ecosystems, and high-quality and up-to-date data (such that reliable recommendations can be adapted as conditions and stocks fluctuate). Alternatively, the effects of uncertainty can be minimized by applying precautionary approaches in the face of limited knowledge [18],[20]. Of the world's 209 EEZs analyzed, 87% have scientific personnel who are qualified (e.g., with Ph.D.- or Masters-level education, or have participated in training courses or relevant conferences) to perform fisheries assessments and provide science-based management advice, approximately 7% use holistic models as the basis of management recommendations (i.e., including a broad set of biological and environmental data on fisheries to enable ecosystem-wide understanding of fisheries drivers and impacts), 61% carry out frequent assessments to ensure the effectiveness of existing management measures, and 17% implement precautionary approaches for at least some species. We summarized all responses that pertain to "scientific robustness" on a linear scale using multidimensional scaling. (Multidimensional scaling is an ordination method that uses the similarities and dissimilarities among responses to reduce the number of variables analyzed. This facilitates the assessment and visualization of patterns from several dimensions into one. Very simplistically, this is analogous to calculating an average of the different scores for each country; see Materials and Methods.) The resulting scale ranged from 0 to 1, and we divided it into four quarters (i.e., from 0 to 0.25, from 0.25 to 0.5, from 0.5 to 0.75, and from 0.75 to 1, with the lowest quarter indicating the worst combination of attributes and the top the best). We found that 7% of all EEZs rank in the top quarter of such a scale (Figure 2, countries depicted in Figure 3A), which account for approximately 9% of the world's fisheries catches and approximately 7% of the world's fished stocks (data are for 2004). Distinguishing between high- and low-income countries using per capita Gross Domestic Product (i.e., 2007 per capita Gross Domestic Product larger or smaller than US $10,000, respectively), we found that high-income countries ranked significantly higher on the scale of scientific robustness (Mann-Whitney U test: $p < 0.00001$).

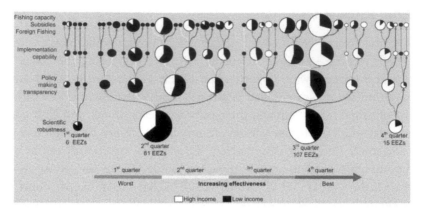

Figure 2. Discrimination of the world's exclusive economic zones (EEZs) according to their management effectiveness. Effectiveness is defined in terms of scientific robustness, policymaking transparency, implementation capability, and extent of fishing capacity, subsidies, and access to foreign fishing. Each attribute was quantified with a set of questions, whose answers were summarized into a single scale using multidimensional scaling (see Materials and Methods). For display purposes, each scale was divided into four quarters aligned from worst- to best-case scenarios (each quarter is color coded as indicated at the bottom of the figure). Our assessment of fishery management effectiveness started with the classification of all analyzed EEZs among the four quarters on the scale of scientific robustness. The EEZs within each of those quarters were then classified among the four quarters on the scale of policymaking transparency, and then those EEZs classified among the quarter of the next attribute, with the subdivision continuing until all EEZs were classified in all attributes. The size of the bubbles is proportional to the number of EEZs classified in each quarter. For purposes of display, subsidies, overcapacity, and fishery access agreements were summarized in a single scale with multidimensional scaling.

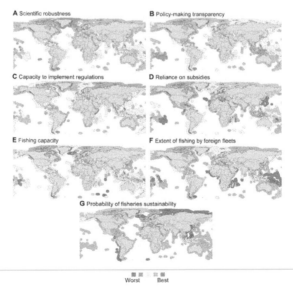

Figure 3. Management effectiveness and sustainability of the world's fisheries. These figures depict the results of experts' opinions on the valuation of scientific robustness (A), policymaking transparency (B), implementation capability (C), subsidies (D), fishing capacity (E) and access to foreign fishing (F). (G) depicts the probability that fisheries in each EEZ are sustainable (Psust) in 2004.

We note that a recent study indicated the success of catch shares, as individual transferable quotas, in preventing fisheries collapses [21]. This strategy has been implemented primarily in the EEZs of New Zealand, Australia, United States, Iceland, Chile, and Peru, which are all countries with robust scientific capabilities (Figure 3A). Our results indicate that the global adoption of individual transferable quotas should be considered with caution given that their underlying success rests on the scientific robustness of the implemented quotas and that few countries meet that condition (Figure 3A).

Policy Transparency

Guidelines to improve the acceptance and compliance with fishing regulations recommend that decisions be based on the best available scientific evidence and follow a transparent and participatory process [8],[14],[22],[23]. Unfortunately, the process of policymaking can be subjected to substantial political pressures, perhaps including corruption. In our survey, management authorities from 92% of the EEZs consider scientific recommendations in formulating policies, and in 87%, all stakeholders are consulted or their opinions considered. Yet in 91% of all EEZs, regulations commonly face economic or political pressures to increase allowable catches or to implement regulations that err on the side of risk rather than caution, whereas a surprising 83% of EEZs are thought to face corruption or bribery. Of all EEZs, 26% rank in the top quarter of a scale of "policymaking transparency," which summarizes, through multidimensional scaling, the attributes of considering scientific advice, participation, pressures, and corruption (countries depicted in Figure 3B). Only 1.4% of all EEZs are in the top quarter on the combined scales of scientific robustness and policymaking transparency (Figure 2), which together accounted for 0.85% of the world's fisheries catch and 1.1% of the world's fished stocks. There were no significant differences between low- and high-income countries with respect to policy transparency. However, the underlying mechanism was different, with low-income countries facing more corruption ($p<0.00001$) and less commonly incorporating scientific advice ($p<0.005$), whereas high-income countries faced slightly more political pressures ($p<0.05$).

Implementation Capability

One of the biggest challenges in fisheries management lies in the implementation and enforcement of regulations [23]. Poverty, unemployment, available infrastructure for control and surveillance, the severity of penalties for violations, and participation in policymaking are all likely influencing the level of compliance with regulations. Proper enforcement through (1) adequate funding and equipment for the managing authorities, (2) patrolling of fishing grounds, and (3) tough penalties for infringements, occurs in 17% of all EEZs. Not surprisingly, no EEZ was free of the effects of poaching (see [24]). On a scale of "implementation capability," which summarizes, through multidimensional scaling, poaching and the different attributes of enforcement, we found that only approximately 5% of all EEZs are in the top quarter of such a scale. Only two relatively small EEZs, those of the Faeroe and Falkland Islands, were in the top quarter for all three indicators of scientific robustness, policymaking transparency, and implementation capability (Figure 2), which combined, accounted for 0.80% of the world's fisheries catch and 0.48% of the world's fished stocks. Better "implementation capability" is frequently more common among high- than low-income countries ($p<0.0001$), which is mainly a consequence of better enforcement ($p<0.00001$) and reduced poaching in the former ($p<0.002$).

Extent of Subsidies, Overcapacity, and Foreign Fishing

When the structure of a management regime is weak, fisheries will be prone to overexploitation due to several factors. Three that have received particular attention are fishing capacity, subsidies, and access to foreign fishing fleets [8],[23],[25],[26]. Open access to fishing (because of lack of effective management) leads to a "race for fish" that commonly increases fleet size and fishing power. This should reduce fish stocks, at which point fishing capacity should stabilize given decreasing profits from reduced catches [8]. Subsidies can override this mechanism by keeping fisheries profitable and encouraging overexploitation [8],[13]. The picture is further complicated by fisheries agreements that allow foreign fleets to catch fish that are not caught by national fleets [25],[26]. Unfortunately, such agreements are commonly made between developing coastal and island states (often with low capacity to assess stocks and to enforce regulations) and developed and heavily subsidized nations [25]. Recent analyses of current agreements indicate a high risk of overexploitation due to several reasons, including selling fishing rights on highly migratory stocks under bilateral agreements, selling access rights without specified catch limits, excessive by-catch, and distortion of reported catches, among others [25],[26]. Such agreements are thought to develop coastal economies through monetary gains and local employment. In certain instances, revenues are also used to generate management plans; their effectiveness, however, is unclear given chronic weaknesses in fisheries governance and management systems [25].

Our assessment of the extent of fishing capacity, subsidies, and access to foreign fishing fleets yielded the following results. We found that fleet sizes are quantified and regulated in 20% of the world's EEZs, although in 93% of EEZs, fishing fleets face some level of modernization to catch fish more efficiently or cheaply. Thus, although fishing capacity may be reduced in terms of fleet size, fishing power may remain constant or even increase due to technological improvements (i.e., fewer improved boats being more effective at catching fish). Effective controls on fleet size were more common among high-income than low-income EEZs ($p < 0.02$), but the former modernized their fleets more often than the latter ($p < 0.00001$). Using multidimensional scaling to summarize the results pertaining to "fishing capacity" (i.e., fleet size controls and fleet modernization), we found high-income EEZs having significantly higher fishing capacity than low-income ones ($p < 0.02$). Fisheries sectors that rely to some degree on subsidies occurred in 91% of the world's EEZs, and more commonly among high- than low-income EEZs ($p < 0.00001$) (see also [27]). Access to foreign fishing is granted in 51% of all EEZs, and is more frequent in low- than high-income EEZs ($p < 0.00001$). In fact, our survey indicated that in 33% of the EEZs that are classified as low income (commonly, countries in Africa and Oceania), most fishing is carried out by foreign fleets from either the European Union, South Korea, Japan, China, Taiwan, or the United States. No single EEZ meets the best standards (i.e., top quarter of the scales) of scientific robustness, policymaking transparency, and implementation capability while being free of the effects of excess fishing capacity, subsidies, or access to foreign fishing (Figure 2).

Extent and Management: Control of Recreational and Small-Scale Fisheries

The notion that industrialized fishing practices are solely responsible for the global fisheries crisis has been challenged by evidence of the significant effects of recreational and small-scale commercial or subsistence fisheries (e.g., [28],[29]). Although less intensive per unit area, small-scale and recreational fisheries can be far more extensive spatially. Small-scale and recreational fisheries are important in 93% and 76% of the world's EEZs, respectively, and small-scale fisheries are increasingly more predominant among low-income EEZs whereas recreational fisheries are more predominant in high-income countries ($p < 0.0001$). Of the world's EEZs, 40% collect at least some data on small-scale fishing, and 13% on recreational fishing; 30% impose regulations on the size of fish caught in small-scale fishing, and 29% do so for recreational fishing, 7% regulate the number of fish caught in small-scale fishing, and 15% do so for recreational fishing, whereas 10% limit the number of fishers in small-scale fisheries, and 3% do so for recreational fishing. These management measures are more frequent in high- than low-income EEZs. Measures to regulate small-scale and recreational fishing are clearly limited and could prove detrimental to food supply and sustainability if they continue to operate outside the control of fisheries management institutions.

Overall Management Effectiveness

To provide a general overview of fisheries management effectiveness, we averaged all scores on the scales of scientific robustness, policymaking transparency, implementation capability, fishing capacity, subsidies, and access to foreign fishing. We excluded the effects of small-scale and recreational fisheries, recognizing that their lack of management would extensively reduce the scores. Only 5% of all EEZs were in the top quarter of this scale (countries depicted in Figure 4), with high-income EEZs having significantly better overall management effectiveness than low-income ones (p<0.00001). A sensitivity analysis indicated that the difference between high- and low-income EEZs was driven mainly by foreign fishing agreements, which disproportionally reduced the average score of low-income EEZs. Excluding foreign fishing access leads to similarly low average scores between high- and low-income EEZs. Similar average scores are, however, explained by different mechanisms, namely excessive fishing capacity and subsidies in high-income EEZs and deficient scientific, political, and enforcement capacity in low-income EEZs.

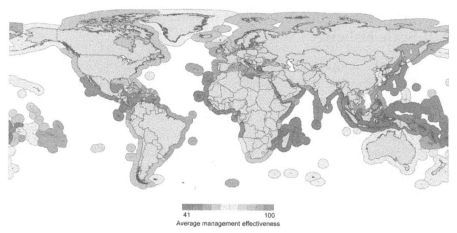

41   100
Average management effectiveness

Figure 4. Overall management effectiveness of the world's exclusive economic zones. This map shows the average, for each surveyed area, of their scores on the scales of scientific robustness, policymaking transparency, implementation capability, fishing capacity, subsidies, and access to foreign fishing.

Effect of Fishery Management on Fisheries Sustainability

One final question that we addressed in this study is to what extent the different attributes of fisheries management analyzed here relate to the actual sustainability of fisheries. We addressed this question using a recently developed method to quantify the probability that ecosystems are being sustainably fished (P_{sust}). This metric assesses the probability that the ratio between the biomass losses due to fishing (i.e., total catch) expressed in primary production equivalents and the primary production of the area in which the catch was taken is sustainable (see Materials and Methods, [30],[31]). We found that this metric is particularly useful to differentiate misinterpretations in landings data when used as an indicator of fisheries status. The metric, for instance, differentiates between countries in which increasing landings (a possible symptom of good fisheries status) are sustainable or not, and between countries in which declining landings (a possible symptom of overfishing or enhanced management [32]) are indicative of the sustainability of fisheries or not. We used classification/regression tree analysis to identify the most likely management attributes that affect the probability of fisheries sustainability; we also included country wealth (i.e., the distinction between high and low income) in the classification tree to analyze differences in fisheries sustainability due to this factor.

Of all management attributes analyzed (i.e., scientific robustness, policymaking transparency, implementation capability, fishing capacity, subsidies, and access to foreign fishing) plus taking into account country wealth, we found that variations in policymaking transparency led to the largest difference in fisheries sustainability. We found that EEZs ranked in the upper best quarter on the scale of transparent policymaking (i.e., EEZs where scientific advice is considered and followed, all parties are consulted and considered, and where corruption and external economic and political pressures are minimal) show the largest probability of having sustainable fisheries compared to EEZs ranked in any of the other three quarters (Figure 5). The probability of sustainability in policy transparent EEZs was 88% compared to 73% in others (Figure 5). We also found that subsidies have an additional negative effect on fisheries sustainability among EEZs with nontransparent policy systems. We found that the probability of fisheries sustainability in nontransparent EEZs was reduced from 78% to 67% due to the effects of even modest subsidies (Figure 5) (i.e., EEZs ranked in the first three quarters on the scale of subsidies or EEZs in which fisheries sectors are dependent minimally to almost entirely on subsidies).

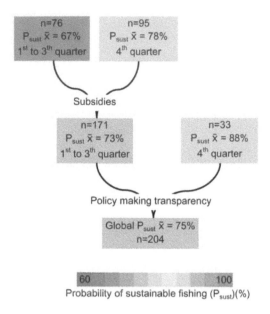

Figure 5. Effect of fishery management on fisheries sustainability. Results of a classification tree aimed to identify the most likely fishery management attributes related to the sustainability of fisheries. In a classification/ regression tree, the factor that maximizes differences in fisheries sustainability is placed at the root of the tree, and the EEZs in each of its quarters are separated into different branches. This method repeatedly tests for significant differences among the EEZs in each branch in the remaining attributes and stops when no significant difference exists in any attribute within the EEZs of any branch (see Materials and Methods). The results shown here include the linking between the probability of fisheries sustainability (Psust) and each of the management attributes analyzed: scientific robustness, policymaking transparency, implementation capability, fishing capacity, subsidies, access to foreign fishing, and country wealth.

The significant effect of policymaking transparency on fisheries sustainability likely relates to the fact that this particular attribute forms the core of the fisheries management process. Firstly, it determines the extent to which scientific advice will be translated into policy, whereas transparent and legitimate participation of involved parties is likely to promote compliance with regulations [22]. Our findings indicate that policymaking transparency is likely to work as a "sustainability bottleneck" through which other positive attributes of fisheries management are filtered. For instance, we found that scientific robustness did not influence the sustainability of fisheries. This may be because, in the process of policymaking, scientific advice may be overridden due to socioeconomic costs and political or corruption pressures. The recent catch quotas for Mediterranean Bluefin tuna (Thunnus thynnus) established by the International Commission for the Conservation of Atlantic Tunas may serve as an example. In this particular case, robust and well-founded scientific advice recommended to maintain catches at 15,000 tonnes per year and to close the fisheries during two spawning months; yet the policy was set at 22,000 tonnes per year, with fishing allowed during critical spawning months. This is a case in which scientific robustness may not necessarily result in sustainability due to significant pressures in the process of policymaking. We also found that variation in implementation capabilities did not have much effect on fisheries sustainability. This result can also be explained by the effect of policymaking transparency. If the policymaking process is participatory and legitimate, it is likely that even poorly enforced systems will move towards sustainability because of voluntary compliance [22]. In contrast, some systems may strongly enforce regulations, but if the regulations were flawed during the process of policymaking, good enforcement may not bring about sustainability either. If the establishment of regulations includes scientific advice and follows a participatory mechanism, it is likely that fisheries will be tightly regulated, regardless of who carries out the fishing, which may also explain the lack of significance of fishing capacity and international fisheries agreements on fisheries sustainability. This is not to say that fishing capacity and foreign fishing access do not have impacts on fisheries sustainability but rather that their effects are moderated by the policymaking process (i.e., fishing capacity and access agreements may have different effects on sustainability in situations that are tightly regulated compared to those that are not). Finally, our results indicate how deficiencies in the process of policymaking can leave fisheries vulnerable to overexploitation due to the effect of subsidies. It is known that subsidies can override possible fishing controls exerted by economic benefits (see section above on subsidies; [8],[13],[27]). We presume, however, that this effect is likely to be more pervasive in nontransparent systems given that fishing remains poorly controlled or regulated and allowed to fluctuate more freely, depending largely on subsidies.

Concluding Remarks

Improvements to fisheries management have been incorporated into international initiatives, which have received broad acceptance (e.g., [14],[15]). Unfortunately, our study shows that there is a marked difference between the endorsement of such initiatives and the actual implementation of corrective measures. The ongoing decline in marine fisheries catches [5],[9],[33]–[36] and the ecological and socioeconomic consequences of a fisheries crisis call for a greater political will of countries worldwide if further fisheries declines and their wider consequences are to be prevented. Effective transfer of improved scientific capacities to policy, achieved through a transparent and participatory process, will be more important than ever in stabilizing our food supply from the sea and preventing unnecessary losses due to management deficiencies. Current projections suggest that total demand for fisheries products is likely to increase by approximately 35 million metric tonnes by 2030 (~43% of the maximum reported catch in the late 1980s)

[3],[4] and by approximately 73% for small-scale fisheries by 2025 [35]. This contrasts sharply with the 20% to 50% reduction in current fishing effort suggested for achieving sustainability [30],[36], and implies that regulators may face increasing pressures towards unsustainable catch quotas. Given that the demand for fish lies outside the control of conventional fisheries management, other national and international institutions will have to be involved to deal with poverty alleviation (inherently improving management) and stabilization of the world's human population (to soften fisheries demand), if pressures on management are to be prevented and sustainability achieved.

Materials and Methods

Conditions Analyzed

We considered factors broadly recognized as critical for the sustainable management of fish stocks (by sustainability, we mean sustainable catches and not social, economic, or institutional sustainability and the like, which at times are also associated with fisheries management and often dominate policy decisions). The factors considered in the present analysis were categorized into those related to the robustness of scientific recommendations, transparency in the process of converting recommendations into actual policy, the capability to enforce and ensure compliance with regulations, and the extent of fishing capacity, subsidies, and access to foreign fishing. Each of these attributes was evaluated with a set of questions whose answers could be categorized in a hierarchical order from worst- to best-case scenarios. In cases where several questions applied to the same attribute, we summarized all responses into a single scale using multidimensional scaling. Multidimensional scaling is an ordination method that uses similarities and dissimilarities among variables to reduce them to a specific number of dimensions. Here, we used the anchored multidimensional scaling method developed by Pitcher and Preikshot [37]. In this method, hypothetical countries are generated with the worst- and best-case scenarios for each question and used as normative extremes of a scale on which real countries are ranked. The approach also incorporates uncertainty using a Monte Carlo simulation tool based on the maximum and minimum possible for each score [38]. A copy of the software is available on request.

Fishery Management Regimes Analyzed

We focused our assessment on fishery management conditions for all ocean realms under the sovereignty of a defined coastal territory. Under the United Nations

Convention on the Law of the Sea [39], the protection and harvesting of coastal resources rest within the 200-nautical mile EEZ of each coastal state. There are, however, exceptions, such as the European Union, whose fisheries regulations are mandated by the Common Fisheries Policy but whose enforcement is the responsibility of the member states; member states also differ in their fishing capability and possibly in their compliance with regulations. Similarly, many countries have overseas territories, which may or may not have autonomous control of the regulation of their fisheries, and consequently, there may be variations in the effectiveness of their management regimes. For instance, Saint Pierre and Miquelon, French Guiana, French Polynesia, French Southern and Antarctic lands, New Caledonia, Saint Martin, Reunion, Guadeloupe, and Martinique all are under the sovereignty of France, which furthermore has direct control over its own Atlantic and Mediterranean coast; yet all of these zones have different management conditions. To consider these differences in fishery management regimes, zones managed under the same entity (e.g., the European Union) or zones in different parts of the world belonging to the same sovereignty (e.g., overseas territories of France, United Kingdom, and United States) were analyzed separately. We also included zones that may not be technically defined or recognized as EEZs under the United Nations (e.g., division among coastal states of the Baltic Sea and Black Sea). In total, 245 such zones exist in the world (see Figure 3), which excludes conflict zones (e.g., the Paracel Islands, Spratly Islands, and Southern Kuriles). Out of those 245 zones, we were unable to gather data for isolated islands under the sovereignty of the United Kingdom (i.e., Ascension, Pitcairn, Saint Helena, South Georgia, and the South Sandwich Islands and Tristan da Cunha) and France (Clipperton Atoll) for which neither contacts nor information was available. We also excluded Monaco and Singapore; interviewees at local authorities (Coopération Internationale pour l'Environnement et de Développement in Monaco and the Agri-Food and Veterinary Authority in Singapore) in both of these countries claimed that although marine fishing occurs, it was minimal and considered insufficient to motivate governmental regulation. The final database contained complete data for 236 zones. Although all data are reported in Figures 3 and 4, the statistics reported in the text were based on 209 inhabited zones for which per capita Gross Domestic Product data exist; that excluded uninhabited and isolated atolls to prevent biases due to the fact that we could not get data for all such areas (i.e., United Kingdom and France, see above).

The Survey

For each of the attributes analyzed (i.e., scientific robustness, policymaking transparency, enforcement capability, fishing effort control, subsidies, and access to foreign fishing), we created a set of questions whose answers could be ranked on a scale from worst- to best-case scenarios. The resulting survey included 23 multiple choice questions and was posted on the Internet (http://as01.ucis.dal. ca/ramweb/surveys/fishery_assessment/) in five different languages (i.e., English, Spanish, French, Portuguese, and German). We searched for contacts (email addresses and phone numbers) of fishery experts for all coastal territories in the world. Our sources of information were reports on scientific and administrative meetings relevant to fisheries, Web pages of nongovernmental organizations, Web pages of fishery management organizations in each territory, and proceedings of international conferences on fisheries. The final directory included contact information for 13,892 people. We sent personalized emails using recommendations of email marketing companies to prevent filtering of emails by local servers and promote participation. The survey started in April 2007 and was completed in April 2008. For zones where we did not receive an email response, we carried out phone interviews with local experts, and both email and phone queries were done until at least one full set of responses was available for each zone. We received 1,188 positive responses including at least one from each country with ocean access. Multiple responses for the same zone were averaged.

Justification of the Approach and Assessment of Responders' Reliability and Validity

Expert opinion surveys have been very popular in social, medical, political, and economic sciences [40], and some examples exist in fisheries studies (e.g., [41]). In fisheries research, expert opinions have been categorized as a "highly reliable" method given that overall, it works as a form of "peer review approach" and, for some crucial issues, is the only knowledge available (see [42]). The approach is also cost-efficient and relatively fast. The collection of empirical data for an analysis of this scale could prove ineffective because country-scale data are patchy, in most cases inaccessible through traditional searching engines, and because old data may not describe current conditions. For these reasons, we chose the survey of local experts to acquire data.

The quality of expert opinion surveys relies on the consistency of responders and their understanding of the issues. These problems are defined as reliability and validity [40], which in statistical terms are analogous to precision and accuracy. The former basically considers the extent to which responders agree in their responses and the latter the extent to which the responses approach the truth. Evaluation of data reliability and validity also allows assessment of the extent of expert biases, which may arise for different reasons (e.g., cultural differences, patriotism, opposition to governmental institutions, etc.). Our assessment of reliability and validity was as follows:

Reliability

To test the extent of consistency among responders, we used data from EEZs for which duplicated responses were received. We performed individual Pearson correlations between each responder and the group of responders (recommended by Fleiss [40]). We also tested the significance of the levels of agreement by comparing the actual levels of agreement among responders with the levels of agreement expected when choices were made randomly (see Figure 1). Analyzing 259 independent responses for 17 EEZs, we found a high level of agreement among responders, with over 72% of the cases showing Pearson correlation coefficients greater than 0.8 (Figure 1A). This was due to the fact that in 67% of the cases, the responders chose exactly the same score for any given question, and in 27%, the nearest choice (Figure 1B). Only in 5% of the cases did the responders differ by more than one choice, and in 0.4%, they chose opposite scores (Figure 1B). The levels of agreement and disagreement were significantly higher and lower, respectively, than those expected by chance (Figure 1B). These high levels of agreement are very likely due to the fact that questions were general and the possible responses relatively broad. Under these conditions, responses by different responders are most likely to converge on similar or closely related scores.

Validity

The survey allowed questions to be left unanswered so that responders could answer only the questions they knew about. Most commonly, responders voluntarily, and at times upon our request, gave contact information for other people better placed to provide missing answers. To address the issue of validity, our survey included a question on the extent to which countries are rebuilding depleted fish stocks, an issue explicitly covered by The United Nations Code of Conduct (Article 7, clause 7.6.10), and evaluated in a survey carried out by Pitcher et al.. [15]. The scores from the two different sources (i.e., expert-based and empirically based) for the countries in common were rescaled from 0 to 1 for comparison, and similarities evaluated using a Pearson correlation. This analysis was based on 28 countries for which empirical data were available and reliable to assign an empirical score. The results of this analysis indicated a strong correlation between expert opinion and empirical data ($r = 0.74$, $p<0.000006$, Figure 1C), although expert opinion tended to overestimate the extent to which countries are rebuilding their depleted fisheries (Figure 1C). Thus, the overall statistics provided here should likely be considered a conservative (more optimistic) view of the actual situation.

Quantification of Fisheries Sustainability

The metric we used to quantify fisheries sustainability has been recently published in two independent publications [30],[31], but not applied to the landings of any country. Here, we provide a brief description of its rationale and calculation, but extended details are provided by Libralato et al.. [31] and Coll et al.. [30].

Fisheries catches represent a net export of mass and energy that can no longer be used within an ecosystem; failure of the ecosystem to compensate for that energy loss implies overexploitation. This notion of overexploitation will require establishing a contrast between the loss of energy in the ecosystem due to a particular catch, the energy at the base of the food web in the area where the catch was taken, and reference points indicating whether the ratio between the energy that is taken (by fishing) and produced (through primary production) is sustainable or not. This concept has been recently incorporated into a metric that aims to quantify the probability that an ecosystem is being sustainably fished (Psust: after [31]). The metric first calculates the amount of Primary Productivity Required (PPR after [43]) to sustain a catch as $PPR = \sum_{s=1}^{s} \frac{W_i}{9} \left(TE^{TL_{i-1}} \right)$, where s is the total number of caught species, Wi is catch weight of each species i, TE is transfer efficiency specific for the ecosystem, and TLi is the trophic level of species i. The metric assumes a conservative 9:1 ratio for the conversion of total weight to carbon [43]. The loss of energy in the ecosystem (i.e., Lindex, after [31]) is calculated by comparing PPR to the primary production at the base of the food web (PP) as $L_{index} = \frac{PPR * TE^{TL_c - 1}}{PP * \ln TE}$, where TLc is the mean trophic level of the catch as calculated from the TL and weight of each species in the catch. PP is parameterized from chlorophyll pigment concentrations and photosynthetically active radiation [30]. The probability that such energy loss is sustainable (i.e., Psust) is calculated by comparing Lindex to reference Lindexes in which overfishing or sustainability have previously been identified. Reference Lindexes were quantified for different regions worldwide using a set of well-documented mass balance models representative of exploited ecosystems and constructed with independent information for each ecosystem [31]. Each of these models is classified as overfished if it meets one or more of the following criteria: (1) biomass of any species falls below minimum biologically acceptable limit, (2) diversity decreases, (3) year-to-year variation in populations or catches increases, (4) resilience to perturbations decreases, (5) economic and social benefits decrease, and (5) nontargeted species get impaired (see [30],[31] and references therein for justification of these criterion). Models were defined as sustainable when the impacts of exploitation did not result in any of the above symptoms. The frequency of sustainable or overfished Lindexes allowed us to calculate the probability of sustainability (Psust) for any particular Lindex

value as $P_{sust}(L_{index}) = \dfrac{N\left(L_{indexes_{sustainable}} > L_{index}\right)}{N\left(L_{indexes_{sustainable}} > L_{index}\right) + N\left(L_{indexes_{overfished}} < L_{index}\right)}$, where N is the num-

ber of models in which Lindexes lead to sustainable or overfishing conditions. Probabilities of fisheries sustainability were calculated for each EEZ in the world using catch data as from the Sea Around Us fisheries database, which contains harmonized data from a variety of sources including the Food and Agriculture Organization (i.e., statistics on fisheries catches from 1950 to 2004; [44]). That database adjusted landings data to account for the fishing of long-distance fishing fleets (i.e., landings that are reported by one country, but fished in a different one). Landings data were also adjusted to include discards [45] and a global estimate of illegal, unreported, or unregulated catches [46],[47].

Linkage between Management Effectiveness and Fisheries Sustainability

Data on fisheries sustainability was quantified for the year 2004 and linked to the effectiveness of fisheries management using a classification/regression tree. A classification tree tests for significant differences in fisheries sustainability among the quarters of each attribute (note that the first and fourth quarters are the extremes of a scale from worst- to best-case scenarios for each attribute; see Figure 2). The attribute that maximizes differences among quarters (i.e., smallest p-value) is placed at the root of the tree and the EEZs in each of those quarters separated in different branches. Subsequently, the EEZs in each branch are tested for significant differences among quarters of the remaining attributes. The attribute that maximizes differences among quarters is placed at the base of the branch and the EEZs in each of those quarters separated in upper branches. The process is repeated until no differences are found within each branch in any remaining attribute. This analysis included all attributes considered in this study: scientific robustness, policymaking transparency, implementation capability, fishing capacity, subsidies, access to foreign fishing, and country wealth (i.e., 2007 per capita Gross Domestic Product larger or smaller than US $10,000, respectively). Given the inflation of Type I errors due to multiple comparisons, significance was set at p<0.01.

Acknowledgements

We would like to thank the numerous people worldwide that participated in the survey, Justin Breen for implementing the survey on the Internet, and Colette Wabnitz, Martin Sperling, Sergio Floeter, Diego Barneche, and Daniella Frensel for translating the survey into different languages. We also thank Cassandra de Young for helpful comments.

Author Contributions

The author(s) have made the following declarations about their contributions: Conceived and designed the experiments: CM RAM KJG BW. Analyzed the data: CM MC SL. Contributed reagents/materials/analysis tools: RUS DZ RW. Wrote the paper: CM MC SL TJP RS DZ RW KJG BW. Collected data: CM.

References

1. FAO (2009) The state of world fisheries and aquaculture 2008. Rome (Italy): FAO Fisheries Department. 162 p.

2. Naylor RL, Goldburg RJ, Primavera JH, Kautsky N, Beveridge MCM, et al.. (2000) Effect of aquaculture on world fish supplies. Nature 405: 1017–1024.

3. Pinstrup-Andersen P, Pandya-Lorch R, Rosegrant MW (1997) The world food situation: recent developments, emerging issues, and long-term prospects. 2020 Vision Food Policy Report. Washington (D. C.): International Food Policy Research Institute. 36 p.

4. Delgado CL, Wada N, Rosegrant MW, Meijer S, Ahmed M (2003) Outlook for fish to 2020: meeting global demand. Food Policy Report. Washington (D. C.): International Food Policy Research Institute. 28 p.

5. Watson R, Pauly D (2001) Systematic distortion in world fisheries catch trends. Nature 424: 534–536.

6. Conover D, Munch SB (2002) Sustaining fisheries yields over evolutionary time scales. Science 297: 94–96.

7. Hutchings JA (2000) Collapse and recovery of marine fishes. Nature 406: 882–885.

8. Beddington JR, Agnew DJ, Clark CW (2007) Current problems in the management of marine fisheries. Science 316: 1713–1716.

9. Worm B, Barbier EB, Beaumont N, Duffy JE, Folke C, et al.. (2006) Impacts of biodiversity loss on ocean ecosystem services. Science 314: 787–790.

10. Myers RA, Baum JK, Shepherd TD, Powers SP, Peterson CH, et al.. (2007) Cascading effects of the loss of apex predatory sharks from a coastal ocean. Science 315: 1846–1849.

11. Pauly D, Watson R, Alder J (2005) Global trends in world fisheries: impacts on marine ecosystems and food security. Phil Trans R Soc B 360: 5–12.

12. World Health Organization (2005) Ecosystems and human well-being: health synthesis. A Report of the Millennium Ecosystem Assessment. Geneva (Switzerland): WHO Press. 53 p.

13. Pauly D, Christensen V, Guenette S, Pitcher TJ, Sumaila UR, et al.. (2002) Towards sustainability in world fisheries. Nature 418: 689–695.

14. FAO (2000) Code of conduct for responsible fisheries. Rome (Italy): FAO. Available: http://www.fao.org/docrep/005/v9878e/v9878e00.HTM. Accessed May 19, 2009.

15. Pitcher TJ, Kalikoski D, Pramod G, Short K (2009) Not honouring the code. Nature 457: 658–659.

16. DeYoung C, editor. (2006) Review of the state of world marine capture fisheries management: Indian Ocean. FAO technical paper #488. Rome (Italy): Food and Agriculture Organization of the United Nations. 458 p.

17. Rosenberg AA, Swasey JH, Bowman M (2006) Rebuilding US fisheries: progress and problems. Front Ecol Environ 4: 303–308.

18. Botsford LW, Castilla JC, Peterson CH (1997) The management of fisheries and marine ecosystems. Science 277: 509–515.

19. Walters CJ, Martell SD (2004) Fisheries ecology and management. Princeton (New Jersey): Princeton University Press. 399 p.

20. Pikitch EK, Santora C, Babcock EA, Bakun A, Bonfil R, et al.. (2004) Ecosystem-based fishery management. Science 305: 346–347.

21. Costello C, Gaines SD, Lyham J (2008) Can catch shares prevent fisheries collapse? Science 321: 1678–1680.

22. Castilla JC, Defeo O (2005) Paradigm shifts needed for world fisheries. Science 309: 1324.

23. Rosenberg A (2007) Fishing for certainty. Nature 449: 989.

24. Agnew D, Pearce J, Pramod G, Peatman T, Watson R, et al.. (2009) Estimating the worldwide extent of illegal fishing. PLosOne 4: e4570. doi:10.1371/journal.pone.0004570.

25. Mbithi Mwikya S (2006) Fisheries access agreements: trade and development issues. ICTSD Natural Resources, International Trade and Sustainable Development Series Issue Paper No. 2. Geneva (Switzerland): International Centre for Trade and Sustainable Development. 58 p.

26. Kaczynski VM, Fluharty L (2002) European policies in Western Africa: who benefits from fisheries agreements? Mar Policy J 26: 75–93.

27. Sumaila UR, Teh L, Watson R, Tyedmers P, Pauly D (2008) Fuel price increase, subsidies, overcapacity, and resource sustainability. ICES J Mar Sci 65: 832–840.

28. Coleman F, Figueira WF, Ueland JS, Crowder LB (2005) The impact of United States recreational fisheries on marine fish populations. Science 305: 1958–1960.

29. Zeller D, Booth S, Davis G, Pauly D (2007) Re-estimation of small-scale fisheries catches for U.S. flag island areas in the Western Pacific: the last 50 years. Fish Bull 105: 266–277.

30. Coll M, Libralato S, Tudela S, Palomera I, Pranovi F (2009) Ecosystem overfishing in the ocean. PlosOne 3: e3881. doi:10.1371/journal.pone.0003881.

31. Libralato S, Coll M, Tudela S, Palomera I, Pranovi F (2008) Novel index for quantification of ecosystem effects of fisheries as removal of secondary production. Mar Ecol Prog Ser 355: 107–129.

32. Mutsert K, Cowan JH, Essington TE, Hilborn R (2008) Reanalyses of Gulf of Mexico fisheries data: landings can be misleading in assessment of fisheries and fisheries ecosystems. Proc Natl Acad Sci USA 106: 2740–2744.

33. Hilborn R, Branch TA, Ernst B, Magnusson A, Minte-Veera C, et al.. (2003) Status of the world's fisheries. Annu Rev Environ Resour 28: 359–399.

34. FAO (2000) The state of world fisheries and aquaculture 2000. Rome (Italy): FAO Fisheries Department. 144 p.

35. Newton K, Cote I, Pilling G, Jennings S, Dulvy N (2007) Current and future sustainability of island coral reef fisheries. Curr Biol 17: 658.

36. Pauly D, Alder J, Bennett E, Christensen V, Tyedmers P, et al.. (2003) The future for fisheries. Science 302: 1359–1361.

37. Pitcher TJ, Preikshot DB (2001) RAPFISH: a rapid appraisal technique to evaluate the sustainability status of fisheries. Fish Res 49: 255–270.

38. Kavanagh P, Pitcher TJ (2004) Implementing Microsoft Excel software for Rapfish: a technique for the rapid appraisal of fisheries status. Fisheries Centre Research Reports. Volume 12 no. 2. Vancouver (Canada): University of British Columbia. Fisheries Centre Research. 75 p.

39. UNCLOS (1982) United Nations Convention on the Law of the Sea of 10 December 1982. Overview and full text. Available: http://www.un.org/Depts/los/convention_agreements/convention_overview_convention.htm. Accessed May 19, 2009.

40. Fleiss JL (1981) Statistical methods for rates and proportions. 2nd edition. New York (New York): John Wiley.

41. Al-Chokhachy R, Fredenberg W, Spalding S (2008) Surveying professional opinion to inform Bull Trout recovery and management decisions. Fisheries 33: 18–28.

42. Sullivan PJ, Acheson JM, Angermeier PL, Fasst T, Flemma J, et al.. (2006) Defining and implementing best available science for fisheries and environmental science, policy, and management. Fisheries 31: 460–467.

43. Pauly D, Christensen V (1995) Primary production required to sustain global fisheries. Nature 374: 255–257.

44. Watson R, Kitchingman A, Gelchu A, Pauly D (2004) Mapping global fisheries: sharpening our focus. Fish Fish 5: 168–177.

45. Kelleher K (2005) Discards in the world's marine fisheries: an update. FAO Fisheries Technical Paper 470. Rome (Italy): FAO Fisheries Department. 134 p.

46. Bray K, editor. (2000) A global review of illegal, unreported and unregulated (IUU) fishing. Expert consulting on illegal, unreported and unregulated fishing. Rome (Italy): FAO Fisheries Department. 53 p.

47. Pitcher TJ, Watson R, Forrest R, Valtysson H, Guénette S (2002) Estimating illegal and unreported catches from marine ecosystems: a basis for change. Fish Fish 3: 317–339.

48. Rice WR (1989) Analyzing tables of statistical tests. Evolution 43: 223–225.

49. Moran MD (2003) Arguments for rejecting the sequential Bonferroni in ecological studies. Oikos 100: 403–405.

Chapter 6

Intense Habitat-Specific Fisheries-Induced Selection at the Molecular Pan I: Locus Predicts Imminent Collapse of a Major Cod Fishery

Einar Árnason, Ubaldo Benitez Hernandez
and Kristján Kristinsson

ABSTRACT

Predation is a powerful agent in the ecology and evolution of predator and prey. Prey may select multiple habitats whereby different genotypes prefer different habitats. If the predator is also habitat-specific the prey may evolve different habitat occupancy. Drastic changes can occur in the relation of the predator to the evolved prey. Fisheries exert powerful predation and can be a potent evolutionary force. Fisheries-induced selection can lead to phenotypic

---

Originally published as Árnason E, Hernandez UB, Kristinsson K (2009) Intense Habitat-Specific Fisheries-Induced Selection at the Molecular Pan I Locus Predicts Imminent Collapse of a Major Cod Fishery. PLoS ONE 4(5): e5529. https://doi.org/10.1371/journal.pone.0005529. © 2009 Árnason et al. https://creativecommons.org/licenses/by/4.0/

changes that influence the collapse and recovery of the fishery. However, heritability of the phenotypic traits involved and selection intensities are low suggesting that fisheries-induced evolution occurs at moderate rates at decadal time scales. The Pantophysin I (Pan I) locus in Atlantic cod (Gadus morhua), representing an ancient balanced polymorphism predating the split of cod and its sister species, is under an unusual mix of balancing and directional selection including current selective sweeps. Here we show that Pan I alleles are highly correlated with depth with a gradient of 0.44% allele frequency change per meter. AA fish are shallow-water and BB deep-water adapted in accordance with behavioral studies using data storage tags showing habitat selection by Pan I genotype. AB fish are somewhat intermediate although closer to AA. Furthermore, using a sampling design covering space and time we detect intense habitat-specific fisheries-induced selection against the shallow-water adapted fish with an average 8% allele frequency change per year within year class. Genotypic fitness estimates (0.08, 0.27, 1.00 of AA, AB, and BB respectively) predict rapid disappearance of shallow-water adapted fish. Ecological and evolutionary time scales, therefore, are congruent. We hypothesize a potential collapse of the fishery. We find that probabilistic maturation reaction norms for Atlantic cod at Iceland show declining length and age at maturing comparable to changes that preceded the collapse of northern cod at Newfoundland, further supporting the hypothesis. We speculate that immediate establishment of large no-take reserves may help avert collapse.

Introduction

Predation is a powerful agent in the ecology and evolution of predator and prey. Prey may select multiple habitats whereby different genotypes prefer different habitats. If the predator is also habitat-specific the prey may evolve different habitat occupancy. Drastic changes can occur in the relation of the predator to the evolved prey. For example the predator may exterminate the prey in the habitat available to the predator and thus lose its prey. The prey selecting the alternative habitat may be released from predation and might even evolve to become a new species free of predation. The stronger the predation and more efficient the predator, the more drastic would be the evolutionary changes. It is, therefore, important to study the way predation acts on genes in species that select multiple habitats by genotype.

Man the hunter has become a mechanized techno-beast, a highly efficient predator. In particular, commercial fisheries searching for fish with computerized fish-finders and airplanes and scooping up fish with several thousand-ton capacities with ships powered by several thousand horsepower engines are a case in point

[1]. Modern fisheries are uncontrolled experiments in evolution [2], [3]. Fisheries target certain phenotypes and, therefore, can be a powerful agent of natural selection. They also frequently are habitat-specific by concentrating on the most accessible locations.

Fisheries-induced selection [4]–[10] can lead to phenotypic changes that influence the collapse and recovery of fisheries [4], [5], [11]. Fisheries-induced selection is primarily discussed as selective mortality directly targeting certain phenotypes such as length- or weight-at-age [4], [6], [7] by size-selective fishing or indirectly targeting phenotypes such as age-at-maturation due to age-at-entry into the fisheries or by location of fishery [4]. However, the genetic determination and heritability of the quantitative traits involved is largely unknown [4], [8], [12]. General estimates of the heritability of life-history traits are low as are estimated selection intensities although fishing mortality can be very high [13], [14]. From these observations emerges the view that fisheries-induced evolution occurs at moderate rates with significant changes only observable at decadal time scales [13]. Nevertheless, the recommendation is that the management adopt short-term conservation/management measures that also meet concerns about long-term evolutionary impact. Thus conservation of old, big fish is promoted as a combined short- and long-term conservation strategy [13]. Evolutionary changes were implicated in the collapse and non-recovery of the northern cod of Newfoundland and in the near collapse of North Sea cod [11], [13], [14]. Following the collapse of the northern cod fishery in Newfoundland [11], the cod fishery at Iceland has remained as one of the worlds major and ostensibly sustainable cod fishery.

The single molecular Pantophysin I (Pan I) locus represents an ancient balanced polymorphism predating the split of Atlantic cod (Gadus morhua) and its sister species [15]. The A and B alleles at the locus differ by multiple nucleotide and six amino-acid substitutions. The locus exhibits local adaptation and is under an unusual mix of balancing and directional selection including current selective sweeps [16], [17]. Pan I variation also has been taken as evidence for population sub-structuring caused by breeding structure [18], [19]. However, the precise roles that fisheries-induced selection and natural selection play in generating that structure are unknown.

Here we test the hypothesis that fisheries can exert powerful habitat-specific selection on a fish species with consequent evolutionary changes. We report exceptionally strong selective changes at the Pan I locus which is involved in habitat selection of depth by genotype in populations of Atlantic cod at Iceland. With a heritability of 100% and indirect but intense selection due to fishing in preferred habitat we observe significant changes on a yearly basis. Thus, fisheries-induced evolution is short term [20]. We identify a threat of collapse due to the selection imposed by the fishery. We consider that if management acted immediately it may be possible to avert collapse. Our study demonstrates the importance of molecular population genetics of targeted loci for studies of fisheries-induced selection and highlights the importance of evolutionary thinking for both short- and long-term management of exploited fish populations.

Results

Allele Frequency and Depth of Sampling

The relationship of allele frequency and depth was highly regular and significant (Figure 1, Table 1) with frequency of the A allele decreasing rapidly to a depth of less than 200 m but staying relatively level in deeper waters. A linear regression equation of allele frequency of A on depth for depth less than 200 m was $P_A = 0.806-0.0044D$ ($t = -14.6$, $P \ll 0.001$) or a 0.44% drop in allele frequency per meter. A generalized linear model (glm) fit gave almost identical results (Figure 1).

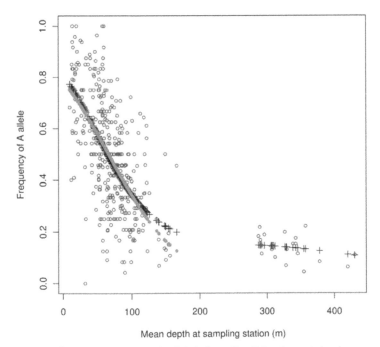

Figure 1. Frequency of Pan I A allele on mean depth (m) of sampling. Points (open circles o) represent frequency at all sampling stations for Atlantic cod in Icelandic Marine Research Institute spring spawning surveys in 2005, 2006, and 2007. Pluses + represent a generalized additive model (gam) smooth fit. Solid dots • represent a generalized linear regression (glm) of allele frequency on depth for depths less than 200 m; glm linear predictor $\eta = 1.297-0.0195$ depth yields an allele frequency intercept of 78.5% and 34.2% at 100 m, a 44.3% change.

Table 1. Pan I allele frequencies and Hardy-Weinberg deviations by 25 m depth classes among Atlantic cod.

Depth	Spring, spawning				Fall, feeding			
	N	p_A	F_{IS}	x^2	N	p_A	F_{IS}	x^2
0–25	345	0.759	0.175	10.57**				
25–50	691	0.656	0.237	38.91***	8	0.875	−0.143	0.16
50–75	1824	0.500	0.129	30.54***	107	0.813	0.139	2.07
75–100	1187	0.393	0.041	2.01	166	0.699	−0.059	0.58
100–125	521	0.315	−0.041	0.88	50	0.640	−0.302	4.56*
125–150	170	0.159	−0.101	1.73	270	0.443	−0.103	2.89
150–175	83	0.289	0.121	1.21	171	0.447	−0.053	0.47
175–200					179	0.243	−0.048	0.41
200–225					182	0.209	0.035	0.23
225–250					283	0.269	−0.007	0.02
250–275					127	0.323	−0.117	1.73
275–300	139	0.169	0.053	0.38	118	0.288	−0.074	0.65
300–325	129	0.163	−0.024	0.07	209	0.220	−0.087	1.59
325–350	178	0.140	−0.163	4.75	261	0.276	−0.189	9.48**
350–375	46	0.120	−0.136	0.85	84	0.286	−0.050	0.21
375–400	24	0.083	−0.091	0.20	153	0.320	−0.111	1.88
400–425	24	0.062	−0.067	0.11	168	0.196	−0.169	4.80*
425–450	23	0.109	−0.122	0.34	145	0.231	−0.068	0.66
450–475					10	0.350	−0.099	0.10
475–500					84	0.244	−0.258	5.61*
500–650					37	0.243	−0.321	3.82*
Sum	5384	0.444	0.195	205.28***	2812	0.342	0.040	4.60*

Atlantic cod at spring spawning and fall feeding grounds at Iceland. Number of individuals, N, and significant deviations from Hardy Weinberg represented by starred x^2 statistics: *: $P<0.05$; **: $P<0.01$; ***: $P<0.001$. Frequency of A allele: p_A. Deviation from Hardy-Weinberg equilibrium: F_{IS}. Test statistic: x^2.

In spring, significant heterozygote deficiency characterized the three top 25 m depth layers, an apparent Wahlund effect [21] due to the convergence in shallow waters of groups of fish that differed in allele frequencies. AA fish were rare in deep water in spring but were found at all depths in fall although they preferred shallow water (Table 1). In fall significant deviations from Hardy-Weinberg were found at various depths representing heterozygote excess in all instances. For fall and spring combined the sign of FIS was most often negative (sign test, $P<0.01$) indicating a general tendency for heterozygote excess.

Population Differentiation in a Two and a Three Dimensional Habitat

Spatial population differentiation exhibited by Pan I [19] is more fully explained by differences in depth among localities although depth and locality are confounded. We found large differences in allele frequencies among localities defined by pooling sampling stations in squares within areas (Figure 2; we pooled to increase sample size). Close inspection of the figure, however, shows that there can be large allele frequency differences among neighboring localities within divisions. These may be described as an inshore/offshore difference, however, depth is a more important explanatory variable. We found apparent spatial differentiation significant in all instances (P = 0.001 based on 1000 permutations in all instances. Table 2). Interestingly, the overall F_{ST} = 0.074 was considerably lower than found for Northeast-Southwest comparison [19]. Also the differentiation among sampling stations within divisions was higher than among divisions (Table 2).

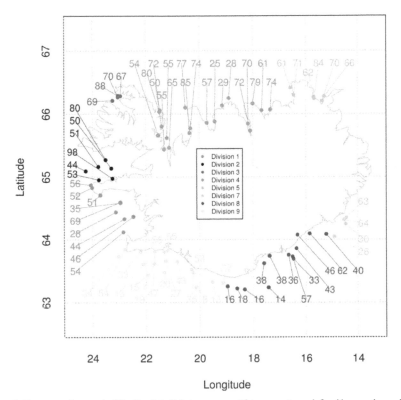

Figure 2. Frequency (percent) of the Pan I A allele in squares within areas. Areas defined by one degree longitude and one half degree latitude (dotted lines) are each split into four equal sized squares (not shown). Sampling stations within subareas are pooled for frequency estimation. Atlantic cod in Icelandic Marine Research Institute spring spawning surveys in 2005, 2006, and 2007. Color coded divisions based on revised METACOD definitions as detailed in paper [19], [61].

Table 2. Hierarchical F statistics among divisions, among sampling stations within divisions and among individuals within stations.

	Division	Station	Individual
Total	0.074	0.192	0.208
Division		0.128	0.145
Station			0.020

Significance P = 0.001 based on 1000 permutations in all instances.

Considering depth of sampling there is clearly a confounding of depth and geographic location. Shallow stations are located in the Northeast and North whereas very deep-water stations are exclusively in the South and Southwest. Thus, the greatest contrast in depth was between the Northeast and South/Southwest. Given the gradient of 0.44% change in allele frequency with 1 m change in depth (Figure 1), depth differences between the Northeast and South/Southwest very likely contributed to the apparent spatial differentiation of Icelandic cod [19]. There were in some instances large differences in allele frequencies among neighboring stations within divisions (Figure 2) particularly in divisions that included sampling stations of different depths. Because depth and location are essentially a factorial or crossed design [22] we cannot use hierarchical or nested F-statistics to test their effects. However, a crossed factor can be tested independently within a level of the other factor [23] and probabilities combined in meta analysis [24]. Differentiation by depth within divisions (Table 3) was relatively high and significant in all instances except division 3. The test combining probabilities was also significant $\left(X^2 = -2\sum \log(P_i) = 97.37, df = 16, P \ll 0.001\right)$. In contrast, considering divisions within depth classes differentiation was considerably lower (Table 4) but nevertheless significant in several instances and overall $\left(X^2 = -2\sum \log(P_i) = 103.20, df = 22, P \ll 0.001\right)$.

Table 3. F statistics among depth classes within divisions.

Depth Class	$F_{Div/Tot}$	$F_{Ind/Tot}$	$F_{Ind/Div}$	$df_{D/T}$	$df_{I/D}$	$df_{I/T}$	G	P
0–25	0.090	0.219	0.142	2	342	345	40.33	0.001
25–50	0.019	0.224	0.209	7	585	593	31.70	0.002
50–75	0.026	0.149	0.127	6	1881	1888	75.37	0.001
75–100	0.054	0.048	−0.006	6	1190	1197	104.40	0.001
100–125	0.058	−0.008	−0.070	4	528	533	50.68	0.001
125–150	0.060	−0.055	−0.122	2	134	137	9.50	0.007
150–175	0.323	0.293	−0.045	1	126	128	35.22	0.001
275–300	−0.007	0.053	0.059	1	137	139	0.01	0.866
300–325	−0.007	−0.025	−0.018	1	127	129	0.45	0.515
325–350	0.006	−0.156	−0.163	1	176	178	1.74	0.189
350–375	0.088	−0.077	−0.180	1	44	46	5.42	0.033
Overall	0.077	0.208	0.142	7	5376	5384	677.39	0.001

Statistics of Division in Total, Individuals in Total and Individuals in Division. G is test statistic and P is based on 1000 permutations in all instances. Divisions are the same as in Table 3.

Overall, therefore, there was greater differentiation by depth than by geographic locality but the two factors remain statistically confounded. Depth of course is a proxy for some biological and environmental factors [25]. For the purposes of this paper the relationship with depth is of great importance. Even if significant spatial variation existed that was not confounded by depth it would not alter implications of the selective effects discussed here.

Table 4. F statistics among divisions within depth classes.

Depth Class	$F_{Div/Tot}$	$F_{Ind/Tot}$	$F_{Ind/Div}$	$df_{D/T}$	$df_{I/D}$	$df_{I/T}$	G	P
0–25	0.090	0.219	0.142	2	342	345	40.33	0.001
25–50	0.019	0.224	0.209	7	585	593	31.70	0.002
50–75	0.026	0.149	0.127	6	1881	1888	75.37	0.001
75–100	0.054	0.048	−0.006	6	1190	1197	104.40	0.001
100–125	0.058	−0.008	−0.070	4	528	533	50.68	0.001
125–150	0.060	−0.055	−0.122	2	134	137	9.50	0.007
150–175	0.323	0.293	−0.045	1	126	128	35.22	0.001
275–300	−0.007	0.053	0.059	1	137	139	0.01	0.866
300–325	−0.007	−0.025	−0.018	1	127	129	0.45	0.515
325–350	0.006	−0.156	−0.163	1	176	178	1.74	0.189
350–375	0.088	−0.077	−0.180	1	44	46	5.42	0.033
Overall	0.077	0.208	0.142	7	5376	5384	677.39	0.001

Statistics of Division in Total, Individuals in Total and Individuals in Division. G is test statistic and P is based on 1000 permutations in all instances. Divisions are the same as in Table 3.

Fishing Pressure

Data on Atlantic cod catches show that, in general, the brunt of the fishing is carried by 5–7 year old fish. By eight years of age a year class is severely reduced and by nine years it is all but fished out. Furthermore, analysis of catch by gear shows that the heavy fishing occurs in shallow water. Catch by long line is to some extent conducted in deeper water (200–300 m) with a shift towards shallower waters in recent years as shown by the higher density of catch at depths less than 100 m and lower density at 200–300 m depths. Although also taking considerable catch in shallow waters, fishing by bottom trawl is distributed over the greatest range of depth.

Bottom trawl which targets greatest depth range brings in more than 40% of the catch. From 1997 to 2007 the catch taken by gear targeting shallow water has increased. Catch with hand line, Danish seine, and gill nets increased up until 2003. However, their catch have decreased since then. Catch with bottom trawl has decreased but catch with long line has increased substantially and after 2005 nearly matches the catch taken by bottom trawl. Thus, fishing mortality is heaviest in shallow water [26] and at least after 2000 fishing targeting shallow water has increased.

Total catch is regulated by government issued quotas of total allowable catch (TAC) which are based on advice from the Marine Research Institute (MRI) of recommended total allowable catch (recommended TAC). Current management strategies of Icelandic cod are based on a catch limitation system where each vessel is allowed a certain share of the TAC. Annual recommended TACs are based on scientific assessment of state of fish stocks and ecosystem condition, but have been reduced lately. For example, the TAC for the 2007/2008 fishing year was 130,000 tons, a 63,000 ton reduction from the previous year. The recommended TAC for the following fishing year was a further 7000 ton reduction but 160,000 was issued. There is a tight correlation of total catch, issued TAC, and recommended TAC. However, total catch almost invariably exceeds issued TACs which in turn are always greater than recommended TAC except for 1996 to 2007 when a catch rule has been in effect.

Annual catch, effort, and catch per unit effort for different gear show a complex interaction with each other and with quotas and TACs (Figure 3). The stock reached an all time low population numbers in 1993–1995 [27]. In the following years the MRI estimated an increase in numbers and increased quotas were issued. Catches of most gear increased and increased effort followed immediately or with a lag (of up to one year) for some gear. Both catch and effort peaked around 2000 and 2001. Catch per unit effort peaked earlier. Catch and effort of both long line and hand line started to increase later than other gear and peaked later or in 2004 and decreased somewhat after that (Figure 3). This, at least partly, accounts for the relative increase of catch by long line in the total catch. As catch diminished after reaching a peak, effort was reduced more rapidly and thus the relative measure of catch per unit effort increased. Using the relative measure of catch per unit effort as an indicator of stock abundance must take total catch into account and whether it is increasing or decreasing [28].

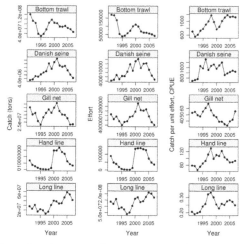

Figure 3. Catch (tons), effort and catch per unit effort, CPUE, at year for different gear. Data are from log book records. Parts of these data are the same as figures 9.3.1. and 9.3.2 in [62]. Units of effort for different gear are described in Methods.

Following the 1995 crash government issued a new catch rule limiting annual quotas to 25% of average fishable biomass. In 2000 and again 2001, with the benefit of hindsight, the MRI re-estimated the population numbers for the previous years. A considerable stock size overestimation and underestimation of fishing mortality was apparent [29], [30], amounting to 25–50%. Issued TACs were based on the overestimated stocks and catches have been 27–40% of the fishable stock, far exceeding the target of 25%. Thus, for example, for 2000 fishing mortality was estimated at 0.86 compared to approximately 0.4 if the catch had been at the 25% target [30]. Consequently the population has experienced increased predation pressure through increased fishing mortality [31].

Part of the overestimation is explained by a lower than predicted mean weight at age [30]. Changes in catchability [28], [31] are, however, the main explanation [30]. There were also changes towards fishing older and larger fish. As an example the gill net fleet changed most of its nets from 7 inch to 9 inch mesh size from 1994–1998 [30].

Probabilistic Maturation Reaction Norms

We applied the principles of estimating probabilistic maturation reaction norms [32]–[34] to evaluate the potential effects of fisheries on changes in life-history traits. We used data on mean maturity, length and age [35] and estimated maturity ogives. From these we estimated the probability of maturing m at length or age (Figure 4). Quantiles of length (and age) at both 50% and 95% probability of maturing show a significant downward trend (slope of linear regression: –0.9 and –2.2 cm per year respectively, P«0.001). Length (and age) at 5% probability of maturing increased (slope of linear regression: 0.4 cm per year, P«0.001).

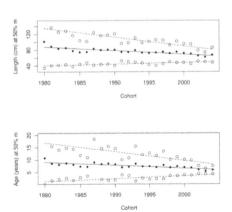

Figure 4. Probabilistic maturation reaction norms: length and age at 50% probability of maturing on cohort. Solid dots and solid lines represent length or age at 50% probability of maturing. Upper and lower open dots and dashed lines represent length or age at 95 and 5% probability of maturing respectively. Lines are linear regression of length or age on cohort. Based on data on mean length, age and maturity ratio from table 3.1.4 in Anonymous [27] (and see [74]).

Genic and Genotypic Frequencies at Age

There were large and highly regular allele and genotypic frequency changes on age within year classes (Figure 5 and Table 5). Year classes are independent realizations of birth and death processes yet changes were in the same direction. Allele and genotypic frequencies in older year classes appeared as continuation of frequencies among more recent year classes (Figure 5). Comparing the 2002 to 1996 year classes, frequencies of A allele on age decreased both within and among year classes ($\overline{\Delta p_A} = -8.4\% \pm 4.8\%$ per year). Similarly, frequencies of AA decreased ($\overline{\Delta f_{AA}} = -7.5\% \pm 5.0\%$), AB increased at young ages but decreased slightly on average ($\overline{\Delta f_{AB}} = -1.7\% \pm 3.0\%$), and BB increased ($\overline{\Delta f_{BB}} = 9.3\% \pm 5.0\%$). There was a reversal with BB becoming the most common genotype by about eight years of age. The 1995 and 1994 year classes had higher A allele frequencies compared to the more recent year classes. However, comparing these two year classes the same pattern of decrease of AA and increase of BB with age held. Evolutionary changes are observable on a yearly basis [13] and thus ecological and evolutionary time scales are congruent.

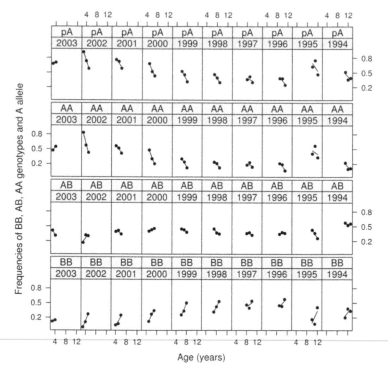

Figure 5. Allelic and genotypic frequencies at age conditioned on year class.

Spring spawning Atlantic cod at Iceland. Frequency of A allele, p_A (top panel row), and frequencies of AA, AB, and BB genotypes, f_{AA}, f_{AB}, and f_{BB} (panel rows 2–4 respectively). Panels represent year classes arranged most recent to older from left to right in each row. Points • represent observed frequencies; lines represent linear regression of frequency on age.

Table 5. ANOVA table of change of genotypic frequency per year within year class.

Source	Df	SS	MS	F	P
Genotype	2	0.28	0.14	16.33	<0.00001
Residuals	54	0.46	0.01		

ANOVA table of change of genotypic frequency per year within year class among the *AA*, *AB*, and *BB* genotypes of the *Pan* I locus in Atlantic cod.

We sampled three consecutive years and thus, except for the most recent and the oldest year classes, a year class entered our samples at three ages. Adjacent year classes have two overlapping age classes. The large observed evolutionary changes predict that the A allele would be decreasing in frequency. This is exactly what we observed as can be seen by comparing allele frequencies at age (Figure 5). In general allele frequencies at age are lower among the more recent year classes with a negative slope of a regression of allele frequency at age on year class. This shows that A allele frequencies have decreased with time.

Fitness Estimation and Prediction of Changes

Genotypic frequencies changed significantly between years within year class (Table 5). Frequency changes can be used to estimate relative fitness of Pan I genotypes. Overall, the genotypic frequencies changed rapidly to about eight years of age (Figure 6) and stayed relatively level after that, an age at which the brunt of fishing of a year class is over. Catch-at-age data show that by eight years of age, a year class is severely reduced and by nine years is almost fished out. Therefore, the selective pressure of the fishery is mostly over by these ages. Taking notice of this fact we took the ratio of the gam predicted frequencies (Figure 6) among 8 year old (post-selection) to 4 year old (entering the fishery) as weights to estimate fitness (Table 6). Relative AA fitness is only 8% and AB 27% showing partial dominance. We used the upper and lower confidence limits to predict best-case and worst-case scenarios. Similar low fitness was obtained using slightly different methods for estimation.

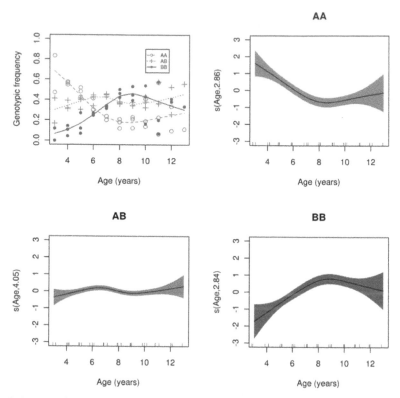

Figure 6. Genotypic frequencies on age in years within year class. Frequencies of AA genotype (red open circles •, dashed line), AB (magenta pluses +, dotted line), and BB (blue filled circles •, solid line). Lines represent a generalized additive model (gam) smooth fit with quasibinomial link (panel A). Panels B, C and D: gam smooth fit of genotypic frequency on age within year class for the AA, AB and BB genotypes respectively; shaded region represents two standard errors above and below fit. Smooth carries estimated degrees of freedom.

Table 6. Predicted genotypic frequencies, weights and fitnesses.

	AA			AB			BB		
Age 4	lower	esti-	upper	lower	esti-	upper	lower	esti-	upper
Age 8	upper	mate	lower	upper	mate	lower	upper	mate	lower
4 years	0.44	0.57	0.68	0.28	0.34	0.41	0.06	0.11	0.19
8 years	0.21	0.18	0.15	0.41	0.38	0.36	0.48	0.44	0.40
U_i	0.48	0.32	0.22	1.44	1.12	0.88	8.68	4.15	2.06
W_i	0.23	0.08	0.03	0.70	0.27	0.10	1.00	1.00	1.00

Genotypic frequencies of 4 and 8 year old predicted by the gam model in Figure 6. Upper and lower are frequencies ±2 standard errors. Weights, U_i, are ratios of frequencies among 8 year old to 4 year old. Ratios of upper to lower and lower to upper frequencies are used for predictions of best-case and worst-case scenarios respectively. The table is arranged accordingly with lower to upper and upper to lower for each genotype. Fitnesses, W_i, are weights scaled to the most fit *BB* genotype.

Plugging the fitness estimates (Table 6) into an equation for allele frequency change under a constant-viability selection model showed that the A allele would be eliminated in 4–5 generations, assuming continued selection of this magnitude. Given a generation time in Atlantic cod of 4.8 years [36], this gives two generations or about 10 years until predicted near disappearance of shallow-water adapted AA fish and four generations or 20 years until disappearance of heterozygous AB fish (Figure 7 panel 1). Using the most optimistic fitness values (Table 6) doubled the time in a best-case scenario (Figure 7 panel 3) with AB heterozygotes still making up some 15% of the population after seven generations. Under a worst-case scenario the most pessimistic values (Table 6) predicted disappearance of AA in one generation and of AB in two generations or 10 years (Figure 7 panel 4).

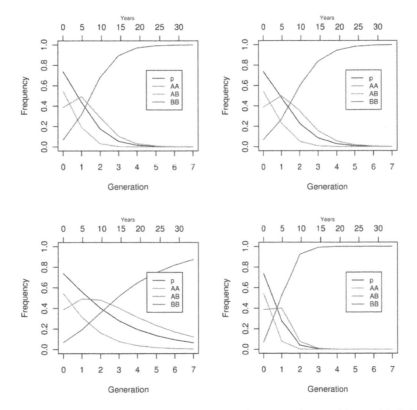

Figure 7. Predicted allele and genotypic frequency changes with a constant-fitness viability model of selection. Top left panel based on fitness estimates from Table 6; lower left panel based on highest fitness estimates from Table 6 (a best case scenario); lower right panel based on lowest fitness estimates from Table 6 (a worst case scenario). Starting frequency of 0.738 assumed based on intercept of gam fit in Figure 6. Years based on a generation time of 4.8 years [36]. Color codes are black for the A allele and red, magenta, and blue for the AA, AB, and BB genotypes respectively.

Discussion

Depth and Population Differentiation

There is an apparent spatial differentiation of Atlantic cod in Icelandic waters with differentiation between the Northeast and Southwest at F_{ST} = 0.261 and F_{ST} = 0.003 for the Pan I and nine microsatellite loci respectively [19]. It is important to examine whether spatial differentiation contributes to the observed selective patterns. The differentiation is clearly driven by the Pan I locus. The very low (but significant) differentiation observed at the microsatellite loci is of questionable biological significance [37] especially considering the fact that two of the nine microsatellite loci used (Gmo8 and Gmo34) are outliers in genetic differentiation [38], [39] being influenced either by direct or hitch-hiking selection. In fact, in a study in northern Norway Gmo34 shows strong linkage disequilibrium with Pan I [39]. The Pan I locus is generally acknowledged to be under selection [15]–[18], [40], [41]. A differentiation of Pan I similar to the Icelandic case [19] also is observed for Arctic and Coastal cod in Norway [18]. Both studies [18], [19] concluded that breeding structure with reduced gene flow was the most likely explanation for the observed spatial differentiation, yet both noticed a relationship of Pan I allele frequency and depth of sampling similar to the above (Figure 1).

The apparent Wahlund effects (deficiency of heterozygotes) in shallow water in spring could be interpreted as signs of population structure by depth similar to that observed between coastal and Arctic cod in Norway [18]. To the extent that such segregation by depth actually occurs, the fisheries-induced habitat-specific selection discussed further below would be more efficient in removing shallow-water adapted fish and the effect would be harder to reverse. Conversely, if, as we suggest, there is near panmixia on breeding grounds, selection would be operating in the face of free gene flow between niches. This would reduce the efficiency of selection, it would take longer to lose the shallow-water adapted fish, and the effect would not be as difficult to reverse. A counter argument is based on the observed heterozygote excess at some depths. Overdominance in fitness is one possible explanation for heterozygote excess. However, it is very difficult to detect selection as a deviation from Hardy-Weinberg equilibrium [42]. If fitnesses form a geometric series genotypic frequencies will be in Hardy-Weinberg before and after selection [43] and directional selection may give rise to spurious overdominance if sampling is done before selection is complete [44]. If overdominance was strong enough to produce a significant heterozygote excess we would expect to see it in our fitness estimation. Such strong overdominance would also be expected to counter the heterozygote deficiency observed in spring. Unequal allele frequencies among male and female parents is another textbook explanation for heterozygote excess [45]. But neither overdominance nor male/female allele frequency differences are satisfactory explanations for the observed patterns for it is hard to see why significant excess would be restricted to certain depths at certain times of year. Negative assortative mating is another textbook explanation for heterozygote excess. The intense fisheries-induced selection observed might lead to secondary selection for apostatic mating that would reduce the effects of the selection among their offspring. Thus there could be an advantage for shallow-water adapted phenotypes to be attracted to phenotypes from the opposite deep-water habitat. Instead we consider the generation of heterozygote deficiency in spring (apparent Wahlund effect) and its disappearance in fall as well as the generation of heterozygote excess in fall and its disappearance in spring as a sign of a dynamical system. We consider the patterns most likely to be due to behavioral responses [46]. Fish fitted with data-storage tags can be classified into deep-water and shallow-water behavioral types that either forage and stay mainly in shallow water or migrate to deeper and colder waters foraging at thermal fronts and going to shallow water only for breeding [47]. These types are correlated with Pan I: AA are shallow-water type, BB are deep-water type that move to shallow waters during breeding and AB are somewhat intermediate although closer to AA than BB [46]. Pan I genotypes thus are functionally related to behavioral types that select their habitat by depth. Under natural conditions the polymorphism may be balanced by shallow-water vs. deep-water niche-variation specialization. If heterozygous AB fish move between the shallow-water and deep-water habitats to a greater extent than homozygous AA and BB fish there could be at any time an excess of heterozygotes at any depth. However, the role of various mating and feeding behaviors in generating heterozygote excess remains to be studied.

Selection at Pan I

The mechanism of balancing selection at Pan I under natural conditions is not known [16] but we suggest specialization to shallow and deep-water niches with free gene flow between the niches as the fish converge on the shallow water breeding grounds. The fitness estimates and relationships observed here, however, very likely tip the balance [16], [17] leading to directional selection in favor of BB deep-water adapted fish. It is possible that some unknown fitness component would overcome the mortality leading to overdominance and balanced polymorphism but there is no evidence for that in our results. The selective pressure is most certainly due to habitat-specific fishing mortality that is heavily directed against shallow water fish [26]. If it continues unabated shallow-water fish will disappear rapidly with consequent collapse of the population and the fishery.

We have not considered here the effects of selection and age-structure [48], [49]. In the absence of selection a population attains a stable age-distribution.

Selection will alter lx and mx schedules and will rip a population out of a stable age-distribution. Demographic processes will tend to restore a stable age-distribution and with continued selection a tug of forces will ensue. Models of intense selection in age-structured populations [49] show that age-structure and age-dependency of selection can either increase or decrease the intensity of selection. In particular, if selective mortality hits reproductive ages (as is true here) age-structure may intensify selection. Therefore, there is hardly reason to think that age-structure will alleviate the threat of collapse. However, this remains to be investigated.

Fisheries-induced selection is most often considered size-selective mortality [12] directly targeting specific phenotypes. Here, however, selection is indirect. Fisheries are to a very large extent conducted in shallow rather than deep water [26] and, therefore, fishing mortality will generate selection against genotypes adapted to shallow water although there is no direct targeting of specific phenotypes. There are general lessons in this for population and conservation genetics that changes in habitat can lead to intense selection even if the mortality is non-selective in the habitat in which it occurs. Sequence variation studies show that the Pan I A and B alleles are ancient predating the split of Atlantic cod from its sister species, the Pacific walleye pollock Theragra chalcogramma [15], [16]. Such studies provide a deep window into the species past evolutionary and selective history. However, here we observe a steep allele frequency gradient with depth and intense fisheries-induced selection. These factors clearly would influence measures of population differentiation such as FST. We, therefore, question the use of Pan I and other loci under such intense selection as markers for analysis of population breeding structure due to reduced gene flow [18], [19], particularly if depth, the confounding of depth and geographic location, and habitat-specific fishing pressures are not controlled for. In particular, we question the practice of combining results from strongly selected loci such as Pan I with variation of supposedly neutral microsatellite loci to study breeding structure [18], [19] due to population isolation. The strongly selected loci will drive the overall measure of differentiation even in the face of considerable gene flow. Fisheries-induced selection can cause differentiation of Pan I among local groups and using that differentiation to argue for local population structure and special management of local populations may be circular.

There is a great need to understand selection and local molecular adaptation and how fisheries can indirectly and inadvertently generate intense selection as in this case. Conservation measures such as conserving large fish [13] would not be enough if they failed to protect genetic variants adapted to local niches.

Fisheries-Induced Changes and Evolution of Life-History Traits

Data on catch per unit effort are difficult to interpret as indicators of population abundance [28]. For the Icelandic cod fishery total catch is highly correlated with TAC and increases and decreases with it. Total catch always exceeds the TAC and thus the fishing mortality is greater than assumed with the recommended TAC. As expected fishing effort increases with catch and TAC and so does catch per unit effort. As catch diminishes for some gear, such as gill net, effort is also reduced and at a low total catch the catch per unit effort may increase again. Other gear behave differently. For long line in particular, which lately is taking a larger share of the total catch, both catch and effort have increased and stayed high. There are signs of changes in catchability [31], [50] and increased fishing pressure and effort as has been observed in the decline and collapse of other cod populations [11], [31].

Rapid changes in maturation preceded the collapse of the northern cod at Newfoundland [6]. Similarly we have indications of maturation changes occurring in Icelandic cod. The caveat is that we have not studied differential maturation of the sexes or potential effects of geographic location (e.g. north vs. south [19]) or environment and we are using averages as data. We interpret the trends in estimated probabilistic maturation reaction norms to mean that the sigmoid (logistic) maturation curves on length (and/or age) are changing. As their inflection points are pushed towards shorter lengths or lower ages they are also changing shape, becoming a stepped function. Probabilistic maturation reaction norms are useful for assessing genetic changes in the presence of environmental variation and phenotypic plasticity [32]. We, therefore, hypothesize that these are selective changes. Small and young fish may be evolving to delay reproduction while larger and older fish evolve to mature earlier and the fish become mature in a narrower window of length (or age). Overall, therefore, we have signatures for Icelandic cod of changes in effort and in life-history traits that are comparable to changes observed in the collapse of other cod stocks.

Olsen et al.. [6] cautioned that "Although eroding maturation reaction norms can thus signal extreme exploitation pressures, they are not to be misinterpreted as signs of imminent stock collapse. But exploitation pressures so strong that they overturn a species' natural pattern of life-history adaptation certainly ought to be cause for concern." Our study certainly appears to meet the criteria for concern. The strength of selection imposed by the fishery in Iceland is extremely high with selection coefficients of 92% and 73% against AA and AB genotypes respectively. This is in the high end in the distribution of known selection coefficients [51],

[52]. The life-history changes coupled with the Pan I changes are perhaps even more dramatic than that documented for the northern cod [6].

Future of Commercial and Fishery

We hypothesize that a collapse of the fishery is imminent if Pan I genotypic frequencies change as predicted. This hypothesis is supported also by changes in life-history. Considering fate of the fishery, deep-water fish are harder to catch and as they increase in frequency the fishery may become commercially in-viable. Fishing mortality in the preferred habitat would then cease before exterminating the A allele and fitness would revert back to natural values. Under that scenario selection pressures will diminish and the A allele may not go extinct completely which presumably would help subsequent recovery [6]. An alternative, and more likely, scenario is contraction of habitat use from tertiary and secondary to the most suitable primary habitat as a density-dependent response similar to the collapse of the northern cod stock in Newfoundland [53] and of the North Sea cod [54]. Fishers will go after smaller and smaller but equally dense clusters of fish in the primary habitat that still allow a profitable commercial fishery until the shallow water fish disappear from all habitats. Fisheries-induced selection at the Pan I locus may have contributed to the collapse of the northern cod and other threatened cod stocks. If fishing mortality causes large decreases in frequencies of the A allele it is unlikely to revert quickly back to previous values after fishing ceases. The fact that we see rapid changes in frequency means that back selection for A from natural causes clearly is much lower than the intense selection against A caused by the fishing mortality. Also, the lack of changes among fish greater than eight years of age (Figure 6) shows no evidence for back selection. Therefore, upon collapse the fishery would take a longer time to recover than it takes to collapse [11]. With AA and AB fish decimated by fishery we can inquire whether BB fish could invade the shallow-water niche and support a commercial fishery. This is unlikely as BB fish are deep-water adapted types [46]. We have found no evidence for historical separation or population structure by depth. However, to the extent that such a structure exists with limitation on interbreeding between shallow water and deep water fish a collapse would be more rapid and its effects would be harder to reverse.

Current management strategies of Icelandic cod are based on a catch limitation system where each vessel is allowed a certain share of the total allowable catch (TAC). Annual TACs are based on scientific assessment of state of fish stocks and ecosystem condition, but have been reduced lately. In addition, special measures for protecting small fish and the ecosystem are implemented. Thus relevant areas may be closed for short periods, if the percentage of small fish or by-catch exceed set limits. Additionally, cod spawning grounds are closed annually at the height of the spawning season in March and April to protect spawning fish. Apparently, however, these measures do not protect shallow-water fish as selection is similarly affecting all year classes that entered our samples (Figure 5). Furthermore, some current management measures may actually intensify selection. For example, during the March/April stop, deep-water BB fish move to shallow waters to spawn [46] (Table 1) whereupon they return to deep water and relative safety from fishing mortality. On re-opening, fishing mortality will hit the AA and AB fish that stay in shallow-water. Thus without knowledge of local adaptation good conservation intentions may exacerbate the problem.

Averting Collapse with No-Take Reserves

We consider that our study meets criteria for concern that the Icelandic cod stock is imperiled. Can anything be done to avert collapse? Upon collapse of the northern cod of Newfoundland the Canadian government imposed a moratorium on fishing [11]. Such a drastic measure if imposed in Iceland doubtless would avert collapse. Alternatively management measures that shifted fishing from shallow-water to deep-water or measures that distributed fishing effort evenly over all depth ranges by controlling fishing by different gear also could possibly help avert collapse. However, we consider that such strategies would be difficult to implement. Alternatively we speculate and suggest that it may be possible to avert collapse by adopting a different strategy of removing selection pressures against shallow-water adapted AA and AB fish. This highlights the use of evolutionary thinking for management and conservation issues. Given that current practices are ineffective in protecting shallow-water adapted fish, we suggest that immediate action is required. We suggest that establishment of large no-take marine reserves that range from the shoreline down to the very deep waters of at least 500 meters or more would protect all genotypes. In the case of Icelandic cod an obvious area is Selvogsbanki and Faxafloi, the main spawning grounds in the Southwest [26], 55. Additional areas would be the shallow-waters in the Northeast which were closed for some years with good results but subsequently re-opened [56]. The advantage of no-take reserves would be to relieve selection pressures against the shallow-water adapted AA and AB fish. Although there are gaps in our knowledge of no-take reserves [57], [58] we predict standard benefits of spillover of adults from prime into secondary and tertiary habitat and export of pelagic eggs and larvae that will ultimately benefit the fishery [59], [60].

Materials and Methods

Sampling and Measurement

To assess temporal and spatial variation in Pan I frequencies we sampled cod measured and aged at all predetermined sampling stations during the Marine Research Institute spring spawning surveys in 2005, 2006, and 2007 in eight of nine divisions revised from definitions in the METACOD project [19], [61]. At each station a set with 12 gill nets of alternating six to nine inch mesh size made with mono-filament or multi-filament yarn was laid out. The sets stayed in place for at least 12 hours. Each net was 50 meters long and a set of 12 thus was 600 m long. The height of nets was 50–60 meshes or about 12–15 m. In the steep-slope deep waters off the South coast a double set of 24 gill nets 1200 m long was laid out. A set of gill nets thus could cover a range of space and depths. We used mean location and mean depth in the analysis. We similarly took stratified random samples of stations taken during the MRI fall ground fish surveys in 2004, 2005, and 2006.

From each net up to 25 fish were taken for measurement of various individual traits. Otoliths were taken for age determination from a single fish from each net and a sample of gill tissue was taken for genetic analysis from these and preserved in 96% ethanol. The year class (cohort) of a fish was determined from the sample year and age read from otolith. Based on our sampling design most year classes entered our sampling for three consecutive ages except the very recent and old year classes which entered for one or two ages.

Commercial catches and effort for 1997–2007 by different gear were obtained from logbook data and from official statistics. Effort of bottom trawl is trawling time in minutes, effort of Danish seine is number of throws, effort of gill net is number of sets of nets, effort of hand line is number of hours at sea, and effort of long line is number of lines times number of hooks per line. Log book results on effort and catch per unit effort for most of the gear have also been presented in figures 9.3.1 and 9.3.2 in [62].

Molecular Analysis

The Pan I locus of Atlantic cod has two alleles, A and B, defined by the presence or absence of a DraI restriction site [63]. We used a proteinase K digestion/chelex 100 method [64] for DNA isolation from tissue. We used primers 3 and 20 [15] to amplify a 489 base pair fragment of the Pan I gene and digested that with DraI to reveal diagnostic bands of the three genotypes on an agarose gel [63]. Altogether we genotyped over 8100 individuals.

Estimating Probabilistic Maturation Reaction Norms

To estimate probabilistic maturation reaction norms [32] we used methods for estimating probabilities of maturing from maturity ogives [33], [34]. We used data on sexual maturity at age in Marine Research Institute spring surveys (Table 3.1.4 in [35]) and mean length and age. We estimated maturity ogives o, probability of being mature at mean length and age, with a generalized linear model (glm) logistic regression: $\log it\left[o(a:l)\right] = \alpha + \beta\left(a:l\right)$ with sample sizes as weights. The data are mean maturity and mean length l at age a and thus fitting a full model [34] is not possible. Instead the interaction term a:l takes into account potential non-linearity of age and length. Following Barot et al.. 33, 34 probability of becoming mature m was estimated as

$$m(a:l) = \frac{o\left(a:\overline{l}\right) - o\left(a-1:\overline{l} - \Delta\overline{l}\right)}{1 - o\left(a-1:\overline{l} - \Delta\overline{l}\right)}.$$

To parameterize the reaction norms we fitted a logistic model of logit(m) on mean length or age and calculated the quantiles of length and age at 5%, 50% and 95% m using the dose.p function of the MASS library under R [65].

Statistics and Fitness Estimation

We mostly used R [66] and various in house functions and packages under R for statistical and genetic analysis. In particular we used the LATTICE package [67], [68] that implements Trellis graphics in R, the MGCV package [69], [70] for fitting generalized additive models (gam) and the HIERFSTAT package [71] for R that implements an algorithm [72] for estimating F-statistics at any level of a nested or hierarchical structure.

Fitness is considered a weight, U, that transforms a genotypic frequency at one age into a frequency at another and higher age. To estimate the weight we took ratios of genotypes frequencies at two ages. We used the generalized additive model (gam) fit (Figure 6), with a quasibinomial link function to model overdispersion, to predict genotypic frequencies and approximate 95% confidence intervals at 4 and 8 years of age (Table 6). We took the ratio of these predictions to estimate the weights. Furthermore, we took the ratios of upper to lower and lower to upper confidence intervals for predictions of best-case and worst-case scenarios. Relative fitnesses, Wi, are the weights scaled to the most fit Pan I BB genotype.

We also took the pooled observed frequencies among 3 and 4 year old (entering the fishery; "pre-selection") and among 8–13 year old (ages at which frequencies do not change much and the brunt of the fishery is over; "post-selection") to estimate the weights and standard errors based on variance of ratios. We used median genotypic frequencies at age within the various year classes to estimate weights as yearly transitions. Assuming independent action of weights in time we multiplied the yearly weights to get an overall weight.

We used our estimated fitness values to plug into a constant-fitness viability model [73] and assuming non-overlapping generations [49]

$$\Delta p_A = \frac{pq\left[p\left(W_{AA} - W_{AB}\right) + q\left(W_{AB} - W_{BB}\right)\right]}{\bar{W}}$$

for predicting allele and genotypic frequency changes at Pan I. The effect of selection and age-structure [48], [49] was not considered here.

Acknowledgements

We thank various members of the Marine Research Institute for help in obtaining samples and members of the University of Iceland population genetics laboratory for help with molecular analysis. We thank Eiríkur Steingrímsson, Jarle Mork, and R. C. Lewontin for comments on the manuscript. We also thank an anonymous reviewer for a thorough critique of the paper.

Author Contributions

Conceived and designed the experiments: EA. Performed the experiments: UBH. Analyzed the data: EA UBH. Contributed reagents/materials/analysis tools: EA KK. Wrote the paper: EA. Participated in the refinement and exposition of ideas: EA UBH KK. Helped design and coordinate the sampling, compiled both individual data and catch data from logbooks: KK.

References

1. Kurlansky M (1997) Cod: A Biography of the Fish That Changed the World. Ottawa, Canada: Alfred A. Knopf.

2. Rijnsdorp AD (1993) Fisheries as a large-scale experiment on life-history evolution—disentangling phenotypic and genetic effects in changes in maturation and reproduction of North-Sea plaice, Pleuronectes platessa L. Oecologia 96: 391–401.

3. Hutchings JA (2004) The cod that got away. Nature 428: 899–890.

4. Law R (2000) Fishing, selection, and phenotypic evolution. ICES J Mar Sci 57: 659–668.

5. Conover DO (2000) Darwinian fishery science. Mar Ecol Prog Ser 208: 303–307.

6. Olsen EM, Heino M, Lilly GR, Morgan MJ, Brattey J, et al.. (2004) Maturation trends indicative of rapid evolution preceded the collapse of northern cod. Nature 428: 932–935.

7. Olsen EM, Lilly GR, Heino M, Morgan MJ, Brattey J, et al.. (2005) Assessing changes in age and size at maturation in collapsing populations of Atlantic cod (Gadus morhua). Can J Fish Aquat Sci 62: 811–823.

8. Kuparinen A, Merilä J (2007) Detecting and managing fisheries-induced evolution. Trends Ecol Evol 22: 652–659.

9. Jørgensen C, Enberg K, Dunlop ES, Arlinghaus R, Bouka DS, et al.. (2007) Managing evolving fish stocks. Science 318: 1247–1248.

10. Conover DO (2007) Nets versus nature. Nature 450: 179–180.

11. Hutchings JA (2000) Collapse and recovery of marine fishes. Nature 406: 882–885.

12. Swain DP, Sinclair AF, Hanson JM (2007) Evolutionary response to size-selective mortality in an exploited fish population. Proc R Soc Ser B 274: 1015–1022.

13. Law R (2007) Fisheries-induced evolution: present status and future directions. Mar Ecol Prog Ser 335: 271–277.

14. Hutchings JA, Fraser DJ (2008) The nature of fisheries- and farming-induced evolution. Mol Ecol 17: 294–313.

15. Pogson GH, Mesa K (2004) Positive Darwinian selection at the Pantophysin (Pan I) locus in marine Gadid fishes. Mol Biol Evol 21: 65–75.

16. Pogson GH (2001) Nucleotide polymorphism and natural selection at the Pantophysin (Pan I) locus in the Atlantic cod, Gadus morhua (L.). Genetics 157: 317–330.

17. Karlsson S, Mork J (2003) Selection-induced variation at the pantophysin locus (Pan I) in a Norwegian fjord population of cod (Gadus morhua L.). Mol Ecol 12: 3265–3274.

18. Sarvas TH, Fevolden SE (2005) Pantophysin (Pan I) locus divergence between inshore v. offshore and northern v. southern populations of Atlantic cod in the north-east Atlantic. J Fish Biol 67: 444–469.

19. Pampoulie C, Ruzzante DE, Chosson V, Jörundsdóttir TD, Taylor L, et al.. (2006) The genetic structure of Atlantic cod Gadus morhua around Iceland:

Insights from microsatellites, the Pan I locus, and tagging experiments. Can J Fish Aquat Sci 63: 2660–2674.

20. Stockwell CA, Hendry AP, Kinnison MT (2003) Contemporary evolution meets conservation biology. Trends Ecol Evol 18: 94–101.

21. Wahlund S (1928) Zuzammensetzung von populationen und korrelationser-scheinungen vom standpunkt der vererbungslehre aus betrachtet. Hereditas 11: 65–106.

22. Johannesson K, Tatarenkov A (1997) Allozyme variation in a snail (Littorina saxatilis)—deconfounding the effects of microhabitats and gene flow. Evolution 51: 402–409.

23. de Meeûs T, Goudet J (2007) A step-by-step tutorial to use HierFstat to analyse populations hierarchically structured at multiple levels. Infect, Genet Evol 7: 731–735.

24. Fisher R (1970) Statistical Methods for Research Workers. Edinburgh: Oliver and Boyd, 14 edition.

25. Case RAJ, Hutchinson WF, Hauser L, Oosterhout CV, Carvalho GR (2005) Macro- and micro-geographic variation in pantophysin (Pan I) allele frequencies in NE Atlantic cod Gadus morhua. Mar Ecol Prog Ser 301: 267–278.

26. Begg GA, Marteinsdottir G (2003) Spatial partitioning of relative fishing mortality and spawning stock biomass of Icelandic cod. Fish Res 59: 343–362.

27. Anonymous (2007) State of marine stocks in Icelandic waters 2007/2008. Prospects for the quota year 2008/2009. Technical Report nr. 138, Marine Research Institute, Reykjavík, Iceland. In Icelandic with English summary.

28. Maunder MN, Sibert JR, Fonteneau A, Hampton J, Kleiber P, et al.. (2006) Interpreting catch per unit effort data to assess the status of individual stocks and communities. ICES J Mar Sci 63: 1373–1385.

29. Anonymous (2000) State of marine stocks in Icelandic waters 1999/2000. Prospects for the quota year 2000/2001. Technical Report nr. 75, Marine Research Institute, Reykjavík, Iceland. In Icelandic with English summary.

30. Anonymous (2001) State of marine stocks in Icelandic waters 2000/2001. Prospects for the quota year 2001/2002. Technical Report nr. 80, Marine Research Institute, Reykjavík, Iceland. In Icelandic with English summary.

31. Myers RA, Hutchings JA, Barrowman NJ (1996) Hypothesis for the decline of cod in the North Atlantic. Mar Ecol Prog Ser 138: 293–308.

32. Heino M, Dieckmann U, Godø OR (2002) Measuring probabilistic reaction norms for age and size at maturation. Evolution 56: 669–678.

33. Barot S, Heino M, O'Brien L, Dieckmann U (2004) Estimating reaction norms for age and size at maturation when age at first reproduction is unknown. Evol Ecol Res 6: 659–78.

34. Barot S, Heino M, OBrien L, Dieckmann U (2004) Long-term trend in the maturation reaction norm of two cod stocks. Ecol Appl 14: 1257–1271.

35. Anonymous (2007) State of marine stocks in Icelandic waters 2006/2007. Prospects for the quota year 2007/2008. Technical Report nr. 126, Marine Research Institute, Reykjavík, Iceland. In Icelandic with English summary.

36. Árnason E (2004) Mitochondrial cytochrome b DNA variation in the high fecundity Atlantic cod: Trans-Atlantic clines and shallow gene-genealogy. Genetics 166: 1871–1885.

37. Waples RS (1998) Separating the wheat from the chaff: Patterns of genetic differentiation in high gene flow species. Heredity 89: 438–450.

38. Nielsen EE, Hansen MM, Meldrup D (2006) Evidence of microsatellite hitchhiking selection in Atlantic cod (Gadus morhua L.): Implications for inferring population structure in nonmodel organisms. Mol Ecol 15: 3219–3229.

39. Westgaard J, Fevolden S (2007) Atlantic cod (Gadus morhua L.) in inner and outer coastal zones of northern Norway display divergent genetic signature at non-neutral loci. Fish Res 85: 306–315.

40. Pogson GH, Fevolden S (2003) Natural selection and the genetic differentiation of coastal and Arctic populations of the Atlantic cod in northern Norway: a test involving nucleotide sequence variation at the Pantophysin (Pan I) locus. Mol Ecol 12: 63–74.

41. Sarvas TH, Fevolden S (2005) The scnDNA locus Pan I reveals concurrent presence of different populations of Atlantic cod (Gadus morhua L.) within a single fjord. Fish Res 76: 307–316.

42. Lewontin RC, Cockerham CC (1959) The goodness-of-fit test for detecting natural selection in random mating populations. Evolution 13: 561–564.

43. Wallace B (1958) The comparison of observed and calculated zygotic distributions. Evolution 12: 113–115.

44. Prout T (1965) The estimation of fitness from genotypic frequencies. Evolution 19: 546–551.

45. Robertson A (1965) The interpretation of genotypic ratios in domestic animal populations. Anim Prod 7: 319–324.

46. Pampoulie C, Jakobsdóttir KB, Marteinsdóttir G, Thorsteinsson V (2007) Are vertical behaviour patterns related to the Pantophysin locus in the Atlantic cod (Gadus morhua L.)? Behav Genet 38: 76–81.

47. Pálsson ÓK, Thorsteinsson V (2003) Migration patterns, ambient temperature, and growth of Icelandic cod (Gadus morhua): evidence from storage tag data. Can J Fish Aquat Sci 60: 1409–1423.

48. Charlesworth B (1994) Evolution in Age-Structured Populations. Cambridge, UK: Cambridge University Press.

49. Galvani AP, Slatkin M (2004) Intense selection in an age-structured population. Proc R Soc Lond Ser B 27: 171–176.

50. Swain DP, Nielsen GA, Sinclair AF, Chouinard GA (1994) Changes in catchability of Atlantic cod (Gadus morhua) to an otter-trawl fishery and research survey in the southern Gulf of St Lawrence. ICES J Mar Sci 51: 493–504.

51. Endler J (1986) Natural Selection in the Wild. Princeton, USA: Princeton University Press.

52. Kingsolver JG, Hoekstra HE, Hoekstra JM, Berrigan D, Vignieri SN, et al.. (2001) The strength of phenotypic selection in natural populations. Am Nat 157: 245–261.

53. Hutchings JA (1996) Spatial and temporal variation in the density of northern cod and a review of hypotheses for the stock's collapse. Can J Fish Aquat Sci 53: 943–962.

54. Blanchard JL, Mills C, Jennings S, Fox CJ, Rackham BD, et al.. (2001) Distribution-abundance relationships for North Sea Atlantic cod (Gadus morhua): observation versus theory. Can J Fish Aquat Sci 62: 2001–2009.

55. Marteinsdottir G, Gunnarsson B, Suthers IM (2000) Spatial variation in hatch date distributions and origin of pelagic juvenile cod in Icelandic waters. ICES J Mar Sci 57: 1182–1195.

56. Schopka SA (2007) Area closures in Icelandic waters and the real-time closure system. A historical review. Technical Report Technical Report nr. 133, Marine Research Institute, Reykjavík, Iceland. In Icelandic with English summary.

57. Sale PF, Cowen RK, Danilowicz BS, Jones GP, Kritzer JP, et al.. (2005) Critical science gaps impede use of no-take fishery reserves. Trends Ecol Evol 20: 74–80.

58. Ewers RM, Rodrigues AS (2008) Estimates of reserve effectiveness are confounded by leakage. Trends Ecol Evol 23: 113–116.

59. Roberts CM, Bohnsack JA, Gell F, P HJ, Goodridge R (2001) Effects of marine reserves on adjacent fisheries. Science 294: 1920–1923.

60. Gell FR, Roberts CM (2003) Benefits beyond boundaries: the fishery effects of marine reserves. Trends Ecol Evol 18: 448–455.

61. Jónsdóttir IG, Campana SE, Marteinsdottir G (2006) Otolith shape and temporal stability of spawning groups of Icelandic cod (Gadus morhua L.). ICES J Mar Sci 63: 1501–1512.

62. ICES (2008) Report of the NorthWestern Working Group (NWWG). Technical Report ICES CM 2008/ACOM:03, ICES Headquarters, Copenhagen, Denmark.

63. Fevolden SE, Pogson GH (1997) Genetic divergence at the Synaptophysin (Syp I) locus among Norwegian coastal and north-east Arctic populations of Atlantic cod. J Fish Biol 51: 895–908.

64. Walsh PS, Metzger DA, Higuchi R (1991) Chelex 100 as a medium for simple extraction of DNA for PCR-based typing from forensic material. BioTechniques 10: 506–513.

65. Venables WN, Ripley BD (2002) Modern Applied Statistics with S. New York: Springer. fourth edition.

66. R Development Core Team (2007) R: A Language and Environment for Statistical Computing. R Foundation for Statistical Computing, Vienna, Austria.

67. Sarkar D (2007) lattice: Lattice Graphics. R package version 0.17-2.

68. Sarkar D (2008) Lattice: Multivariate Data Visualization with R. New York: Springer.

69. Wood S (2006) Generalized Additive Models: An Introduction with R. London: Chapman and Hall/CRC.

70. Wood S (2008) Fast stable direct fitting and smoothness selection for generalized additive models. J R Stat Soc Ser B 70: 495–518.

71. Goudet J (2005) HIERFSTAT, a package for R to compute and test hierarchical F-statistics. Mol Ecol Notes 5: 184–186.

72. Yang R (1998) Estimating hierarchical F-statistics. Evolution 52: 950–956.

73. Hedrick PW (2005) Genetics of Populations. Sudbury, Massachusets: Jones and Bartlett Publishers. third edition.

74. Björnsson H, Sólmundsson J, Kristinsson K, Steinarsson BÆ, Hjörleifsson E, et al.. (2007) Stofnmæling botnfiska á Íslandsmidum (SMB) 1985–2006 og stofnmæling botnfiska ad haustlagi (SMH) 1996–2006. undirbúningur, framkvæmd og helstu nidurstödur. Technical Report nr. 131, Hafrannsóknastofnunin. Marine Research Institute, Reykjavík, Iceland. In Icelandic with English summary.

Chapter 7

Analysing Ethnobotanical and Fishery-Related Importance of Mangroves of the East-Godavari Delta (Andhra Pradesh, India) for Conservation and Management Purposes

F. Dahdouh-Guebas, S. Collin, D. Lo Seen, P. Rönnbäck,
D. Depommier, T. Ravishankar and N. Koedam

ABSTRACT

Mangrove forests, though essentially common and wide-spread, are highly threatened. Local societies along with their knowledge about the mangrove

Originally published as Dahdouh-Guebas, F., Collin, S., Lo Seen, D. et al. Analysing ethnobotanical and fishery-related importance of mangroves of the East-Godavari Delta (Andhra Pradesh, India) for conservation and management purposes. J Ethnobiology Ethnomedicine 2, 24 (2006). https://doi.org/10.1186/1746-4269-2-24. © 2021 BioMed Central Ltd. (http://creativecommons.org/licenses/by/2.0)

also are endangered, while they are still underrepresented as scientific research topics. With the present study we document local utilization patterns, and perception of ecosystem change. We illustrate how information generated by ethnobiological research can be used to strengthen the management of the ecosystem. This study was conducted in the Godavari mangrove forest located in the East-Godavari District of the state Andhra Pradesh in India, where mangroves have been degrading due to over-exploitation, extensive development of aquaculture, and pollution from rural and urbanized areas (Kakinada).

One hundred interviews were carried out among the fisherfolk population present in two mangrove zones in the study area, a wildlife sanctuary with strong conservation status and an adjacent zone. Results from the interviews indicated that Avicennia marina (Forsk.) Vierh., a dominant species in the Godavari mangroves, is used most frequently as firewood and for construction. Multiple products of the mangrove included the bark of Ceriops decandra (Griff.) Ding Hou to dye the fishing nets and improve their durability, the bark of Aegiceras corniculatum (L.) Blanco to poison and catch fish, and the leaves of Avicennia spp. and Excoecaria agallocha L. as fodder for cattle. No medicinal uses of true mangrove species were reported, but there were a few traditional uses for mangrove associates. Utilization patterns varied in the two zones that we investigated, most likely due to differences in their ecology and legal status. The findings are discussed in relation with the demographic and socio-economic traits of the fisherfolk communities of the Godavari mangroves and indicate a clear dependency of their livelihood on the mangrove forest.

Reported changes in the Godavari mangrove cover also differed in the two zones, with significantly less perceptions of a decrease in the protected area, as compared to the adjacent non-protected area. A posteriori comparisons between sequential satellite imagery (retrospective till 1977) and respondents that were at least 15 years back then, revealed a mangrove decrease which was however perceived to different extents depending on the area with which the fishermen were familiar. While local needs had not been incorporated in the existing policy, we created a framework on how data on ethnobotanical traditions, fishery-related activities and local people's perceptions of change can be incorporated into management strategies.

Background

Mangrove forests fulfill a number of well-documented and essential ecological functions in tropical and subtropical regions. They generate a variety of natural resources and ecosystem services that are vital to subsistence economies and sustain local and national economies [e.g., [1-6]]. Mangroves provide breeding, spawning, hatching and nursery grounds for both coastal and offshore fish and shellfish stocks [3,7-13]. They also serve as a physical buffer between marine and terrestrial communities [e.g., [14-17]]. For local peoples, mangrove supply wood and products are harvested directly within the mangrove forest. Rapid population growth and increase utilization of mangrove habitats threatens these communities. Developing sustainable management policies that also consider the subsistence requirements of local people, is a high priority (e.g., [18,19]), particularly in India. Socio-economic or socio-ecological studies on mangroves are becoming more and more used [e.g., [20]]. However, so far, few ethnobiological surveys in mangroves have been conducted, in particularly for the general documentation of mangrove ethnobiology [e.g., [2,4,21]], the retrospective study of ecosystem changes (e.g., [22-24]), and for the investigation of management issues prior to the adoption of a particular policy [e.g. 25-27]. The same is true for the ethnobiological aspects of the seagrass (28) and coral reef ecosystems (29), which are often adjacent to mangroves.

Mangrove cover in India is estimated to be around 6,700 km2 (30), of which 80% occurs in extensive deltaic mangrove formations along the east coast, and in the Andaman and Nicobar Islands [31]. In the State of Andhra Pradesh, a long coastline in the Districts of Krishna, Godavari East and Godavari West host natural mangrove forest along with Casuarina equisetifolia Forest & Forest plantations. The Indian mangrove flora comprises 50 species (incl. mangrove associates) and is dominated by Avicennia and Rhizophora spp., except for the Godavari wetlands, where Rhizophora is poorly represented [32].

The Godavari Delta, like many other deltaic systems in India, has been highly altered by human activity [32]. Since at least 1893, mangroves in this area have been subjected to heavy exploitation for fuelwood. Mangrove forests suffered heavily under various working plans until the 1978 Coringa Wildlife Sanctuary was created in the northern part of the Godavari mangrove [33,34]. The Forest Service permitted wood harvest in selected mangrove blocks. These areas were clear-cut, with the hope that the mangrove forests would regenerate naturally. Residents in nearby towns used the mangroves for agriculture, salt production and aquaculture. The Coringa Wildlife Sanctuary and other areas in the Godavari Mangrove Forest were subjected to heavy felling and feral cattle grazing, resulting in large scale depletion of the Godavari mangroves [33]. The forest is still degrading under increasing anthropogenic pressure from rural and urban areas near the city of Kakinada [35]. Causes for Godavari mangrove degradation includes conversion to aquaculture ponds, pollution, eutrophication and siltation of Kakinada Bay and its rivers, anthropogenically induced river flow change and erosion, seasonal hydrological changes, and over-exploitation by villagers [36-38]. The latter cause has lead to the current ban on wood extraction [39].

Although the current statutory provisions prohibit removal of wood, grazing of animals and establishing other activities such as shrimp farms, the Godavari mangroves are being used in an unsustainable manner [40]. Therefore, together with the Forest Department (FD) and non-governmental organisations (NGO's), the M.S. Swaminathan Research Foundation (MSSRF) initiated the Coastal Wetlands: Mangrove Conservation and Management-Project in 1997 [41]. This project empowers local people to develop subsistence policies and provides resources that serve as alternatives for mangroves (e.g. gas stove instead of firewood). It is within this framework that the present study fits.

To organise participatory activities community-based organisations formed the Eco-Development Committee (EDC) and the Vana Samrakshana Samithi (VSS). A subcommittee called Mangrove Restoration and Management Committee was created to ensure locals' participation in the restoration project (Personal communication: Forest Department, Wildlife Conservation Rajahmundry, 2001). Handouts in Telugu (the local language) about the project were published, community meetings were held, Mangrove Clubs were formed, and illustrations on the importance of the mangroves were painted on the walls of the demonstration villages to increase local awareness [42].

The goals of the present study, carried out in a wildlife sanctuary and an adjacent non-sanctuary area, are, to acquire information on traditional uses of the mangrove ecosystem from the fishermen communities in these two areas of the Godavari mangroves, to acquire information on local perception of change, and to show how these ethnobiological data in sites with different protection status can be used to improve conservation and management of the area.

Methods

Description of the Study Site

The 33,263 ha Godavari mangrove wetlands are located between 16°30'-17°00'N and 82°10'-80°23'E in the East-Godavari District (Figure 1). Situated at the mouth of the 1,330 km long Godavari River (India's second longest), the Godavari mangrove forest is the second largest mangrove area on India's East Coast. It includes 15 'true mangrove species' sensu Jayatissa et al. [43] and Dahdouh-Guebas et al.. [16]. The most important species are Avicennia marina (Forsk.) Vierh., Avicennia officinalis L., Excoecaria agallocha L., Aegiceras corniculatum (L.) Blanco, Sonneratia apetala Buch.-Ham., Ceriops decandra (Griff.) Ding Hou, Rhizophora apiculata Blume and Rhizophora mucronata Lamk. [44]. Mangrove nomenclature is following Tomlinson [45], whereas that of other species is following Mabberley [46].

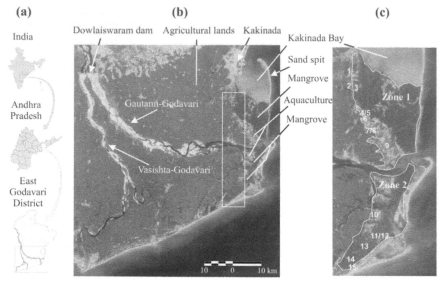

Figure 1. Study site. (a) Location map of India showing the state of Andhra Pradesh and the East-Godavari District (redrawn from NRSA [82]). The small black rectangle indicates the area in (b). (b) Satellite image of the Godavari Delta taken in March 1999. Adjacent to the study area, the white rectangle (ca. 320 km²) indicates the area used to extract demographic data (% fishermen) stored in the database of the South Indian Fertility Project (French Institute of Pondicherry, 2001). (c) Map of the study area investigated with the two zones and the 15 villages therein (numbered as in table 1).

Table 1. Criteria and their sources for the relative distinction of the Godavari mangroves into two zones, and villages studied in each zone (village numbers correspond with those in figure 1). The n-values between brackets indicate the number of questionnaires per village used in this analysis (total = 100). The asterisk refers to Appendix 1, which provides the legal text.

CRITERIA	ZONE 1	ZONE 2
Separation into Reserve Forests with restricted activities (1947 Indian Forest Law part C §20)	yes	yes
Wildlife Sanctuary* [33]	PRESENT: Coringa Wildlife Sanctuary	ABSENT
Prohibitions*	felling of trees and any type of extraction	felling of trees and collection of green wood
Restrictions on entry*	only civil servants or people living inside allowed	also people not inhabiting the sanctuary allowed
Mangrove density	higher	lower
Mangrove species richness	lower	higher
Implementation of Forest Department regulations	to a high degree	to a lesser degree
Presence of Forest Department personnel	strongly present	less present
Accessibility of villages adjacent to the mangrove	very accessible	less accessible
Rehabilitation program/mangrove plantations	present	absent
Aquaculture	present	present
New aquaculture ponds	less present	strongly present
Villages sampled	1. Chollangipetta (n = 7) 2. Kotthuru (n = 6) 3. Ramannapalem (n = 6) 4. Peddha Bodduvengatapalem (n = 5) 5. Chinna Bodduvengatapalem (n = 5) 6. Chinna Valasala (n = 6) 7. Peddha Valasala (n = 7) 8. Laksmipathipuram (n = 6) 9. Gadimoga (n = 7)	10. Balusutippe (n = 9) 11. Molletimoga (n = 6) 12. Kothapallem (n = 8) 13. Pora (n = 8) 14. Pandi-Pallam (n = 10) 15. Neellarevu (n = 4)

A major part of the Godavari mangroves is separated from the Bay of Bengal by Kakinada Bay. Two major shifts in the main course of the Godavari River and the formation of a sand spit have occurred since the construction of the Cotton Barrage at Dowlaiswaram in 1852 (Figure 1). Until the 1930s, the Godavari flowed northwards, opening into Kakinada Bay. Between the 1930s and the 1970s, its course gradually shifted southwards. Since the 1970s the Godavari River flows eastwards. These shifts can be explained by a combination factors including the flatness of the alluvial zone, variations in river flow, and frequent cyclonal activity in the area [47].

Sampling Design and Methodology

We divided the Godavari mangrove area in 2 distinct zones based on a priori sample criteria (Table 1). The most important criterion was the differential legal protection status : Zone 1 comprised the Coringa Wildlife Sanctuary, whereas Zone 2 was a non-sanctuary area (Figure 1). We sampled the local population of nine villages in Zone 1 and six villages in Zone 2. The Hindu fishermen communities inhabiting these villages belong to the Agnikula Kshatriya caste. Their common language is Telugu, a Dravidian language, which is largely spoken in Andhra Pradesh [48]. Additional details regarding the socio-cultural background of the sampled communities can be found in Suryanarayana [48].

In each village, we randomly selected 4 to 10 households for interviews. A total of 55 households completed questionnaires in Zone 1 and 45 in Zone 2. We took the household as a sampling unit and we interviewed as described in Dahdouh-Guebas et al.. [4: p 516]. We conducted interviews in Telugu, with the assistance of two English-Telugu bilingual translators native to the East-Godavari District. We assessed the mangrove knowledge of respondents with ethnobotanical questions, aided by a botanical photographic catalogue showing the tree physiognomies, leaves, fruits, flowers and seeds of each mangrove species. The rest of the semi-structured questionnaire contained both multiple choice and open-ended questions, which covered ethnobotanical and fishery-related issues, local perception of change in the mangroves, as well as personal socio-economic questions for each household (Appendix II). The questionnaire had not the aim to analyse gender issues or other within-household differentiation on the level of resource use. The survey was complemented with visual observations, and the collection of secondary data from both governmental organisations and NGOs. All fieldwork was carried out in October and November 2001.

There were no direct statistics available about the percentage of fisherfolk that we interviewed. According to the demographic data of 2000 obtained from the Mandal Offices of Tallarevu and Katrenikona a total of 34,625 people inhabited the villages surveyed, all of which had access to electricity, and, apart from Pora and Neellarevu, all villages contained a school. From the database of the South Indian Fertility Project at the French Institute of Pondicherry we calculated that the percentage of total active population (i.e. not schooling, not retired or not unemployed, although we acknowledge that it is possible that those classified as schooling, unemployed or retired would still be involved in fishing, catching or collecting in the mangrove, probably as an important coping strategy) in the villages adjacent to the survey area (Figure 1) constituted 36.4 % of the total population in that area, and that 15.7 % of this total active population were fishermen. There were no available data about active population within our study area, so we assumed that the proportion of total active population in our study area was not lower than the above figure for the adjacent villages. However, it is very likely that the study area had a higher proportion of fishermen, particularly in Zone 2. Considering a maximal proportion-of-fishermen-range between 15.7% and 100 %, and assuming that all members of the active population are married and divided into households with 2 parents, our survey then covered between 1.57 % and 10.11 % of the fishermen households in the study area, which is a demographically sound sampling basis.

Statistical Analyses

To analyse the questionnaire data statistically we used the χ^2-test or the related G-test [49] when confronting various classes. These tests were most preferable as we were dealing with qualitative response classes. We did between-zone comparisons of means using t-tests. We did combinatory statistical analyses involving age by splitting the age classes in two equal groups and by confronting the upper with the lower age classes (see results). In the retrospective questions (past decade), we omitted answers from respondents below 25 years of age from further analysis, because younger cohorts could not realistically answer these questions (e.g. youngsters of 25 in 2001 were just born in 1977, see results for a posteriori comparisons with retrospective remotely sensed imagery).

Results

Demography

The age of the fishermen interviewed ranged from 16 to 55 years old, of which 88 % was native to the villages. The main income of all the respondents originated from fishing, and ranged from less than 2,000 to more than 10,000 Indian Rupees (INR) annually (Table 2) – during the fieldwork 1 € = 43.48 INR -. A majority of the fisherfolk lived in a simple kutcha house (Figure 2a; Table 2) and possessed little extra items (e.g. farm animals, bicycle, TV). Considering this sampling homogeneity, and considering that the number of interviews per wealth class per village was low for most wealth classes we did not go into their statistical comparisons.

Figure 2. Photographs illustrating ethnobiological relationships and impacts on the mangrove. (a) House construction of a kutcha (roughly meaning 'low class'). (b) House construction of a pucca (roughly meaning 'high class'). (c) Traditional use of mangroves as fuelwood. (d) Fishermen holding a tray with pieces of Ceriops decandra bark used for dyeing fishing nets. They also show two freshly dyed nets and in the background previously dyed nets are hung to dry. (e) Herdsman milking his feral water buffalo that is consuming Avicennia alba twigs. (f) Sorting of Avicennia spp. seedlings in a mangrove nursery. (g) Although the cause of the destruction of the mangroves on the foreground is natural (cyclone 07B), the irony of this photograph is that in the background fishermen are fishing for species that are dependent on the mangrove otherwise functioning as breeding, spawning, hatching and nursery grounds. (h) Shrimp farm ponds established along Kakinada Road near Gadimoga (Zone 1) at the expense of mangrove forest. (i) Publicity in favour of shrimp farming, showing the (short-term) economic gains that may result from this activity (golden bracelet). (Photographs by Sarah Collin, Deirdre Vrancken and Nico Koedam).

Table 2. Annual income in Indian Rupees (INR) and other assets available to the 100 fisherfolk households interviewed.

ASSETS		# HOUSEHOLDS
Annual income :		
	< 2,000 INR	5
	2,000 – 5,000 INR	47
	5,000 – 10,000 INR	26
	> 10,000 INR	9
	no answer	13
Agricultural land		2
Coconut trees (*Cocos nucifera* L.)		29
Neem trees (*Azadirachta indica* A. Juss.)		1
Moringa trees (*Moringa* spp. Adans.)		4
Smallstock with goats		9
Livestock with buffaloes		3
Nava (= boat) :		
	own property	40
	shared property	27
	rented	24
	motorised	2
Bicycle		15
TV		23
Gas stove		4
Electricity		51
House type :		
kutcha = wood and mud hut, palm roofing (Fig. 2a)		65
semi-*kutcha* = tached hut, palm roofing		15
semi-*pucca* = tiled house		1
pucca = concrete house (Fig. 2b)		19

Ethnobiology

Respondents referred to the general mangrove forest as mada adavi, meaning Avicennia forest. When inquiring about the exact meaning of 'mangrove', 56 respondents referred to the vegetation, 44 to the entire ecosystem (fauna, flora), 27 to the windbreak protecting their villages against cyclones and floods, and 8 to the direct resource (firewood, building wood, fodder).

The level of knowledge for the 13 true mangrove species encountered in this study was subdivided into 4 categories, each corresponding to a minimum number of species recognised : low (< 5 species), fair (5–7 species), good (8–10 species) and very good (> 10 species). Of all respondents, 83% had a good or very good knowledge (Table 3). When combining this level of knowledge with the age of the respondents we saw that, although there are obvious differences between the single age classes per se, there was no significant trend of mangrove knowledge with age (upper versus lower four age classes; $\chi 2 = 0.027$; df = 1; p > 0.1). The level of knowledge varied across mangrove species and according to the zone the respondents lived in. Zone 1 respondents were less likely to recognize A. marina, A. officinalis, Ceriops decandra, Lumnitzera racemosa Willd., Rhizophora mucronata and Xylocarpus granatum König, and more likely to recognise Avicennia alba Blume, Bruguiera gymnorrhiza (L.) Lamk. and Sonneratia apetala in the same zone ($6.920<\chi 2<53.875$; df = 1; $2.14*10-13 < p < 0.03$). There were no significantly different levels of knowledge between the zones for Aegiceras corniculatum, Excoecaria agallocha, Rhizophora apiculata and Sonneratia caseolaris (L.) Engler ($0.000<\chi 2< 2.296$; df = 1; $0.477< p < 1.000$). Another striking observation was that in Zone 2 Avicennia alba and the mangrove associate Hibiscus tiliaceus L. were unknown.

Table 3. Combinatory analysis of the level of knowledge of true mangrove species and the age of the respondents. Methods and results on the statistical analysis are given in the text.

AGE CLASS	# RESPONDENTS (= 100%)	LEVEL OF KNOWLEDGE (%)			
		bad	fair	good	very good
16 – 20	10	0	0	70.0	30.0
21 – 25	12	8.0	25.0	33.5	33.5
26 – 30	17	0	30.0	35.5	35.5
31 – 35	22	0	4.0	67.0	29.0
36 – 40	13	0	30.0	35.0	35.0
41 – 45	11	0	0	54.5	45.5
46 – 50	11	0	18.0	64.0	18.0
51 – 55	4	0	25.0	25.0	50.0
TOTAL	100	1.0	16.0	45.0	38.0

The respondents commonly referred to the 'use' of mangroves as a fishing area (89 %), in which they penetrate on average 10 km in Zone 2 and 15 km in Zone 1 (t = 2.25; df = 88; p < 0.05). On average, they visited mangroves for fishing 15 times per month in Zone 1 and 23 times per month in Zone 2 (t = -5.60; df = 68; p < 0.001).

Some of the uses of mangroves are illustrated in Figure 2. Among the wood and non-wood mangrove uses, a majority of the households reported the personal use of mangrove wood for fuel (Figure 2c) and construction (Figure 3). Within the construction class, respondents distinguished between poles (36% of construction use), roof beams (35 %), fences (26 %) and shelters (3 %). In addition to true mangrove species, 41% of the fishermen harvested other species for fuel, including Borassus flabellifer L., Cocos nucifera L., Casuarina equisetifolia and Prosopis spicigera, or they used sun-dried cow dung or a gas stove. However, since the true mangrove species had nearly ideal calorific values, the villagers found it difficult to use alternative resources. Likewise, 57% of the fishermen used Borassus flabellifer L., Bambusa arundinacea (Retz.) Willd. and Casuarina equisetifolia as alternative construction species.

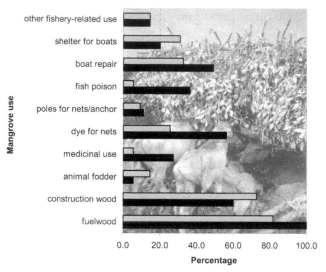

Figure 3. Percentage use of different mangrove use classes in Zone 1 (black) and Zone 2 (grey) amongst the 100 interviewed households (n_{zone1} = 55; n_{zone2} = 45). The background photograph shows Avicennia branches used as fodder for feral water buffaloes. (Photograph by Deirdre Vrancken).

There was no significant difference between the two zones for fuelwood use, but there was a significant difference in the frequency and in the distance that respondents travelled to collect it. On average, inhabitants of Zone 1 travelled 17 km 11 times per month, while those in Zone 2 travelled 27 km 5 times per month (frequency : t = -4.46; df = 55; p < 0.001, distance : t = 3.40; df = 72; p < 0.002). Zone 2 inhabitants also used significantly more mangrove as building wood ($\chi2$ = 9.065; df = 1; p < 0.01). Among the other uses (Figure 3), there were also significantly higher uses of true mangroves species or mangrove associate species for medicine ($\chi2$ = 5.792; df = 1; p < 0.02), dye for nets ($\chi2$ = 4.398; df = 1; p < 0.05) and fish poison ($\chi2$ = 10.705; df = 1; p < 0.01) in Zone 1 as compared to Zone 2. There were however no significant trends in mangrove use with age (0.004<$\chi2$< 1.822; df = 1; n.s.). We also did not find differences in mangrove use between the income classes for which enough data were available (class 2,000–5,000 INR and class 5,000–10,000; see Table 2). Therefore income was not further analysed as a socio-economic factor in the light of the results presented in this paper.

Mangrove uses by species are reported in Table 4. Mangrove associates Thespesia populnea (L.) Solander ex. Correa and Clerodendron inerme (L.) Gaertn.

were also used, as the most used species for boat repair (21%) and as one of the least used species for fodder (1%) respectively. Mangrove bark was employed as a dye plant (Figure 2d). Between 1 and 2 kg of Ceriops decandra bark was boiled in water to create a red dye to color and increased the durability of fishing nets. This was done once or twice per month, depending of the village. The bark of Aegiceras corniculatum was converted into a paste and used as a fish poison. Some villagers also reported medicinal use of the mangrove associates Caesalpinia bonduc (L.) Roxb., Clerodendron inerme, Dalbergia spinosa Roxb., Derris trifoliata Lour. and Hibiscus tiliaceus, but no consistent data were obtained. The shopkeeper of an Ayurvedic shop in Kakinada reported that Avicennia marina was used as a drug against diarrhoea and dysentery, but an Ayurvedic manufacturer in Udoppa, 30 km north of Kakinada, could not confirm this. We explored local Ayurvedic literature about the topic and refer to Nadkarni [50] and Jain and Defilipps [51] for detailed information.

Table 4. Tree and shrub species of the Godavari mangroves and their reported multiple uses by the fishermen of the riverine villages.

BOTANICAL NAME	VERNACULAR NAMES IN TELUGU*	fuelwood (Fig. 2c)	constructi on wood	fodder (Fig. 2e)	boat repair	poles for nets/anchor	other uses
Aegiceras corniculatum	*Guggilam, Dudumara*	32	2	0	0	0	fish poison
Avicennia alba	*Gundu mada, Vilava mada*	<1	0	0	0	0	no reports
Avicennia marina	*Tella mada*	100	60	11	9	2	no reports
Avicennia officinalis	*Nalla mada*	85	36	4	<1	0	no reports
Bruguiera gymnorrhiza	*Thuddu ponna, Uredi*	1	19	0	0	1	no reports
Ceriops decandra	*Gatharu, Thogara*	3	16	<1	0	0	dye/tannin for fishing nets (Fig. 2d)
Excoecaria agallocha	*Tilla, Tella, Chilla*	21	14	9	3	3	no reports
Lumnitzera racemosa	*Thanduga, Kadavi, Kadivi, Than*	17	25	1	13	0	no reports
Rhizophora apiculata	*Uppu ponna, Kaaki ponna*	2	26	0	0	5	no reports
Rhizophora mucronata	*Uppu ponna*	2	26	0	0	5	no reports
Sonneratia apetala	*Kalinga (Chinna), Kyalanki*	1	<1	1	0	0	no reports
Sonneratia caseolaris	*Kalinga (Peda), Kandia*	1	<1	1	0	0	no reports
Xylocarpus granatum	*Chenuga*	<1	<1	0	2	0	no reports

* Complemented or corrected by the nomenclature given by Pròsperi et al. [81].

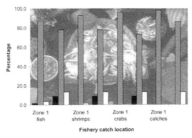

Figure 4. Reported perception on the changes in fish, shrimp and crab catch, and of catches in general between 1991 and 2001, in both zones (n_{zone1} = 55; n_{zone2} = 45). Black = increase; grey = decrease; white = no change. The background photograph shows crab and fish sale at a local market. (Photograph by Nico Koedam).

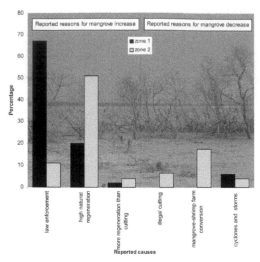

Figure 5. Reported causes for the reported increase and decrease in mangrove cover (n = 100). The background photograph shows the mangrove habitat for fish and shellfish destroyed by a cyclone. (Photograph by Nico Koedam).

In the mangrove communities, penaeid shrimps were the most important commercial catch by value (53%), followed by fish (32%), crabs (15%) and shrimp seed (1%). However, finfish catches were as important as penaeid shrimps by weight. More than 90% of the catch was sold, with no significant differences between fish, shrimps or crabs, or between zones.

Local Perception on Dynamics and Regulations

Seventy six percent of the fishermen of Godavari mangroves reported that the mangrove vegetation had increased over time, and they shared the opinion that this trend would continue. Seventy percent also indicated that the mangrove associate Acanthus ilicifolius L. had increased in vegetation cover. Among the reported reasons for the reported increase were the implementation and enforcement of Forest Department rules, a high natural regeneration, and a natural regeneration that exceeds the incidence of cutting (Fig 5). Local respondents reported that when the mangrove trees were cut, stumps would regenerate by producing new shoots. Illegal cutting of trees, mangrove conversion to aquaculture, and destructive weather phenomena (cyclones, storms) were the major reported responses for a reported decrease (Figure 5). There was no significant age trend in the proportion of people that reported an increase or decrease of the mangrove area (χ^2 = 0.025; df = 1; n.s.). Answers for the category of fishermen that were at least 15 years old in 1977, indicated that the answers for this category was not significantly different from the full set of data (G<0.70; df = 2; p > 0.1 n.s.). This extra test was necessary because we compared the perception of the fishermen with data based from satellite images of January 1977 (Landsat MSS), March 1993 (Landsat TM) and March 1999 (IRS LISS III) in de Solan (2001) and in VUB [39] a posteriori. Contrary to fishermen's perceptions, this revealed a decrease in mangrove vegetation cover. We confronted these results with the area acquainted with by the fishermen, by comparing the average distance that respondents travel for fishing (from interviews), with the remotely sensed changes that occurred within a buffer around their village with that distance as a radius. Applying GIS-technology (geographical information systems), we examined the changes in the mangrove within a 250 m margin of creek or sea separate from changes in the interior of the forest. We found that for all villages mangrove decrease largely occurred in the interior of the forest, and that colonisation (or planting; see Figure 2f) of new mangroves almost exclusively occurred along the water edge of creeks.

There was no significantly different view between people with a good to very good knowledge on mangrove species and people with a bad to fair knowledge ($\chi^2 = 1.830$; df = 1; n.s.). There was, however, a clear geographical trend. The proportion of people reporting a decreasing mangrove cover was significantly larger in Zone 2 than in Zone 1 ($\chi^2 = 7.238$; df = 1; 0.001<p < 0.01).

Fishermen unanimously reported that the catches have declined over the past 10 years (Figure 4), but the report of this decrease was significantly higher in Zone 1 as compared to Zone 2 (4.865<χ^2< 10.277; df = 1; 0.001< p < 0.05). The causes to which the fishermen attributed this decrease cover both ecosystem-related and fishery-related issues (Figure 6).

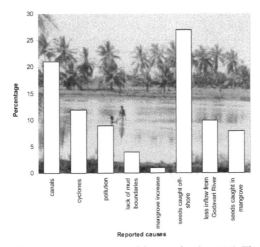

Figure 6. Reported causes for the reported decrease in fishery catches (n = 100). The background photograph shows the collection of shrimp seed near Gadimoga in Zone 1 (Photograph by Sarah Collin).

Ninety five percent of the fishermen were aware of the Forest Department regulations. The remaining 5% entirely were from Zone II where the implementation of the rules was less pronounced, the number of Forest Department personnel was lower and the accessibility to the villages was poorer (cf. Table 1). Out of the 95% of fishermen that knew the rules, 97% accepted the rules because, as one respondent stated, "a ban on cutting means an increase in mangrove cover, which is directly beneficial for the livelihood of the villagers." But 35% of the fishermen disliked the fact that the cutting was illegal, since fuelwood was used daily for cooking and other household purposes. The high fines when caught while cutting or collecting green wood were not appreciated: 32 INR (1 € = ca. 44 INR in 2001), which maybe doubled or even increased by a five-fold, for one load 10 to 20 kg.

Discussion

Mangrove Etymology

The term 'mangrove forest', 'mada adavi' in Telugu, refers to the genus Avicennia (mada), but it is unclear whether it is the genus that adopted the name 'mangrove' because of its high abundance in this local forest (implying that in other regions, where other genera are more abundant, people would refer to the 'mangrove forest' with other names), or it is the forest in general that was named after this genus (implying that even in areas with other generic abundances people would still refer to the mangrove with the 'Avicennia' genus). Although less logic, the latter was observed in the Teacapan-Agua Brava Lagoon in Mexico, where people regularly referred to Laguncularia racemosa (L.) Gaertn.f., the locally most abundant species, as 'mangle rojo', which commonly indicates the regionally more abundant Rhizophora mangle L. [52]. It remains however very informative to analyse the etymology of the species or genus names, which provides insight on their popularity (knowledge by local people), ethnobotany and ecology. 'Tella mada' (Avicennia marina) thus means 'White Avicennia', a species which in English is commonly known as the 'Grey mangrove'. 'Nalla mada' (Avicennia officinalis) means 'Black Avicennia', a vernacular name which in English is reserved for Avicennia germinans (L.) Stearn.. 'Chinna' and 'Peda' are adjectives and mean respectively 'small' and 'large', used in the Sonneratia (Table 4) because the first species does not reach the heights of the second. 'Guggilam' refers to the tree Aegiceras corniculatum whereas 'Guggilupu' refers to its fruit. Also for climbing mangrove associates similar etymologies exists, such as 'Tiga' literally meaning 'creeper' and used for Derris trifoliata (Nalla tiga).

Socio-Demographic and Economic Traits

With an average annual income of about 3,500 INR, fishermen are considered to be among the poorest communities in society in India [see also [40]]. Most fishermen families (65%) live in kutcha's, the simplest among the four common house types (Table 2). Although this house type has been used a standard-of-living indicator, our study reveals that this may be inaccurate, since only 25% of people that earn between 2,000 and 5,000 INR annually claim to live in a kutcha house.

Ethnobotany and Fisheries

It is clear from the results that although the vegetation is of prime importance (cf. ethnobotanical uses, fisheries ground), the fishermen interpret the broader concept, function and service of the mangrove. Therefore we suggest to adopt the new term 'anthroposystem', defined as an ecosystem in which the traditional user is a subsistent ecosystem element.

Respondents do not distinguish between Rhizophora mucronata and R. apiculata (both Ponna or Uppu ponna), but they do distinguish between Avicennia marina (Tella mada) and A.officinalis (Nalla mada), with A. alba (Vilava mada) also less known (15% of respondents). Significant differences were observed between the knowledge in Zone 1 and Zone 2. This could be due to differences in abundances in the two zones. Data suggest that residents of Zone 2 visit mangrove areas more frequently than do those of Zone 1. This may lead to their greater familiarity with the species.

Although there are relatively few studies on the human uses of mangroves; publications on mangroves from Kenya [4,53,54], Tanzania [55], Vietnam [56], Mexico [24] and the Philippines [6,21], all report that construction and fuelwood are the primary uses of mangrove species. In the Godavari delta, Avicennia spp. and Rhizophora spp. are used in a mixture as poles and beams for hut building, and to construct fences and shelters (this study), but one report also highlighted the rare Xylocarpus to be exploited for its valuable timber [57]. In West Bengal, Bruguiera gymnorrhiza and Heritiera littoralis Dryand. have been reported as particularly valuable timber [58]. In Kenya, Rhizophora is favourised for house construction because of their ability to grow long and straight [4], but in the Godavari mangroves this genus is not as densely represented and rarely reaches appropriate sizes for hut building [32,44].

Although the long-standing traditional relationship with feral water buffaloes is important in the livelihood of the local people (Figure 2e) [59], and buffaloes have been observed foraging the mangrove, almost none of the fishermen (13%) admit letting their cattle graze in the forest. They claim to cut Avicennia spp. and

Excoecaria agallocha leaves and bring them to the village where the cattle roam around.

Although some medicinal use of true mangrove species has been documented [2,4,60], no such medicinal use was reported in the present survey (there were however a few examples of medicinal uses for mangrove associate species). This was contrary to our expectation that was based on the legacy of Ayurvedic and plant medicine in India. We do report the use Ceriops decandra bark, to color and preserve fishing nets. This traditional way of better preserving fishing nets was very relevant in the past when fishing nets were manufactured in cotton [48]. Even though most fishing nets now are made of nylon, 47% of the fishermen interviewed, continue to dye them with the red Thogara paste.

Local Perception on Dynamics and Regulations

There are scores of ethnobiological publications on resource utilisation, and sustainability [e.g., [62,62]]. However, the use of ethnobiological surveys in current and retrospective assessment and monitoring of natural resource status and of ecosystem change in tropical coastal ecosystems, though very promising, is novel [22,63-65]. The majority (76%) of respondents reported that the Godavari mangrove cover has increased compared with the past and they share the opinion that this trend will continue in the future. However, the Godavari mangroves have not been spared by man and have been subjected to heavy exploitation to meet local demands of fuelwood in the past [32,33]. They are still degrading due to a combination of various physical, biological and especially anthropogenical factors [36,41].

Some areas of the Godavari mangroves have been lost by conversion to shrimp farms and erosion [39,47]. During the 22-year period covered by the satellite images, a relative progression of the mangrove in the northwest into Kakinada Bay and a relative regression in the eastern parts can be noted as well [39,47]. The mangrove areas clear-felled by various working plans of the Forest Department in the past [33], were still present on the 1977 satellite image but these open areas have been regenerating successfully [47]. These observations indicate that what people perceive is not always actually being recorded with remote sensing technology. Rather than contradictory, ethnobiological data and remote sensing are complementary, and discrepancies should be interpreted in a sound framework [24,63]. The discrepancy also could be due to the fact that fishermen are acquainted with a relatively small and non-random portion of the area. Being familiar with the water edge only may be the reason why most respondents report a positive feeling about the status of the mangrove. Second, the respondents' distinctions between true mangrove species and mangrove associates may also have biased our a posteriori confrontation with remote sensing data. They reported for instance the dramatic expansion of the mangrove associate Acanthus ilicifolius, but this type of distinction of herbaceous plants is possible only with imagery with submeter spatial resolution, such as IKONOS [66]. In addition, expansion of so-called mangrove species, which in reality are mangrove associated species, may lead to misinterpretation and may mask cryptic ecological degradation in mangrove ecosystems and jeopardise functionality [16,64]. This illustrates once more that remote sensing and ethnobiological surveys are complementary and should be interpreted as such.

Fishermen reported that increased mangrove vegetation resulted from natural regeneration of cut-down stumps. However, only few species (e.g., Avicennia marina, Avicennia officinalis and Excoecaria agallocha) produce stump sprouts [67].

As in many areas world-wide [68-70], Andhra Pradesh has witnessed a shrimp farm industry explosion from 6,000 ha in 1990 to as much as 84,300 ha in 1999, representing more than half of the total shrimp culture area in India [20]. Often this occurs at the expense of mangroves, which function as feeding and nursery grounds [71].

Shrimp farm operations were cited as a cause for the reported decline in fish catches. A small percentage (9%) of the fishermen of the Godavari mangroves attribute aquaculture effluents as the main cause of declining harvests, but also other sources of pollution are likely to contribute [cf. [35]]. The devastating tropical cyclone 07B (6th November 1996) with its typhoon wind speeds of 212.4 km.hr-1 [72], killed 848 persons, damaged 594,000 houses, destroyed 496 ha of crops, and killed 13,507 livestock and 2,079,000 chickens and other poultry [73]. Yet only 12% of the fishermen (Figure 2g) report this to be a significant cause of declining yields. Fishermen also attribute the decline to the creation of drainage canals.

Apart from change in the mangrove area, fishermen also reported a number of fishery-related causes to declining catches. Up to 27% of the mangrove fishermen accuse their peers of overharvesting shrimp larvae, juvenile and adults (offshore trawling), leading to the decrease in catch within the mangroves. This argument is confirmed by Rönnbäck et al.. [20], who report that the coastal waters surrounding the Godavari River are especially rich in penaeid shrimp resources and that trawl catches are dominated by newly emigrated juvenile and sub-adult life stages. The aquaculture-related fisheries for wild shrimp seed and broodstock support major operations in the area, but are fraught with bycatch problems [20]. Surprisingly, none of the respondents reported fish and shellfish habitat loss as a major reason behind declining catches, even though they are well aware of the role of mangroves in supporting fish and shellfish populations. Possibly the respondents were afraid of criticising aquaculture activities, and of direct conflicts with this sector.

Use of Ethnobiological Data in Management Policy

Both scientific and societal elements should form the basis of an efficient conservation and management scheme. Such elements include biological monitoring from remote sensing [e.g., [19]], ecological economics [e.g., [5]], ethnobiological traditions and perceptions (this study), and even eco-religious approaches. With respect to the latter, Palmer and Finlay [74] paraphrases the message of the Bhagavad Gita is 'conserve ecology or perish' – *The Bhagavad Gita* is the dialogue between the Hindu Lord Sri Krishna, the supreme personality of Godhead, and his intimate disciple and Prince of India Arjuna, and is considered the essence of Vedic knowledge. These types of religious texts, which are well-known by the people, have proved to be determining elements to turn failing management policies into success [74]. Too often government policies are based only on monodisciplinary scientific studies, or, worse, just assumptions. Another shortcoming of management plans in other countries is that lack of economically acceptable alternatives for mangrove resource utilisation cause dependency [4]. Apart from scientific data many more elements can and must be used to optimalise a policy. More precisely, the policy should be at the heart of the ecologic, economic and socio-cultural reality of the communities involved. Local people are often forced to adapt to a legal conservation framework without receiving alternatives to traditional uses, or without in-depth comparative analysis of the advantages and disadvantages of the alternatives provided. Without incorporating or sufficiently studying the elements and issues of local stakeholders (food, housing, religion,...), we expect conservation and management strategies to fail.

The different views on mangrove increase or decrease from people in Zone 1 and Zone 2, can be explained by the different legal status of both zones. The implementation of the regulations set up by the Forest Department is better organised in Zone 1, which is declared as the Coringa Wildlife Sanctuary [33]. Interestingly, our results show that people inhabiting this protected area perceive more mangrove increase, whereas people inhabiting Zone 2 (which is not a sanctuary) report significantly more often a mangrove decrease. Such responses can be integrated in future management as indicators for the success of the policy with respect to mangrove conservation (Figure 7). However, the acceptance of a ban on cutting does not guarantee the social success of forest legislation, as fines are too high, and restricted access to natural resources has been reported to increase poverty in India [75]. In Kakinada, the Forest Department provides welfare measures to the villagers living around the mangrove areas to reduce their dependency on mangrove's natural resources [32]. At present, land-based alternatives for fire-wood, construction, fencing and fodder are provided. For the Godavari, the pro-vision of gas stoves by the Forest Department and several NGOs is an alternative for the use of mangrove wood as fuelwood. Unfortunately wood is free, gas is not. Only 4% of the respondents of Zone I possess a gas stove. Providing such an alter-native, or providing alternative wood species in artificial plantations as firewood, only works if the special characteristics of the smokeless mangrove species is taken into account.

The main cause of mangrove decrease reported by the fishermen differs across villages. In Pora (Zone II), located at the end of a long strip of mangrove area be-ing converted to shrimp ponds, 17.5% of the respondents give this conversion as the main cause. The fishermen of Peddha Valasala (Zone I) and Neellarevu (Zone II) report the natural cause of cyclones and tropical storms. Around Peddha Va-lasala, there are only small mangrove patches and a continuous stretch of land is occupied by shrimp farms without any mangrove protection [76]. Neellarevu is located on an island in the mangrove forest, completely isolated from other vil-lages. Tropical cyclone 07B (6th November 1996), although devastating for the coastal villages, has been a revelation for the communities living in areas where natural mangrove forests protected residents from the cyclone fury. The answers received in Neellarevu and Peddha Valasala, where flooding damaged many of the semi-kutcha and semi-pucca houses (Figure 2a, b), claiming a high death toll (loc. cit.), and where previously many patches of surrounding mangrove forest were destroyed (Figure 2g), are in line with a created awareness amongst the coastal vil-lages to preserve this unique mangrove ecosystem as a natural dyke [15,40,77].

Currently, also the very conversion from mangrove forest to shrimp farm ponds, possibly under political patronage [cf. [68]], and the publicity about the (short-term) economic gains involved (Figure 2i), are in strong contrast with the mangrove forest management policy. In other areas on the Indian subcontinent, mangroves are cleared to build tourist resorts which "are supposed to be located right at the beach front" (pers. obs.). This once more neglects the power of man-groves to buffer ocean surges such as from cyclones or tsunamis, and it still remains uncertain to which extent a death toll of more than a quarter million people from a single ocean surge (tsunami of 26 December 2004) will have an effect on global coastal zone management in general, and on the enforcement of local mangrove management policy in particular [16]. Although local inhabitants foraging in the Godavari mangrove at the time of the tsunami disaster testify to have survived thanks to the mangrove (pers. comm. K. Ilangovan, French Institute of Pondich-erry, January 2005), most attention from media and global organisations focused on an early warning system to announce such events without attention for early warning of mangrove degradation [cf. [64,66]].

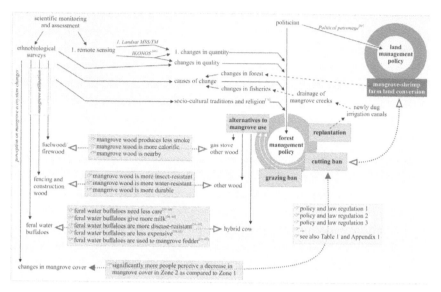

Figure 7. The use of ethnobiological survey data in management policy. The scheme shows forest management actions (central green circle with boxes), and what these actions are primarily based on (elements preceded by a number '1'). It also illustrates where ethnobiological elements could be used to improve the management (elements without a number). Contradictions or conformities between the management actions and the ethnobiological findings are given by the grey dotted arrows (contradiction = open arrow, conformity = closed arrow), and the boxes overlaying them provides a bulleted list with details. Unveiling such contradictions using ethnobiological surveys can help improve the policy. There is also one indication of conflict amongst policies (forest management policy versus land management policy), and impacts involved in the management are given as black dashed arrows. CWS = Coringa Wildlife Sanctuary. Superscripted letters refer to literature references.

Based on the present study, we made a synthesis of the elements that are used in forest management policy to find out that they did not successfully address the needs of the local communities (Figure 7). As elaborated also in the previous sections, we found for instance that quantitative information on extent of the mangrove as detected from classical remote sensing technology is primarily used to define management rules, whereas qualitative information assessed through other remote sensing tools or ethnoscientific surveys in particular, provide a better ecologic and socio-economic basis for a management policy [66,78]. We extracted the elements of our ethnobiological survey, as well as some elements from scientific literature, that point out contradictions between the policy or the alternatives provided by the government on one hand, and the effects of the laws or the evaluation of the alternatives by the local people on the other hand (Figure 7). In an Indonesian case-study Armitage [79] suggested that there is a need to formulate, propose, implement and monitor strategies that contest existing policy narratives and challenge entrenched economic interests and power relationships. It is clear that ethnobiological data, as collected and used in the present study, can be used to display contradictions and to adapt and improve the management. Although the present findings are detailed and provide a good reference on the ethnobotanical aspects of the Godavari mangroves, this type of study should be repeated in 5–10 years to assess traditional use dynamics. This would provide also useful information on the perceptions of the local fisherfolk that can be integrated in existing mangrove management plans, but also on the success of the forest management policy in the elapsed years.

Conclusion

Tropical coastal populations, particularly in developing countries, can be highly dependent on the mangrove ecosystem for multiple purposes [2,4,80]. This statement can be elucidated by the results presented in this study, which shows that 90% of the respondents state that the Godavari mangroves are 'very important' for their livelihood. Firstly, the mangroves form a natural protection against cyclones and floods, which is realised more in villages 'facing the cyclones at the frontline.' Secondly, the mangrove ecosystem provides them with direct natural resources, such as fuel and construction wood, fodder for the cattle and fishery-related activities. Avicennia marina, a dominant species in the Godavari mangroves, is most frequently used as firewood, for construction purposes and as fodder for cattle. The bark of Ceriops decandra is prepared traditionally to enhance the durability of the fishing nets. No medicinal use of the mangroves was reported in contrast with other areas [2,4]. Reported changes in the evolution of the Godavari mangrove cover show to be differential in two zones that differed in legal protection status, with significantly less perceptions of a decrease in the protected area, as compared to the adjacent non-protected area. Whereas, the results of our survey research indicated that elements essential to their lifestyle, have not been incorporated in the existing policy, we illustrate how data on ethnobotanical traditions, fishery-related activities and local people's perceptions of change can point out contradictions and discrepancies with the current management policy, and can therefore be used to improve the policy.

Competing Interests

The author(s) declare that they have no competing interests.

Acknowledgements

From the French Institute of Pondicherry we thank D. Grandcolas for her useful socio-geographic input, G. Okrukaimani and Chandra Shekar for the translations, P. Grard and J. Prósperi for the botanical photographs, and S. Oliveau for the calculations of the demographic data from the database of the South Indian Fertility Project. We are particularly indebted to Nirmalla, Sirisha and their mother for making the field stay in Kakinada so pleasant. The first author is a Post-doctoral Researcher from the Fund for Scientific Research (FWO-Vlaanderen). The research was also funded by the European Commission (Contract ERB IC18-CT98-0295), and the Flemish Inter-University Council (VLIR), and it was awarded prizes by the Flanders Marine Institute (VLIZ Encouragement Award Marine Sciences) and by the International Society of Ethnobiology (9th International Congress on Ethnobiology on Social Change and Displacement, in collaboration with 45th Annual Meeting of the Society for Economic Botany, and the 8th International Congress of the International Society of Ethnopharmacology, 13–18 June, Canterbury, U.K.). This research is within the objectives of the International Geosphere-Biosphere Programme (IGBP), Past Global Changes (PAGES) Focus 5 : Past Ecosystem Processes and Human-Environment Interactions.

References

1. Rollet B: Les utilisations de la mangrove. Journal d'Agriculture Tropicale et de Botanique Appliquée T 1975, XXII(7–12):1149.

2. Bandaranayake WM: Traditional and medicinal use of mangroves. Mangroves and Salt Marshes 1998, 2:133–148.

3. Rönnbäck P: The ecological basis for economic value of seafood production supported by mangrove ecosystems. Ecological Economics 1999, 29:235–252.

4. Dahdouh-Guebas F, Mathenge C, Kairo JG, Koedam N: Utilisation of mangrove wood products around Mida Creek (Kenya) amongst subsistence and commercial users. Economic Botany 2000, 54(4):513–527.

5. Rönnbäck P: Mangroves and Seafood Production. The ecological economics of sustainability. PhD Thesis 2001.

6. Walters BB: Patterns of local wood use and cutting of Philippine mangrove forests. Economic Botany 2005, 59(1):66–76.

7. Robertson AI, Duke NC: Mangroves as nursery sites: comparisons of the abundance and species composition of fish and crustaceans in mangroves and other nearshore habitats in tropical Australia. Marine Biology 1987, 96:193–205.

8. Robertson AI, Blaber SJM: Plankton, epibethos and fish communities. In Tropical Mangrove Ecosystems. Edited by: Robertson AI, Alongi DM. Washington D.C.: American Geophysical Union; 1992:173–224.

9. Baran E, Hambrey J: Mangrove conservation and coastal management in southeast Asia : what impact on fishery resources ? Marine Pollution Bulletin 1998, 37(8–12):431–440.

10. Baran E: A review of quantified relationships between mangroves and coastal resources. Phuket Marine Biology Centre Research Bulletin 1999, 62:57–64.

11. Barbier EB: Valuing the environment as input: review of applications to mangrove-fishery linkages. Ecological Economics 2000, 35:47–61.

12. Barbier EB: Habitat-fishery linkages and mangrove loss in Thailand. Contemporary Economic Policy 2003, 21(1):59–77.

13. Mumby PJ, Edwards AJ, Arlas-González JE, Lindeman KC, Blackwell PG, Gall A, Gorczynska MI, Harborne AR., Pescod CL, Renken H, Wabnitz CCC, Llewellyn G: Mangroves enhance the biomass of coral reef fish communities in the Caribbean. Nature 2004, 427:533–536.

14. Massel SR, Furukawa K, Brinkman RM: Surface wave propagation in mangrove forests. Fluid Dynamics Research 1999, 24(4):219–249.

15. Badola R, Hussain SA: Valuing ecosystem functions: an empirical study on the storm protection function of Bhitarkanika mangrove ecosystem, India. Environmental Conservation 2005, 32(1):85–92.

16. Dahdouh-Guebas F, Jayatissa LP, Di Nitto D, Bosire JO, Lo Seen D, Koedam N: How effective were mangroves as a defence against the recent tsunami? Current Biology 2005, 15(12):R443–447.

17. Kathiresan K, Rajendran N: Coastal mangrove forests mitigated tsunami. Estuarine, Coastal and Shelf Science 2005, 65:601–606.

18. Cormier-Salem MC: The mangrove: an area to be cleared ... for social scientists. Hydrobiologica 1999, 413:135–142.

19. Dahdouh-Guebas F, Ed: Remote sensing and GIS in the sustainable management of tropical coastal ecosystems. Environment, Development and Sustainability 2002, 4(2):93–229.

20. Rönnbäck P, Troell M, Zetterström T, Babu DE: Mangrove Dependence and Socio-Economic Concerns in Shrimp Hatcheries of Andhra Pradesh, India. Environmental Conservation 2003, 30:344–352.

21. Primavera JH: The yellow mangrove : its ethnobotany, history of maritime collection, and needed rehabilitation in the central and southern Philippines. Philippine Quarterly of Culture and Society 2000, 28:464–475.

22. Kovacs JM: Perceptions of environmental change in a tropical coastal wetland. Land Degradation & Development 2000, 11:209–220.

23. Walters BB: People and mangroves in the Philippines : fifty years of coastal environmental change. Environmental Conservation 2003, 30(2):293–303.

24. Hernández-Cornejo R, Koedam N, Luna AR, Troell M, Dahdouh-Guebas F: Remote sensing and ethnobotanical assessment of the mangrove forest changes in the Navachiste-San Ignacio-Macapule lagoon complex, Sinaloa, Mexico. Ecology & Society 2005, 10(1):16.

25. Kaplowitz MD: Assessing mangrove products and services at the local level: the use of focus groups and individual interviews. Landscape and Urban Planning 2001, 56(1–2):53–60.

26. Omodei Zorini L, Contini C, Jiddawi N, Ochiewo J, Shunula J, Cannicci S: Participatory appraisal for potential community-based mangrove management in East Africa. Wetlands Ecology and Management 2004, 12:87–102.

27. Walters BB, Sabogal C, Snook LK, de Almeida E: Constraints and opportunities for better silvicultural practice in tropical forestry: an interdisciplinary approach. Forest Ecology and Management 2005, 209:3–18.

28. Wyllie-Echeverria S, Arzel P, Cox PA: Seagrass conservation : lessons from ethnobotany. Pacific Conservation Biology 2000, 5:329–335.

29. Rosa IML, Alves RRN, Bonifácio KM, Mourão JS, Osório FM, Oliveira TPR, Nottingham MC: Fishers' knowledge and seahorse conservation in Brazil. Journal of Ethnobiology and Ethnomedicine 2005, 1:12.

30. Spalding M, Blasco F, Field F: World Mangrove Atlas. Okinawa: The International Society for Mangrove Ecosystems; 1997.

31. Jagtap TG, Chavan VS, Vistawale AG: Mangrove ecosystems of India: a need for protection. Ambio 1993, 22(4):252–254.

32. Varaprasado Rao N: Mangroves in A.P. with special reference to Coringa Wildlife Sanctuary. Rajahmundry: Forest Department; 1997.

33. Mittal SR: Management plan for Coringa Wildlife Sanctuary. Rajahmundry: Forest Department; 1993.

34. Blasco F, Aizpuru M: Classification and evolution of the mangroves of India. Tropical Ecology 1997, 38(2):357–374.

35. Azariah J, Azariah H, Gunasekaran S, Selvam V: Structure and species distribution in Coringa mangrove forest, Godavari Delta, Andhra Pradesh, India. Hydrobiologia 1992, 247:11–16.

36. Ravishankar T, Ramakrishna D, Ramasubramanian R, Srinivasa Rao N, Sridhar D: Community participation in mangrove restoration management. In Discussion papers on the national Training Workshop on Rehabilitation of

Degraded Mangrove Forests: Practical Approaches. Kakinada: M.S. Swaminathan Research Foundation; 2001.

37. Hema Malini B, Nageswara Rao K: Coastal erosion and habitat loss along the Godavari delta front – a fallout of dam construction. Current Science 2004, 87(9):1232–1236.

38. Tripathy SC, Ray AK, Sarma VV: Water quality assessment of Gautami-Godavari mangrove estuarine ecosystem of Andhra Pradesh, India during September 2001. Journal of Earth System Science 2005, 114(2):185–190.

39. VUB- Vrije Universiteit Brussel: Assessment of mangrove degradation and resilience in the Indian subcontinent: the cases of Godavari Estuary and South-west Sri Lanka. Summary of Scientific Results. Third Annual Report (Contract ERB IC18-CT98-0295). Brussels; 2001.

40. Sri Manoranjan B: Resource management for higher sustainable production in coastal areas. Rajahmundry : Forest Department; 2001.

41. MSSRF – M.S. Swaminathan Research Foundation: Coastal wetlands: mangrove conservation and management – Andhra Pradesh. Kakinada.

42. MSSRF – M.S. Swaminathan Research Foundation: Coastal wetlands: mangrove conservation and management. Summary of socio-economic benchmark survey – Matlapalem, AP (Godavari Mangroves). Kakinada. 1998.

43. Jayatissa LP, Dahdouh-Guebas F, Koedam N: A review of the floral composition and distribution of mangroves in Sri Lanka. Botanical Journal of the Linnean Society 2002, 138:29–43.

44. Satyanarayana B, Raman AV, Dehairs F, Kalavati C, Chandramohan P: Mangrove floristic and zonation patterns of Coringa, Kakinada Bay, East Coast of India. Wetlands Ecology and Management 2002, 10:25–39.

45. Tomlinson PB: The Botany of Mangroves. Cambridge: Cambridge University Press; 1986.

46. Mabberley DJ: The Plant-Book. Cambridge: Cambridge University Press; 1987.

47. de Solan B: La mangrove de la Godavari (Andhra Pradesh): montrer son évolution et la comprendre. Une analyse par télédétection et SIG. In Thesis. Rennes: Ecole Supérieure Agronomique de Rennes; 2001.

48. Suryanarayana M: Marine fisherfolk of North-East coastal Andhra Pradesh. Calcutta: Anthropological Survey of India; 1977.

49. Sokal RR, Rohlf FJ: Biometry. The principals and practice of statistics in biological research. New York: W.H. Freeman and Co; 1981.

50. Nadkarni AK: Dr. K.M. Nadkarni's Indian Materia Medica. Third edition. Bombay: Popular Prakashan Pvt. Ltd.; 1976.

51. Jain SK, DeFilipps RA: Medicinal Plants of India. Algonac: Reference Publications; 1991.

52. Kovacs MJ: Assessing mangrove use at a local scale. Landscape and Urban Planning 1999, 43:201–208.

53. Obade TP: Anthropogenically induced changes in a Kenyan mangrove ecosystem explained by application of remote sensing and Geographic Information Systems (GIS). In MSc Thesis. Brussels: Vrije Universiteit Brussel; 2000.

54. Obade P, Dahdouh-Guebas F, Koedam N, De Wulf R, Tack JF: GIS-based integration of interdisciplinary ecological data to detect land-cover changes in creek mangroves at Gazi Bay, Kenya. Western Indian Ocean Journal of Marine Science 2004, 3(1):11–27.

55. Kajia Y: Assessment of the effects of rice cultivation in the mangrove forests of the Rufiji Delta (Mainland Tanzania). In MSc Thesis. Brussels: Vrije Universiteit Brussel; 2000.

56. Stolk ME: Patterns of mangroves use in Hoanh Bo District, Quang Ninh Province, Northern Vietnam. In MSc Thesis. Brussels: Vrije Universiteit Brussel; 2000.

57. Raju JSSN: Xylocarpus (Meliaceae): A less-known mangrove taxon of the Godavari estuary, India. Current Science 2003, 84(7):879–881.

58. Pernetta JC: Marine protected areas needs in the South Asian Seas Region. India. A Marine Conservation and Development Report. Volume 2. Gland: IUCN; 1993.

59. Dahdouh-Guebas F, Vrancken D, Ravishankar T, Koedam N: Short-term mangrove browsing by feral water buffaloes: conflict between natural resources, wildlife and subsistence interests ? Environmental Conservation, in press.

60. Premanathan M, Arakiki R, Izumi H, Kathiresan K, Nakano M, Yamamoto N, Nakashima H: Antiviral properties of a mangrove plant, Rhizophora apiculata Blume, against human immunodeficiency virus. Antiviral Research 1999, 44:113–122.

61. Gadgil M, Berkes F, Folke C: Indigenous knowledge for biodiversity conservation. Ambio 1993, 22:151–156.

62. Sillitoe P, Dixon P, Barr J: Indigenous Knowledge Inquiries: A Methodologies Manual For Development. London: Intermediate Technology Publications; 2005.

63. Dahdouh-Guebas F, Van Pottelbergh I, Kairo JG, Cannicci S, Koedam N: Human-impacted mangroves in Gazi (Kenya): predicting future vegetation based on retrospective remote sensing, social surveys, and distribution of trees. Marine Ecology Progress Series 2004, 272:77–92.

64. Dahdouh-Guebas F, Hettiarachchi S, Lo Seen D, Batelaan O, Sooriyarachchi S, Jayatissa LP, Koedam N: Transitions in ancient inland freshwater resource management in Sri Lanka affect biota and human populations in and around coastal lagoons. Current Biology 2005, 15(6):579–586.

65. Rist S, Dahdouh-Guebas F: Ethnosciences – A step towards the integration of scientific and traditional forms of knowledge in the management of natural resources for the future. Environment, Development & Sustainability 2006, in press.

66. Dahdouh-Guebas F, Van Hiel E, Chan JC-W, Jayatissa LP, Koedam N: Qualitative distinction of congeneric and introgressive mangrove species in mixed patchy forest assemblages using high spatial resolution remotely sensed imagery (IKONOS). Systematics and Biodiversity 2005, 2(2):113–119.

67. MSSRF – M.S. Swaminathan Research Foundation: Discussion papers on the national Training Workshop on Rehabilitation of Degraded Mangrove Forests: Practical Approaches. Kakinada. 2001.

68. Foell J, Harrison E, Stirrat RL: Participatory approaches to natural resource management – the case of coastal zone management in the Puttalam District. Summary findings of DFID-funded research 'participatory mechanisms for sustainable development of coastal ecosystems' (Project R6977). Brighton: School of African and Asian studies; 1999.

69. Naylor RL, Goldburg RJ, Mooney H, Beveridge M, Clay J, Folke C, Kautsky N, Lubcheno J, Primavera J, Williams M: Nature's subsidies to shrimp and salmon farming. Science 2000, 282:883–884.

70. Dahdouh-Guebas F, Zetterström T, Rönnbäck P, Troell M, Wickramasinghe A, Koedam N: Recent changes in land-use in the Pambala-Chilaw Lagoon complex (Sri Lanka) investigated using remote sensing and GIS : conservation of mangroves vs. development of shrimp farming. Environment, Development and Sustainability 2002, 4(2):185–200.

71. Chandra Mohan P, Rao RG, Dehairs F: Role of Godavari mangroves (India) in the production and survival of prawn larvae. Hydrobiologia 1997, 358(1/3):317–320.

72. Dillon CP, Andrews MJ: 1996 annual tropical cyclone report. Guam: Naval Pacific Meteorology and Oceanography Center; 1997.

73. DHA – Department of Humanitarian Affairs: India cyclone. DHA-Geneva Information report N°2. Geneva: United Nations; 1996.

74. Palmer M, Finlay V: Faith in conservation: new approaches to religions and the environment. Washington D.C.: International Bank for Reconstruction and development/The World Bank; 2003.

75. Reddy SRC, Chakravarty SP: Forest dependence and income distribution in a subsistence economy: evidence from India. World Development 1999, 27(7):1141–1149.

76. AFPRO – Action for Food Production: Baseline study for training in sea safety development program in East-Godavari district, Andhra Pradesh, India. Hyderabad: Department of Fisheries, Government of Andhra Pradesh; 1998.

77. Pearce F: Living sea walls keep floods at bay. New Scientist 1996, 150(2032):7.

78. Dahdouh-Guebas F, Ed.: Bridging the gap between natural resources and their human management for the future using ethnosciences. Environment, Development and Sustainability,, in press.

79. Armitage D: Socio-institutional dynamics and the political ecology of mangrove forest conservation in Central Sulawesi, Indonesia. Global Environmental Change 2002, 12:203–217.

80. Ewel KC, Twilley RR, Ong JE: Different kind of mangrove forests provide different goods and services. Global Ecology and Biogeography Letters 1998, 7:83–94.

81. Pròsperi J, Ramesh BD, Grard P, Depommier D: Mangroves V.I.0 β. A Multimedia identification system. Pondichéry/Matara: University of Ruhuna/ Vishakpatnam: Andhra University/Paris: Centre de coopération internationale en recherche agronomique pour le développement; 2002.

82. NRSA – National Remote Sensing Agency: Land Use Planning Atlas of East Godavari District, A.P. Hyderabad. 1994.

Chapter 8

Lack of Cross-Scale Linkages Reduces Robustness of Community-Based Fisheries Management

Richard Cudney-Bueno and Xavier Basurto

ABSTRACT

Community-based management and the establishment of marine reserves have been advocated worldwide as means to overcome overexploitation of fisheries. Yet, researchers and managers are divided regarding the effectiveness of these measures. The "tragedy of the commons" model is often accepted as a universal paradigm, which assumes that unless managed by the State or privatized, common-pool resources are inevitably overexploited due to conflicts between the self-interest of individuals and the goals of a group as a whole. Under this paradigm, the emergence and maintenance of effective community-based efforts that include cooperative risky decisions as the establishment of marine reserves could not occur. In this article, we question these assumptions and show

Originally published as Cudney-Bueno R, Basurto X (2009) Lack of Cross-Scale Linkages Reduces Robustness of Community-Based Fisheries Management. PLoS ONE 4(7): e6253. https://doi. org/10.1371/journal.pone.0006253. © 2009 Cudney-Bueno, Basurto. https://creativecommons. org/licenses/by/4.0/

that outcomes of commons dilemmas can be complex and scale-dependent. We studied the evolution and effectiveness of a community-based management effort to establish, monitor, and enforce a marine reserve network in the Gulf of California, Mexico. Our findings build on social and ecological research before (1997–2001), during (2002) and after (2003–2004) the establishment of marine reserves, which included participant observation in >100 fishing trips and meetings, interviews, as well as fishery dependent and independent monitoring. We found that locally crafted and enforced harvesting rules led to a rapid increase in resource abundance. Nevertheless, news about this increase spread quickly at a regional scale, resulting in poaching from outsiders and a subsequent rapid cascading effect on fishing resources and locally-designed rule compliance. We show that cooperation for management of common-pool fisheries, in which marine reserves form a core component of the system, can emerge, evolve rapidly, and be effective at a local scale even in recently organized fisheries. Stakeholder participation in monitoring, where there is a rapid feedback of the systems response, can play a key role in reinforcing cooperation. However, without cross-scale linkages with higher levels of governance, increase of local fishery stocks may attract outsiders who, if not restricted, will overharvest and threaten local governance. Fishers and fishing communities require incentives to maintain their management efforts. Rewarding local effective management with formal cross-scale governance recognition and support can generate these incentives.

Introduction

Coastal fishing communities are increasingly exposed to global market pressures, making them more vulnerable to "roving bandits" who can deplete local fishing stocks and move on to other areas to do the same [1], seriously threatening ecosystems and the people who depend on them to survive [2], [3], especially those located in developing countries [4]. To overcome the threat of roving bandits and overexploitation of fisheries, international financial organizations and some national governments are investing huge sums to foster the establishment of marine reserves and community-based management (CBM) [5], [6].

The research community, however, is divided regarding the potential effectiveness of CBM for developing sustainable fisheries [7]. The capabilities of managing coastal fisheries locally, although well documented [8]–[11] have often been ignored or criticized, viewed as relics that are irrelevant to contemporary situations [12], [7]. Indeed, local fisheries are rapidly appearing and—just as rapidly—disappearing in response to emerging global markets and overfished stocks

[1], leaving little time to develop effective customary management practices with which to avoid local overexploitation. Many fishery officials and scholars still accept "the tragedy of the commons" model [13] that assumes that due to conflicts between the self-interest of members of a group and the goals of a group as a whole, common-pool resources need to be managed by the State or privatized to avoid overexploitation. Under this paradigm, the emergence and maintenance of effective community-based efforts that include costly and risky decisions as the establishment of marine reserves would not occur, particularly in recently organized fisheries.

In this study, we question these assumptions and show that the realities of commons dilemmas can be complex and scale dependent. Recently organized fisheries have the potential to develop effective community-based management practices that include the establishment of marine reserves. However, we also show that CBM can collapse when local communities lack linkages to higher levels of governance that help legitimize their organizational efforts [14]. We illustrate the potential for rapid rise and fall of communal self-governance in young fisheries through an effort for CBM of a network of marine reserves in Northwest Mexico.

Based on extensive ecological and social studies conducted prior, during, and after reserve establishment, we observed the evolution of CBM efforts in a recently organized inshore fishery of the Gulf of California, Mexico. These efforts - which included the establishment of a marine reserve network and locally enforced harvesting rules - led to a substantial and documented increase in local resource abundance [15]. The network includes an offshore reserve surrounding an island and two coastal reserves, providing protection to roughly 30% of a fishing sector's fishing grounds (Fig. 1). Reserves were created by a cooperative of 22 commercial divers of Puerto Peñasco, a fishing and tourism hub located in the northeastern portion of the Gulf of California, as a means to protect and enhance mollusk stocks, particularly rock scallops (Spondylus calcifer) and black murex snails (Hexaplex nigritus), two staple resources of commercial divers [16]–[18]. The divers' cooperative has been harvesting benthic shellfish from rocky reefs and adjacent sandy areas for approximately 30 years [19]. The cooperative is comprised of no more than two generations of divers (age range 24–58 yrs) who arrived from communities south of Puerto Peñasco in the late 1970's [19]. While the fishing cooperative in its current structure was formally established in 2001, some members belonged to a similar cooperative before that and had been harvesting benthic resources since the late 1970s. Some of these older fishers were leaders in the new cooperative and played an important role in creating management measures.

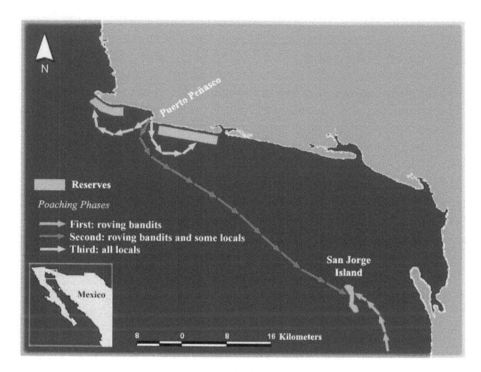

Figure 1. Community-based marine reserve network of Puerto Peñasco and sequential phases of poaching. During the first phase (green), divers from other locations, "roving bandits," poached on the San Jorge Island reserve. This was followed by local rule breaking and some divers from Puerto Peñasco poaching on the Island (second phase, pink arrows). During the third phase (yellow), all members of the Puerto Peñasco diving cooperative broke their local rules and coastal reserves were targeted. It took less than two months for rules to be broken by all local divers after entrance of roving bandits.

Local divers created and enforced the reserves while working closely with researchers from a local non-governmental conservation organization (Centro Intercultural de Estudios de Desiertos y Océanos) and an academic institution (the University of Arizona) to design and implement a monitoring program for their fisheries (Cudney-Bueno worked with these institutions and the cooperative of divers between 1997–2004, helping facilitate fishers' organizational meetings and leading the monitoring program with commercial divers). Fishers approached these institutions in 1998 to help conduct population assessments and biological studies of their fisheries. Close collaboration between fishers, researchers, and local institutions emerged, leading to the subsequent design, establishment, and enforcement of the network of marine reserves four years after this collaboration began [19]. The network was designed by fishers based on their knowledge of local currents, of differences in densities of mollusks, and recollection of previous abundant sites that the cooperative wanted to see rebound. Researchers worked with fishers to help bridge their local knowledge with experimental design and establish a monitoring program to measure changes within and outside the reserves [19]. Reserve establishment occurred without waiting for official government recognition at a time when other efforts for the establishment and management of marine reserves in the region had been highly conflictive [20].

Undoubtedly, the presence of researchers and an NGO played an important role in facilitating the establishment of reserves. However, we do not elaborate on the conditions or the relative importance of factors that led to the establishment of reserves since this is discussed elsewhere [19] and is forthcoming work. Instead, we describe the main components of the CBM system, its effects on local resources and cooperation, as well as the processes and factors leading to a downfall in local governance and subsequent effects on resources protected. We discuss these results in the context of the "tragedy of the commons" paradigm and emerging worldwide efforts to foster CBM of fisheries.

Results

Local Management of the Commons

Community-based management of the marine reserve network relied primarily on a suite of simple rules and means of enforcement, leadership of key individuals, meeting venues that allowed for social and ecological feedbacks, and capitalizing on the region's physical and environmental characteristics.

Formal and Informal Rules and Sanctions

Fishers designed, monitored, and enforced three main forms of rules: resource-based rules (snail fishing banned June and July; fishing banned within reserves), monitoring rules (mandatory participation in and financial contributions for monitoring) and administrative rules (mandatory: participation in cooperative meetings, monthly financial contributions to the cooperative, and timely provision of paperwork for cooperative). These rules and their sanctions were built primarily on foundations of trust and reciprocity and concerns for the group's well being. Hence, the most effective and usual form of enforcement relied on variations of peer pressure and public shame. This, in essence, could ultimately threaten the rule-breaker's reputation and his social bonds and norms—also known as social capital [21]—with the rest of the members of the group. These were de facto sanctions with no legal standing under the statutes of the cooperative. While other formal sanction types were developed, they were either largely avoided, often changed, or were applied last (Table 1). On occasions, local government officials provided enforcement support that was based entirely on the rapport built between fishers and officials, as reserves were yet to be formalized by the government.

Table 1. Formal and informal sanctions by rule type devised by local fishers.

Rule Type	Sanction Types	Formal (F)/Informal (I)
Resource-based rules	On site warnings and verbal confrontations	I
	Peer pressure/public shame during meetings	I
	Threats of temporary confiscation of boat	F
	Threats of expulsion from the cooperative	F
Monitoring rules	Peer pressure/public shame during meetings	I
	Extend rule breaker's monitoring responsibilities	F
Administrative rules	Peer pressure/public shame during meetings	I
	Threat of expulsion from cooperative after 3 faults	F
	Temporary confiscation of boat	F
	Expulsion from the cooperative	F

To exemplify this, 100% of fishers interviewed said that they trusted that other fellow fishers for the most part respected the reserves. Similarly, when asked the open ended question "In what way would breaking cooperative rules affect you?" all answers fell into three categories: 1) personal guilt and sense of betrayal to the group, 2) concern of the rest of the group's opinion about one's actions, and 3) concern over the possibility of losing trust and friendship.

If someone within the group cheated, the first approach of members of the cooperative was to tap into the personal guilt associated with the event. Often, it was only necessary to bring the case to the attention of the group without singling out specific people. "Cheaters" assumed that at least someone else likely knew who the culprit was. This way, the informer's reputation was also protected and he would not be labeled as an accuser. "Accuser" is one of the worst labels a commercial diver can have, largely because it can undermine his network ties and reliance on these ties when in need of any help. During interviews, when given a choice to express what would be worse, for the group to label you as an accuser or as a cheater, practically all divers found it impossible to make a choice. They were both seen as equally detrimental.

Role of Leadership and Cooperative Meetings

When direct accusations in front of the group did take place, however, these were carried out by the more elder or experienced divers who had already gained high levels of respect within the group. These individuals played a pivotal role during meetings. They gave credibility to agreements and helped maintain, although often contentious, a respectful meeting atmosphere. They were also the main players involved in confronting cheaters directly on site when found breaking any given rule.

Cooperative meetings encompassed a key component for the evolution of cooperation and maintenance of checks and balances. Between summer 2002 and summer 2004, 15 meetings were held, all with an attendance of at least 80% of members. More than acting as a means to discuss various issues pertaining to administration, these meetings provided the main venue to maintain the checks and balances of the system and its functionality. They provided a forum for the development of trust, the generation of rules and sanctions, and allowed for collective feedbacks from biological knowledge gained while commercial diving and/or monitoring. This, in turn, reinforced among the group the perceived benefits of the reserves and played a key role in dismissing poaching allegations and re-enforcing group strength. For instance, it was common for rumors of poaching to develop and quickly spread within the group. However, these rumors were usually dismissed during cooperative meetings. Given that fishers were directly involved in the monitoring process, with designated individuals repeatedly monitoring the same areas jointly with academic researchers, there was a strong sense of individual appropriation towards each monitored site and of other group members' respect towards the reports or opinions of these individuals. Knowing that poaching allegations were often false and that resources were in good health would in turn re-enforce the unity and strength of the group and trust in its members. In short, burden of proof regarding the state of reserves and fishing areas fell largely on fishers themselves.

Role of the Physical and Environmental Layout

Knowing if, when, and where a poaching event took place was facilitated by the region's physical and environmental layout. It is simple to know where a diver fishes on a daily basis. The group is small and highly communicative, allowing for the quick spread of information. Coastal reserves are found close to port and fishing activities within them can be easily detected either from shore or from fishing areas. In addition, diving patterns within any given month are constrained by environmental factors, particularly tidal currents and visibility. For instance, during monthly spring tides, when tidal currents are strongest, divers are largely constrained to dive within the reefs south of port. During neap tides, they target offshore areas and the reefs north of port.

In the case of San Jorge Island, which is found farther offshore and is harder to patrol, enforcement relied on more active means. These means, however, were geared towards patrolling entrance of outsiders and not of members of the cooperative.

It is quickly known when a diver from Puerto Peñasco goes to the island as this trip demands extra preparation and usually involves overnight stays. Cooperative members would sometimes carry out trips to the island during neap tides with the sole purpose of seeing if anyone was there. However, on three occasions when credible rumors emerged about outsiders poaching, Puerto Peñasco divers also gained the support of the local Navy and fisheries offices to assist in patrolling and enforcement operations. This collaboration was based on the rapport built between divers and local government officials throughout the years rather than as a mandate, as reserves were yet to be formalized at a federal level.

The unique environmental characteristics of the region also facilitated local divers' efforts to discourage settlement of outsiders. A case in point is when a prominent Puerto Peñasco buyer hired divers from another region to work for him at low wages and increase his revenues. Local divers advised them to fish in areas and times marked by intense currents and low visibility. These divers never developed the skills to dive in the region and left soon after.

Outcomes Strengthen Cooperation

This relatively informal governance system was highly effective. Regular underwater monitoring visits to reserve sites revealed minimal evidence of fishing activity within reserves. Finding evidence of rock scallop fishing is facilitated by the fact that the right valve remains attached to the rock after the scallop muscle has been removed and its bright white color contrasts with the rest of the reef. In the case of black murex snails, these are only harvested when they form large summer breeding aggregation mounds [18]. We saw the same aggregations repeatedly in reserve sites. Similarly, we confirmed only 13 poaching events in at least 2,000 fishing trips conducted between Summer 2002-end of Spring 2004. Hence, rule compliance was high and accomplished primarily through means of social pressures rather than through heavy policing by external officials.

Only two years after the establishment of the reserve network, populations of black murex and rock scallops had increased markedly on the San Jorge Island reserve (Fig. 2) [also see 15], with relative densities of up to 160 individuals per 100 m2 that exceed any others reported for the Gulf of California [22], [23]. As has been shown in other cases [24], positive effects were also seen in fishing areas adjacent to the reserves [15]. Relative densities of juvenile rock scallop had increased by up to 40.7% within coastal reserves and by 20.6% in fished areas, and changes were also evident for black murex, with more than a three-fold increase in density of juveniles within fished areas [15]. This increase in density of juveniles, combined with predictions from larval transport models and field oceanography data, indicate enhanced recruitment via protection of larval sources within the reserve network [15].

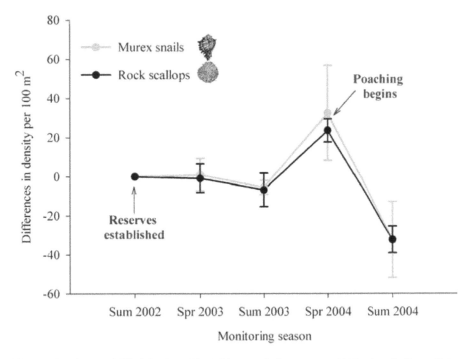

Figure 2. Rapid rise and fall of San Jorge Island fishery stocks from summer 2002, when the Puerto Peñasco community-based reserve network was established, to the end of summer 2004, three months after roving bandits poached on the island.

The graph depicts differences in relative densities (S.E. bars included) from one monitoring season to another for the main species harvested: murex snails (Hexaplex nigritus) and rock scallops (Spondylus calcifer).

Data from divers' catches of rock scallop showed an increase in average mass of 19.9% (F2, 897 = 10.78; p<0.0001, 1-way ANOVA) in the two years since reserve establishment (Fig. 3A). Similarly, average mass of black murex increased by 74.74% in reserves (F2, 220 = 77.75; p<0.001, 1-way ANOVA) and by 35% in fishing areas (F2, 421 = 23.80; p<0.001, 1-way ANOVA) (Fig. 3B).

Rapid feedback from fishing resources allowed fishers to expect future benefits of the group's various initiatives. In interviews conducted prior to providing results on monitoring efforts, over 85% of fishers reported benefits from the reserves, wanting to continue the reserves into the future.

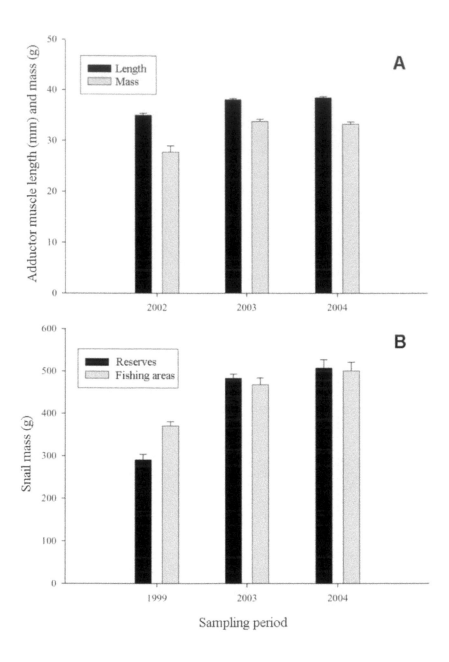

Figure 3. Changes in length and mass of rock scallops (Spondylus calcifer) and black murex snails (Hexaplex nigritus) before and after reserve establishment.

(a) Comparison of the average adductor muscle length and mass of rock scallops from fishing areas (2002 = Spring, two months before reserve establishment). Data from reserves was not obtained as animals would have needed to be sacrificed. (b) Comparison of the average live mass of black murex snails from reserve and fishing areas.

The Fall of Cooperation

Shortly after these initial positive outcomes, local governance faced severe external challenges in the form of: 1) lack of recognition at higher levels of government of the local arrangements, 2) abrupt changes in local government leadership, such as the replacement of the chief of the fishing agency office in Puerto Peñasco, who was a local himself, and (3) new fishing pressure from outsiders. News about these community-based management efforts and the abundance of resources at the reserves spread quickly at a regional scale. "Roving bandits" from more than 300 km away (along the coastline, eight hours travel by boat) began fishing the island. Since reserves and territorial use rights were not formally recognized by the Government, local fishers did not have the right to expel others from their reserves and de facto fishing grounds. Doing so could put them and the cooperative at risk. Furthermore, they no longer had any support of local government officials, which reduced even more any options available to cope with the problem. Various cooperative members confronted outsiders verbally on at least 3 different occasions and physical confrontation was seriously considered, an option that was soon dropped as it would have had serious legal consequences on the cooperative and those involved. Unable to deter poaching, those local divers who historically had fished on the island the most opted to fish there before outsiders finished reaping the benefits of their own investment. What followed was a rapid cascading effect on fishing resources and locally-designed rule compliance. The island quickly became a free for all. In one month, mollusk populations were reduced in half (Fig. 2). As the island was being harvested by most cooperative members, local divers who tend to fish closer to shore also began to target coastal reserves even without direct pressure from roving bandits in these areas. There were no incentives to continue protecting coastal reserves knowing that fellow fishers were obtaining large catches offshore and that prices would likely fall due to market saturation. Coastal reserves also became an opportunity to rapidly harvest in a setting with limited competition from other fellow fishers. In less than two months since roving bandits first entered the area, all reserves had been targeted by every member of the cooperative at least once (Fig. 2 and Table 2).

Social capital that initially allowed for a rapid evolution of cooperation and self-governance for resource management now facilitated overexploitation and rule breaking (Table 2). Once most members of the cooperative had broken some rule, accountability had been eroded and fishers were no longer willing to hold their peers accountable for poaching due to strong social ties with them in other dimensions outside of fishing. Key cooperative members, for instance, stopped attending meetings in order to avoid encounters with specific people, at times family related. Between June 2004 and December 2005, of six cooperative meetings held, only one had the minimum quorum necessary to make decisions recognized under the bylaws of the cooperative (50%+1 members).

Table 2. Rules and levels of compliance before and after entrance of roving bandits.

Rule Type	Compliance Time A (Before)	Compliance Time B (After)
Resource-based rules		
Snail fishing banned May-July	1	5
Fishing banned within reserve network	1	5
Monitoring rules		
Participation in monitoring	1	5
Financial support for monitoring	1	5
Administrative rules		
Participation in all meetings	1	4
Monthly financial contribution	2	5
Providing paperwork necessary for the cooperative	2	2

†Compliance levels based on percentage of fishers known to have broken the rule at least once: 1 = very low (<10%), 2 = low (10–40%), 3 = moderate (41–60%), 4 = high (61–90%), 5 = very high (>90%). Time A = June 2001–May 2004, Time B = first six months (June–November 2004) after entrance of roving bandits.

Discussion

A combination and interaction of three main factors led to the initial downfall of cooperation within this CBM system: lack of government recognition, changes in local government leadership, and entrance of roving bandits. Once rule-breaking had become prevalent, these factors were further exacerbated by the existing strong social ties among cooperative members, ties that under other conditions had been conducive to CBM. Although we cannot quantify which factor weighed more over the other, not having formal government recognition of management guidelines and of fishers' de facto fishing grounds clearly hampered any means to cope with roving bandits, clouding any possible incentive to cooperate. The importance of this recognition in helping to foster local cooperation is further exemplified with recent developments. In summer 2006, the Mexican Government granted a fishing concession to the Puerto Peñasco diving cooperative, providing its members exclusive access rights to fish rock scallop within what were before de facto fishing grounds [19]. Rights to the concession demand adherence to a government recognized management plan, which includes the establishment of seasonal and area closures, as well as total allowable catches [19]. Divers are once again conducting subtidal monitoring, are active participants in stock assessments to define their annual quota, have defined new rules for overall governance of the cooperative, and meetings are regularly held at full quorum [personal communication, Iván Martínez, Centro Intercultural de Estudios de Desiertos y Océanos].

The downfall of this CBM effort could be simplistically attributed to a local tragedy of the commons [13]. Realities and outcomes of commons dilemmas can, however, be much more complex and scale-dependent. We have shown that cooperation for community-based management of a local commons, in which marine reserves form a core component of the system, can indeed emerge and evolve rapidly in a fishery with limited collaborative experience. Locals can find an effective set of rules to self-govern their marine resources and with which sustainable resource exploitation is more likely to take place. This cooperation in turn can be reinforced by perceived rapid positive outcomes of local management interventions. However, as effective a CBM system may be at a local scale, it is likely that this effectiveness may only last as long as the system remains buffered from external pressures.

It is important that CBM efforts that incorporate marine reserves are initially implemented in systems where responses can be measured rapidly and where there is an existing social base for reserve establishment. Reserves reduce the total fishing area, initially render an economic cost to fishers, and complicate management of risk by reducing the physical spaces available to choose from in accordance to variations in environmental conditions and the state of their resources. Stakeholder participation in monitoring, where there is a rapid feedback of the system's response, can play a key role in reinforcing cooperation. In this regard, some sessile or semi-sessile fisheries with rapid growth rates can be good candidates to invest and promote the emergence of social capital and reserve establishment. They can provide a fast feedback to fishers and a setting that can be tinkered with [19], [25], eventually leading to developing some of the factors associated with ways of organizing activities that are sustainable for the long term [26].

Nevertheless, even if CBM efforts are effective within the local biophysical and social context, we show that cooperation and strong social capital alone are not enough to sustain their efficacy. Fishers and fishing communities need to be granted formal government recognition of their locally-devised management structures when they appear to be effective. Higher levels of governance have the ability to create incentives for the emergence of local cooperation leading to sustainable resource use. One way of providing incentives for successful CBM is by rewarding such efforts with formal cross-scale governance recognition and support.

In an increasingly globalized economy, the existence of isolated and buffered fishing communities has largely been lost. Yet, as we show, effective CBM that includes costly decisions like the establishment of marine reserves can emerge even in these settings. Not granting appropriate forms of territorial use rights nor formally recognizing and giving viability to effective local management structures and arrangements, as simple or complex as these may be, could threaten a community's existing foundations for sustainable use of fishery resources. In short, without effective cross-scale institutional arrangements in place that provide robustness to a CBM system, just as cooperative behavior can arise it can also fall along with the biological resource base intended to be managed.

Materials and Methods

Our research followed a mixed method approach that combined qualitative and quantitative research in the social and biophysical sciences, including the development of larval dispersal models to assess reserve effects within and outside of reserves. Comprehensive results and analyses of the biophysical research are provided elsewhere [15]. At the broadest level, our findings build on social and ecological research with commercial divers of Puerto Peñasco (conducted by Cudney-Bueno) before (1997–2001), during (2002) and after (2003–2004) the establishment of marine reserves. This timeframe entails more than 600 days living in Puerto Peñasco, participation in 147 commercial diving trips, and attending 30 meetings of the Sociedad Cooperativa Buzos de Puerto Punta Peñasco, Puerto Peñasco's divers' fishing cooperative.

Research was based on principles of participatory research [27], where stakeholders are actively involved in research and decision-making. Fishers formed part of the research process by having designed and established their marine reserves and monitored the state of their fishery resources within these. Before monitoring began, commercial divers were trained by R. Cudney-Bueno. All biological monitoring was conducted by academic researchers in collaboration with divers.

Following is a summary of the methods used to address the effects and evolution of the Puerto Peñasco community-based marine reserve initiative.

Social Qualitative and Quantitative Research

Ethnographic research on the Puerto Peñasco diving fishery began five years before the establishment of marine reserves, which allowed us to address social dynamics prior to and after the establishment of reserves. Between Summer 2003–2004, Cudney-Bueno conducted fieldwork specifically targeted to address a) if current collective action for the establishment of marine reserves developed quickly and with no or very limited previous experience to define and/or establish collective management decisions, and b) the conditions that facilitated and led to the establishment of community-based management efforts. Through oral histories, we searched for previous cooperative efforts and key past events or situations that could have shaped fishers' interests in adopting more conservation-oriented measures. Oral histories also allowed us to single out and understand relevant issues that may not be as clearly or obviously identified with the use of directed questions. Full results and analyses of this ethnographic research go beyond the scope of this paper and are in preparation.

We complemented our qualitative research with structured interviews. Throughout March and at the beginning of April 2004, we conducted structured interviews with 18 fishers, representing 82% of the members of the diving cooperative of Peñasco. These interviews primarily addressed perceptions of fishers as to the effects and efficacy of their management efforts, factors affecting the evolution of cooperation within the cooperative such as the building of trust among cooperative members. We conducted all interviews at fishers' homes.

Having had the time to build sufficient rapport and trust with local fishers, it became possible to gain a comprehensive understanding of the diving fishery, how divers define and enforce rules and regulations, record the presence or absence of poaching events, and note if conflict resolution and consensus-building processes were facilitated or halted.

Quantitative Estimation of Population Parameters

We estimated changes in relative densities of rock scallop (Spondylus calcifer) and black murex snail (Hexaplex nigritus) in reserve and fishing sites for two consecutive years beginning in Summer 2002, one month preceding reserve establishment. These species were selected for being the main species targeted by the commercial diving fishery and representing the main reason leading to the establishment of the reserves.

The region monitored encompassed the reefs of San Jorge Island and those found near the fishing town of Puerto Peñasco (within 3 km from highest tide line) in the eastern part of the northern Gulf of California, Mexico. This region extends from 31,22,18.1 N; 113,39,09.4 W to 31,15,03.8 N; 113,20,48.1 W (see Fig. 1).

We subdivided the region into 5 sampling areas: a) two coastal reserves, Las Conchas and Sandy; b) two coastal fishing areas, Los Tanques and La Cholla; c) one offshore island reserve, San Jorge Island (Fig. 1). We conducted density counts of juveniles and adults in 58 100 m2 permanent quadrants distributed randomly across these 5 sampling areas. To reduce heterogeneity associated with depth, we restricted all sampling to depths ranging from 40–65 ft. This also reduced health risks associated with diving and facilitated overall monitoring as we were able to remain underwater for longer periods of time. A comprehensive account of methods used to analyze reserve effects on recruitment within and outside reserves - which combines results from monitoring data, outputs of larval transport models, and field oceanography - is provided elsewhere [15].

Quantitative Estimation of Changes in Size and Mass of Harvested Species

We estimated changes in the size and mass of adult (harvested) black murex and rock scallop. For black murex, we collected specimens (n = 244) from breeding aggregations of reserve and fishing sites before the establishment of reserves (Summer 1999) and after their establishment (Summer 2003 and 2004). All snails from reserves were returned to the collecting site, whereas snails from fishing sites were obtained from fishers' catches [for details on sampling black murex see 16].

For rock scallops, we estimated changes in average length and mass of the adductor muscle, the part of the animal that is commercialized and that fishers return to port. Since the only way of obtaining samples of the adductor muscle is by killing the animal, we limited our samples and analyses to fishing sites. All samples were constricted to Spring (post reproduction) to avoid variations in weight and size caused by glycogen accumulation in the muscle pre and post reproduction [17]. We obtained a total of 1081 samples of rock scallop from Spring 2002 (pre-reserve establishment) to 2004 [see sampling details in 17]. Both black murex and rock scallop data were analyzed using a 1-way ANOVA framework and Tukey's HSD multiple comparison test to determine pairwise relationships.

Acknowledgements

We thank the divers of the Sociedad Cooperativa Buzos de Puerto Punta Peñasco, the staff of the Centro Intercultural de Estudios de Desiertos y Océanos and O. Morales and J. Rupnow for overall support and participation in field monitoring. P. Turk-Boyer provided invaluable assistance and enriched our discussions throughout this study. M. Moreno assisted in the development of Figure 1. J. Donlan, E. Ostrom, and W. W. Shaw provided helpful insights through the development of this manuscript. This is a scientific contribution of the PANGAS Project, www.pangas.arizona.edu.

Author Contributions

Conceived and designed the experiments: RCB. Performed the experiments: RCB. Analyzed the data: RCB XB. Contributed reagents/materials/analysis tools: RCB. Wrote the paper: RCB XB.

References

1. Berkes F, et al.. (2006) Globalization, roving bandits, and marine resources. Science 311: 1557–1558.

2. Pauly D, Christensen V, Dalsgaard J, Froese R, Torres F (1998) Fishing down marine food webs. Science 279: 860–863.

3. Jackson JBC, et al.. (2001) Historical overfishing and the recent collapse of coastal ecosystems. Science 293: 629–638.

4. Berkes F, Mahon R, McConney P, Pollnac R, Pomeroy R (2001) Managing small-scale fisheries. Alternative directions and methods. Ottawa, Canada: International Development Research Centre.

5. Mansuri G, Rao V (2004) Community-based and driven development: A critical review. The World Bank Research Observer 19: 1–39.

6. Balmford A, Gravestock P, Hockley N, McClean CJ, Roberts CM (2004) The worldwide costs of marine protected areas. Proceedings of the National Academy of Sciences 101: 9694–9697.

7. Pauly D (1997) Small-scale fisheries in the tropics: Marginality, marginalization and some implications for fisheries management. In: Pikitch EK, Huppert DD, Sissenwine MP, editors. Global trends: Fisheries management. Bethesda, Maryland: American Fisheries Society Symposium. pp. 40–49.

8. Johannes RE (2002) The renaissance of community-based marine resource management in Oceania. Annual Review of Ecology and Systematics 33: 317–340.

9. Ostrom E (1990) Governing the commons: The evolution of institutions for collective action. New York, NY: Cambridge University Press.

10. Dyer CL, McGoodwin JR (1994) Folk management in the world's fisheries, lessons for modern fisheries management. Niwot, CO: University Press of Colorado.

11. Basurto X (2005) How locally designed access and use controls can prevent the tragedy of the commons in a Mexican small-scale fishing community. Journal of Society and Natural Resources 18: 643–659.

12. Finlayson AC (1994) Fishing for truth: A sociological analysis of northern cod stock assessments from 1977–1990. St. John's, Newfoundland: Institute of Social and Economic Research, Memorial University of Newfoundland.

13. Hardin G (1968) The tragedy of the commons. Science 162: 1243–1248.

14. Lejano RP, Ingram H (2007) Place-based conservation: Lessons from the Turtle Islands. Environment 49: 18–29.

15. Cudney-Bueno R, Lavín MF, Marinone SG, Raimondi PT, Shaw WW (2009) Rapid effects of marine reserves via larval dispersal. PLoS ONE 4(1): e4140. doi:10.1371/journal.pone.0004140.

16. Cudney-Bueno R, Rowell K (2008) The black murex snail, Hexaplex nigritus (Mollusca, Muricidae), in the Gulf of California, Mexico: II. Growth, longevity, and morphological variations with implications for management of a rapidly declining fishery. Bulletin of Marine Science 83: 299–313.

17. Cudney-Bueno R, Rowell K (2008) Establishing a baseline for management of the rock scallop, Spondylus calcifer (Carpenter 1857): Growth and reproduction in the upper Gulf of California, Mexico. Journal of Shellfish Research 27: 625–632.

18. Cudney-Bueno R, Prescott R, Hinojosa-Huerta O (2008) The black murex snail, Hexaplex nigritus (Mollusca, Muricidae), in the Gulf of California, Mexico: I. Reproductive ecology and breeding aggregations. Bulletin of Marine Science 83: 285–298.

19. Cudney-Bueno R, Bourillón L, Sáenz-Arroyo A, Torre-Cosío J, Turk-Boyer P, Shaw WW (2009) Governance and effects of marine reserves in the Gulf of California, Mexico. Ocean and Coastal Management 52: 207–218.

20. Turk-Boyer PJ (2007) Growing a conservation community: the CEDO Story. In: Felger RS, Broyles B, editors. Dry borders: Great natural reserves of the Sonoran Desert. Salt Lake City: The University of Utah Press. pp. 548–559.

21. Putnam RD (2000) Bowling alone: America's declining social capital. Journal of Democracy 6: 65–78.

22. Baqueiro E, Massó JA, Guajardo H (1988) Distribución y abundancia de moluscos de importancia comercial en Baja California Sur. Instituto Nacional de la Pesca, México, Serie de Divulgación 11: 1–32.

23. Villalejo-Fuerte M, Arellano-Martínez M, Ceballos-Vázquez BP, García-Domínguez F (2002) Reproductive cycle of Spondylus calcifer Carpenter, 1857 (Bivalvia: Spondylidae) in the Bahía de Loreto National Park, Gulf of California, Mexico. Journal of Shellfish Research 21: 103–108.

24. Roberts CM, Bohnsack JA, Gell F, Hawkins JP, Goodridge R (2001) Effects of marine reserves on adjacent fisheries. Science 294: 1920–1923.

25. Basurto X (2008) Biological and ecological mechanisms supporting marine self-governance: The Seri callo de hacha fishery in Mexico. Ecology and Society 13: 20. http://www.ecologyandsociety.org/vol13/iss2/art20.

26. Ostrom E (2005) Understanding institutional diversity. Princeton: Princeton University Press.

27. Bernard HR (2001) Research methods in anthropology: Qualitative and quantitative approaches. California, USA: Alta Mira Press.

Chapter 9

A Step Towards Seascape Scale Conservation: Using Vessel Monitoring Systems (VMS) to Map Fishing Activity

Matthew J. Witt and Brendan J. Godley

ABSTRACT

Background

Conservation of marine ecosystems will require a holistic understanding of fisheries with concurrent spatial patterns of biodiversity.

Methodology/Principal Findings

Using data from the UK Government Vessel Monitoring System (VMS) deployed on UK-registered large fishing vessels we investigate patterns of fisheries activity on annual and seasonal scales. Analysis of VMS data shows that regions of the UK European continental shelf (i.e. Western Channel and Celtic Sea, Northern North Sea and the Goban Spur) receive

---

Originally published as Witt MJ, Godley BJ (2007) A Step Towards Seascape Scale Conservation: Using Vessel Monitoring Systems (VMS) to Map Fishing Activity. PLoS ONE 2(10): e1111. https://doi.org/10.1371/journal.pone.0001111. © 2007 Witt, Godley. https://creativecommons.org/licenses/by/4.0/

consistently greater fisheries pressure than the rest of the UK continental shelf fishing zone.

Conclusions/Significance

VMS provides a unique and independent method from which to derive patterns of spatially and temporally explicit fisheries activity. Such information may feed into ecosystem management plans seeking to achieve sustainable fisheries while minimising putative risk to non-target species (e.g. cetaceans, seabirds and elasmobranchs) and habitats of conservation concern. With multilateral collaboration VMS technologies may offer an important solution to quantifying and managing ecosystem disturbance, particularly on the high-seas.

Introduction

For global commercial fisheries to maintain a sustainable future [1], [2] there is a need to develop and implement ecosystem management plans that enable managed exploitation of fish stocks while mitigating against bycatch [3]–[7]. These goals are most likely to be achieved through the development of spatially explicit models on the distribution of fisheries activity, commercially desirable fish stocks and non-target species and habitats.

Knowledge regarding the spatial ecology of non-target species of conservation concern (e.g. cetaceans, elasmobranchs, turtles and seabirds) is ever-growing from boat and aerial surveys [8], an increasing array of electronic tagging and tracking methods [9], [10], plus molecular and other forensic techniques [11], [12]. Analyses of capture records from vessels carrying independent observers have both elucidated the ecology of non-target species but also provided effort-corrected and temporally and spatially relevant insights into the magnitude of impacts of different gear types [13]–[15].

Creating a generalised, yet spatially and temporally explicit, understanding of fisheries effort with which to evaluate potential capture of target stocks and minimise putative risk to non-target species and habitats is far from trivial. Information on the at-sea distribution and behaviour of fishing vessels may be obtained from routine and opportunistic surveillance by enforcement agencies using boats and planes, but these approaches lack spatial and/or temporal coverage. Catch-book data can be used but are subject to potential biases in reporting [5]. Vessel Monitoring Systems (VMS) deployed by several nations on large commercial fishing vessels [16] could however provide patterns of fisheries activity as they have good temporal and spatial coverage and are catch-book and vessel-master independent.

In the Europe Union, VMS operates on larger vessels of Member States fishing fleets (≥15 m overall length). Such vessels employ a range of fishing techniques to exploit demersal and pelagic fish species (e.g. dredging, beam trawling, pair-trawling, gill netting and longlining). These techniques have their respective degrees of selectivity for both their intended catch species but also non-target species and variable impacts on habitats. For example, small cetacean bycatch is commonly associated with bottom set gill-netting and pair trawling [17], whereas dredging is more harmful to benthic habitats [18].

Here we investigate the utility of data from the UK VMS to describe patterns of at-sea space use by large UK-registered fishing vessels. Such data may ultimately inform seascape scale conservation by feeding into marine spatial planning activities [19] that should ensure sustainable persistence of commercial fisheries and effective mitigation of putative risk to species and habitats of conservation concern.

Results

Mapping of VMS data highlights considerable heterogeneity in space use (Figure 1a). Regions of the UK continental shelf and the European continental shelf-edge (i.e. Western Channel and Celtic Sea, Northern North Sea and the Goban Spur) receive appreciable fisheries pressure. Shelf habitats (≥25 m and ≤150 m depth, 85% of the UK declared fishing zone), received 64.1% of fisheries activity. Shelf-edge habitats (≥150 m and <250 m depth), which are not exclusively within the UK declared fishing zone, received 16.6% of fisheries activity.

To validate the presented fishing patterns (Figure 1a) we mapped sea fisheries statistics for landings of demersal and pelagic fish (Figure 1b), by area of capture, landed by UK-registered vessels during 2004 (presented in ICES statistical reporting boxes) [20]. When comparing these figures to the mean annual pattern of fisheries activity (Figure 1a) we see there is a statistically significant correlation (Spearman rank order correlation rs = 0.6, $P<0.05$) between the levels of fishing activity and declared fish landed. Also insightful is the general correlation of fisheries hotspots with the magnitude of the number of vessels registered in proximate harbours (represented by the filled circles); for example, Newlyn in the southwest and Peterhead and Northern Ireland in the northeast and northwest respectively (Figure 1b).

It is highly likely that VMS data plots fishing activity with a much greater degree of precision than inferences that could be made from catch-book data. Is this high resolution picture predictable across years and across seasons as would be needed for efficient design of spatially explicit management? When we spatially map coefficient of variation (CV) among years (Figure 1c) and across months (Figure 1d) it is clear that hotspots of fisheries activity are consistent through time.

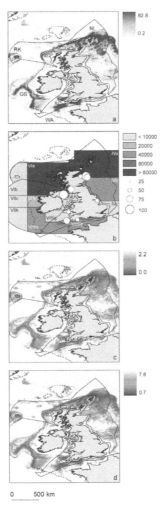

Figure 1. a) Mean annual spatial distribution of fisheries activity derived from VMS records using a simple speed filter.

The color scale indicates the mean annual number of VMS derived data points within 9 km² pixels, solid line circumscribes the UK declared fishing zone, broken line is 200 m depth contour. Regional labels: Western Channel (WA), Goban Spur (GS), Rockall (RK) and Northern North Sea (NI). b) Tonnes of fish (demersal and pelagic) landed by UK registered vessels from the shown ICES statistical reporting boxes. Total number of vessels registered at main UK fishing ports greater than 17 metres in overall length (filled circles). All vessels for Northern Ireland have been mapped to Belfast. c) Coefficient of variation of the mean annual distribution of fisheries activity, lighter colors indicate areas of greatest variability in space-use, darker areas indicate regions of consistent space-use on annual time-scales. d) Coefficient of variation of the mean monthly distribution of fisheries activity, lighter colors indicate areas of greatest variability in space-use, darker areas indicate regions of consistent space-use on monthly time-scales.

Discussion

VMS was initially conceived to assist in the monitoring and control of fisheries activities and was legislated prior to changes in EU common fisheries policy [21], which emphasised a greater focus on understanding the effects of fishing at an ecosystem level. We show that VMS, while not designed to understand putative risk to marine ecosystems, can aid EU Member State's obligations under the Common Fisheries Policy and Habitats Directive to manage ecosystem impacts of fisheries. VMS mapping generates a spatially and temporal explicit view of fisheries activity at a far greater resolution than catch-book statistics. VMS data have great potential to highlight areas where the success of ecosystem management plans may be investigated.

The importance of the identified centres of fisheries activity (i.e. Western Channel and Celtic Sea, Northern North Sea and the Goban Spur) can be explained from biological and physical oceanographic perspectives. These are regions where seafloor topography and currents set up physical features that act to support upwelling, enhanced mixing, input of nutrient rich waters, or aid the development and maintenance of frontal systems that aggregate biological matter [22], [23]. These features support primary and secondary productivity, the resulting energy of which is transferred to higher trophic levels within regional food webs. Such factors highlight why fisheries and many marine megavertebrate species seeking prey occupy similar habitats.

With the increased resolution of spatio-temporal patterns of fisheries a step improvement in knowledge of the spatial distribution of species and habitats of conservation concern is required. This requirement has been met, in part, by UK and EU funded research on small cetaceans [8], [24] and seabirds [25]. There are however statistical problems preventing the data from such studies being used as a full correlative data layer to compare with patterns of fisheries activity (i.e. merging species specific distribution and abundance data produced from differing survey methodologies; pers. comm. Simon Northridge–NERC Sea Mammal Research Unit, UK). More recently SCANS II, funded through the EU-LIFE program and participating EU Member States, has aided a more quantitative understanding of the spatial distribution and abundance of cetaceans [26]. Seasonal patterns of distribution and abundance are however still lacking and given the seasonal nature of fisheries such information is required to gain a coherent understanding of putative risk.

Although the VMS approach is a step forward in aiding the development of ecosystem management plans, there are a number of important caveats that must be considered in the interpretation of our findings, which suggest future directions for research. The fisheries activity maps are indicative of the spatial and temporal distribution of large UK-registered fishing vessels only. The patterns are therefore biased towards more offshore fishing activity and represent only a subset of the UK fleet. In addition, we only present data from the UK-registered fleet and not from other EU Member States operating in UK waters. The lack of these data does not detract from the utility of VMS data in providing a spatially and temporally explicit understanding of fisheries activity. Their absence does, however, highlight the need for integration with VMS data from other Member State vessels operating in UK domestic waters. A synoptic European view of fisheries activity will be essential for understanding the relationship between fisheries and migratory target and non-target species as they move seasonally between the waters of distant Member State.

The absence of metadata in the UK VMS on vessel gear type required us to use assumptions on movement speeds that most likely characterise fishing behaviour across several fishing methods employed by larger fishing vessels. In using a narrow range of speeds we believe we have been parsimonious in our estimation of when a vessel might be engaged in fishing. The common factor that a fishing vessel travels at slower speeds during fishing, gear deployment and retrieval, be it demersal or pelagic gear, provides a characteristic, albeit coarse, signal upon which to partition data. Expanding and contracting the width of the speed filter has the effect of widening or constricting the observed spatial patterns; what remain consistent are the identified centres of fisheries activity. Identification of these areas, their spatial range and their seasonality, provides important information for spatial management plans that could seek to manage fish stock extraction while mitigating risk to non-target species and habitats.

Not all fisheries techniques pose the same degree of risk to species and habitats of conservation concern, yet this lack of metadata does not prevent a coarse spatial interpretation of the putative risk posed to these groups as gear types, with their associated risks, are commonly deployed in known depths of water over particular habitat types. Moreover, non-target species adopt fairly predictable habitat utilisation patterns and physical habitats that represent areas of increased biodiversity can be mapped [27]. In deeper off-shore waters, such as those of the continental shelf-break, fishing vessel activity most likely represents pelagic techniques such as mid-water trawling and purse-seining. In shallower waters, fisheries activity will increasingly involve demersal techniques including bottom trawls and dredging. In the absence of robust metadata it may however, be possible to use behavioural rules on turning angles, bathymetry in the area of operation and information on movement patterns to help assist in more accurately characterising and spatially placing fishing behaviour. The development and implementation of electronic logbook system for fisheries [28], [29] may make a substantial contribution in European waters; providing spatially explicit information on gear deployment, duration of fishing and capture of target and non-target species.

Recent work to describe trawl intensity received by the seabed [30], [31] highlights additional uses of VMS for ecosystem management. Such approaches help describe the amount of disturbance an area receives. When integrated with knowledge of benthic habitat type [27] and derived habitat sensitivity, VMS data might provide better ways to manage the seabed and the fish stocks they support. VMS may also have utility in assisting the designation and subsequent measurement of the effectiveness of Marine Protected Areas that function to conserve both target stock spawning biomass and non-target species and habitats. VMS could assist in optimally selecting such areas.

Notwithstanding the caveats, the simple and coherent patterns of habitat occupation by fishing vessels presented here suggest that fishing activity could be managed on a more finely resolved spatial and temporal basis. Furthermore, with multilateral collaboration VMS technologies may offer an important solution to quantifying and managing ecosystem disturbance particularly on the high-seas, which has become evermore important as fisheries move into deeper [32] and more distant waters.

Materials and Methods

Vessel Monitoring System

The Vessel Monitoring System (VMS) is an automated method of recording the location of fishing vessels at sea. The system consists of a tamper-proof installation onboard fishing vessels registered in the UK and was introduced under European Commission legislation (EC 686/97). Each unit consists of a global positioning satellite (GPS) receiver; a satellite transmitter and a power backup that will last approximately 72 hours [33]. From the year 2000, these units were mandatory for fishing vessels greater than 24 metres overall length, from 2004 they were mandatory for vessels greater than 18 metres length and from 2005 for vessels greater than 15 metres overall length. VMS units are required to report 99% of all locations accurate to within 500 metres [33], [34]. VMS units operating in UK waters report location and ancillary data (i.e. speed and heading), via satellite communication, on a 2-hour duty cycle to the UK Fisheries Monitoring Centre (FMC). The FMC may request the location of a fishing vessel at any time from the VMS unit. VMS units can also be tasked to increase the reporting frequency within certain regions or within the waters of other EU Member States.

VMS Dataset

VMS data were obtained from the UK Sea Fisheries Inspectorate in 2005 (now the Marine and Fisheries Agency of the Department for Environment, Food and

Rural Affairs). This dataset contained 5,788,188 records. Each record contained geographic coordinates in decimal degrees (World Geodetic System 1984 format) an accompanying time stamp in UTC and a vessel identification number. All received data were anonymous with respect to their vessel registration numbers, dimensions and administrative ports. The mean number of VMS records per year was 840,182±60,346 SD (range 756,863 to 926,363). Filters were applied to the VMS dataset to remove: a) erroneous geographic records outside the range 90°S to 90°N, –180°W to 180°E, b) records outside the 5 year study period, set to be 01-01-2000 to 31-12-2004, and c) records with elevations greater than 50 metres above sea-level as determined from the TerrainBase digital elevation model [35]. The number of vessel identification numbers appearing in the dataset declined annually (from 422 in 2000 to 334 in 2004); however, new identification numbers were introduced annually to the dataset during the study period (2001 n = 23, 2002 n = 26, 2003 n = 20 and 2004 n = 32).

Route Reconstruction

Fishing trips were reconstructed as follows: a 5 km buffer zone was constructed around the coastline of Europe, this was used to determine when vessels were leaving or nearing ports. All records belonging to a vessel were assigned a logical flag (1 or 0) to indicate whether they were inside or outside this coastal buffer zone. The start and finish of a fishing trip was determined when a vessel moved out of and back into the zone with respect to time. Records occurring within the buffer zone were discarded. A speed filter was applied to remove improbable locations; this process removed locations necessitating travel speeds greater than 100 km hr-1 (~55 knots) between time adjacent locations. The filter was triggered on 1,015 trips and removed 6,891 records.

Potential trips were discarded if they contained ≤3 VMS records, or were ≤6 hours in duration or had transmission breaks ≥5 days; removing 28,800; 12,121 and 168,549 records respectively (in total 3.6% of the original dataset). It is likely that these filters remove some legitimate fishing trips of short duration and may underestimate near-shore fishing effort. However, they were required to minimising the degree of visual supervision needed to manage this large dataset while maximising retention of VMS data. Post filtering the dataset contained 56,434 fishing trips.

The modal frequency of record transmission was 2 hours. To ensure temporal consistency among data, all trips were re-sampled where necessary to a 2 hour±15 minutes frequency using great circle, speed-appropriate, principles. This process maximised the retention of transmitted records, only filling temporal gaps where necessary and resulted in a 14% reduction from pre-treated data, making available

3,635,855 data points. The mean net change in the number of data points following this temporal alignment process for each trip was –8.9; 28,320 trips experienced a net addition, receiving an average of 10±19 data points, 13,986 trips experienced a net reduction, losing an average of 56±198 VMS records; 14,776 trips experienced no adjustment in their temporal frequency.

Vessel Behaviour

A speed rule was used to distinguish fishing from steaming or near-stationery movement. It was necessary to construct derived speeds for all VMS records as prior to 1-1-2006 transmission of speed and heading was not mandatory [34]. Derived speeds represent the speed of movement between time adjacent records within a fishing trip. We compared transmitted vessel speeds available from 40,681 fishing trips (3,126,213 VMS records) to comparative derived speeds to ensure that these speeds were closely mirrored. The process identified 78.9% of fishing trips yielded statistically significant positive correlations between transmitted and derived speeds (Pearson correlation coefficient; $p \leq 0.05$; mean $r2 = 0.6$ for all fishing trips). The speed filtering process assigned 1,710,725 data points (47% of available data) as representing fishing activity.

The upper and lower speed thresholds for determining fisheries activity were influenced by the frequency distribution of vessel speeds, and from published values [30], [31], [36]. As the UK VMS database retains incomplete data on vessel gear type and vessels can change their gear seasonally it was necessary for the speed rule to encompass many types of fisheries activities, for example beam trawling, gill netting and longlining. The lack of metadata prevents VMS data from being partitioned by gear type. Fishing activity was therefore assigned to all vessels travelling at speeds ≥3.0 and ≤10.0 km h–1, (~1.5 to 5.5 knots). While this approach is a coarse manner in which to filter the data, the assigned limits circumscribes the speeds at which larger vessels move while undertaking fisheries activities.

Mapping Fisheries Activity

Fisheries activity was gridded at a spatial resolution of 9 km2 (3 km by 3 km pixel) by summing the number of VMS derived data points coincident to each pixel over monthly and annual scales.

Acknowledgements

We thank K. Porter (Department for Environment, Food and Rural Affairs) for assistance in obtaining VMS data. We thank S. Northridge (Sea Mammal

Research Unit, UK) and J. Reid (Joint Nature Conservation Council, UK) for constructive criticism during the analytical phase of this work. We thank L. Hawkes for comments on drafts of the manuscript.

Author Contributions

Conceived and designed the experiments: MW BG. Performed the experiments: MW. Analyzed the data: MW. Contributed reagents/materials/analysis tools: MW. Wrote the paper: MW BG. Other: Co-devised the analysis: BG.

References

1. Pauly D, Christensen V, Guenette S, Pitcher TJ, Sumaila UR, et al.. (2002) Towards sustainability in world fisheries. Nature 418: 689–695.

2. Zeller D, Pauly D (2005) Good news, bad news: global fisheries discards are declining, but so are total catches. Fish and Fisheries 6: 156–159.

3. Hall MA, Alverson DL, Metuzals KI (2000) By-catch: Problems and solutions. Marine Pollution Bulletin 41: 204–219.

4. Lewison RL, Crowder LB (2003) Estimating fishery bycatch and effects on a vulnerable seabird population. Ecological Applications 13: 743–753.

5. Lewison RL, Crowder LB, Read AJ, Freeman SA (2004) Understanding impacts of fisheries bycatch on marine megafauna. Trends in Ecology & Evolution 19: 598–604.

6. Myers RA, Worm B (2005) Extinction, survival or recovery of large predatory fishes. Philosophical Transactions of the Royal Society B 360: 13–20.

7. Votier SC, Furness RW, Bearhop S, Crane JE, Caldow RWG, et al.. (2004) Changes in fisheries discard rates and seabird communities. Nature 427: 727–730.

8. Hammond PS, Benke H, Berggren P, Borchers DL, Buckland ST, et al.. (1995) Distribution and abundance of the harbour porpoise and other small cetaceans in the North Sea and adjacent waters. Final report to the European Commission under contract LIFE 92-2/UK/27.

9. Sims DW, Southall EJ, Richardson AJ, Reid PC, Metcalfe JD (2003) Seasonal movements and behaviour of basking sharks from archival tagging: no evidence of winter hibernation. Marine Ecology-Progress Series 248: 187–196.

10. Croxall JP, Silk JRD, Phillips RA, Afanasyev V, Briggs DR (2005) Global circumnavigations: Tracking year-round ranges of nonbreeding albatrosses. Science 307: 249–250.

11. Bearhop S, Thompson DR, Phillips RA, Waldron S, Hamer KC, et al.. (2001) Annual variation in Great Skua diets: The importance of commercial fisheries and predation on seabirds revealed by combining dietary analyses. The Condor 103: 802–809.

12. Bowen BW, Bass AL, Soares L, Toonen RJ (2005) Conservation implications of complex population structure: lessons from the loggerhead turtle (Caretta caretta). Molecular Ecology 14: 2389–2402.

13. Northridge S, Morizur Y, Souami Y, Canneyt Ov (2006) PETRACET: Project EC/FISH/2003/09. Final report to the European Commission 1735R07D. Lymington, UK: MacAlister Elliott and Partners Ltd.

14. Carranza A, Domingo A, Estrades A (2006) Pelagic longlines: A threat to sea turtles in the Equatorial Eastern Atlantic. Biological Conservation 131: 52–57.

15. Phillips RA, Silk JRD, Croxall JP, Afanasyev V (2006) Year-round distribution of white-chinned petrels from South Georgia: relationships with oceanography and fisheries. Biological Conservation 129: 336–347.

16. Molenarr EJ, Tsamenyi M (2000) Satellite-based vessel monitoring systems international legal aspects and developments in state practice. United Nations.

17. Department for Environment Food and Rural Affairs (2004) Explanatory memorandum to the south-west territorial waters (prohibition of pair trawling) order 2004-No. 3397.

18. Gilkinson KD, Gordon DC, MacIsaac KG, McKeown DL, Kenchington ELR, et al.. (2005) Immediate impacts and recovery trajectories of macrofaunal communities following hydraulic clam dredging on Banquereau, eastern Canada. ICES Journal of Marine Science 62: 925–947.

19. Department for Environment Food and Rural Affairs (2005) Charting progress: An integrated assessment of the state of the UK seas.

20. Marine Fisheries Agency (2005) United Kingdom Sea Fisheries Statistics 2004. London.

21. Council of the European Union (2002) Council Regulation (EC) No 2371/2002 of 20 December 2002 on the conservation and sustainable exploitation of fisheries resources under the Common Fisheries Policy.

22. Le Fevre J (1986) Aspects of the Biology of Frontal Systems. Advances in Marine Biology 23: 163–299.

23. Huthnance JM, Coelho H, Griffiths CR, Knight PJ, Rees AP, et al.. (2001) Physical structures, advection and mixing in the region of Goban spur. Deep Sea Research Part II: Topical Studies in Oceanography 48: 2979–3021.

24. Reid JB, Evans PGH, Northridge SP (2003) Atlas of cetacean distribution in north-west European waters. Peterborough, UK: Joint Nature Conservation Committee.

25. Stone C, Webb A, Barton C, Ratcliffe N, Reed T, et al.. (1995) An atlas of seabird distribution in north-west European waters. Peterborough, UK: Joint Nature Conservation Committee.

26. Hammond PS (2007) Final report on Small Cetacean Abundance in the European Atlantic and North Sea (SCANS-II). Department for Environment, Food and Rural Affairs.

27. Connor DW, Gilliland PM, Golding N, Robinson P, Todd D, et al.. (2006) UKSeaMap: the mapping of seabed and water column features of UK seas. Peterborough, UK: Joint Nature Conservation Committee.

28. Gallaway BJ, Cole JG, Martin LR (2003) Description of a simple electronic logbook designed to measure effort in the Gulf of Mexico shrimp fishery. North American Journal of Fisheries Management 23: 581–589.

29. Commission of the European Communities (2004) Proposal for a Council regulation on electronic recording and reporting of fishing activities and on means of remote sensing. No 724-2004.

30. Deng R, Dichmont C, Milton D, Haywood M, Vance D, et al.. (2005) Can vessel monitoring system data also be used to study trawling intensity and population depletion? The example of Australia's prawn fishery. Canadian Journal of Fisheries and Aquatic Sciences 62: 611–622.

31. Mills CM, Townsend SE, Jennings S, Eastwood PD, Houghton CA (2007) Estimating high resolution trawl fishing effort from satellite-derived vessel monitoring system data. ICES Journal of Marine Science 64: 248–255.

32. Pauly D, Alder J, Bennett E, Christensen V, Tydemers P, et al.. (2003) The Future for Fisheries. Science 302: 1359–1361.

33. Department for Environment Food and Rural Affairs (2005) Satellite Monitoring: UK fishing boats satellite-tracking device specifications.

34. Department for Environment Food and Rural Affairs (2005) Satellite Monitoring: Notes for Guidance.

35. National Geophysical Data Centre (2007) http://www.gfdl.noaa.gov/products/vis/data/datasets/TerrainBase.html.

36. Murawski SA, Wigley SE, Fogarty MJ, Rago PJ, Mountain DG (2005) Effort distribution and catch patterns adjacent to temperate MPAs. ICES Journal of Marine Science 62: 1150–1167.

Chapter 10

Rapid Effects of Marine Reserves via Larval Dispersal

Richard Cudney-Bueno, Miguel F. Lavín, Silvio G. Marinone, Peter T. Raimondi and William W. Shaw

ABSTRACT

Marine reserves have been advocated worldwide as conservation and fishery management tools. It is argued that they can protect ecosystems and also benefit fisheries via density-dependent spillover of adults and enhanced larval dispersal into fishing areas. However, while evidence has shown that marine reserves can meet conservation targets, their effects on fisheries are less understood. In particular, the basic question of if and over what temporal and spatial scales reserves can benefit fished populations via larval dispersal remains unanswered. We tested predictions of a larval transport model for a marine reserve network in the Gulf of California, Mexico, via field oceanography and repeated density counts of recently settled juvenile commercial mollusks before and after reserve establishment. We show that local retention of larvae within a reserve network can take place with enhanced, but spatially-explicit, recruitment to local fisheries. Enhancement occurred rapidly (2 yrs), with up to

Originally published as Cudney-Bueno R, Lavín MF, Marinone SG, Raimondi PT, Shaw WW (2009) Rapid Effects of Marine Reserves via Larval Dispersal. PLoS ONE 4(1): e4140. https://doi.org/10.1371/journal.pone.0004140.

a three-fold increase in density of juveniles found in fished areas at the down-stream edge of the reserve network, but other fishing areas within the network were unaffected. These findings were consistent with our model predictions. Our findings underscore the potential benefits of protecting larval sources and show that enhancement in recruitment can be manifested rapidly. However, benefits can be markedly variable within a local seascape. Hence, effects of marine reserve networks, positive or negative, may be overlooked when only focusing on overall responses and not considering finer spatially-explicit responses within a reserve network and its adjacent fishing grounds. Our results therefore call for future research on marine reserves that addresses this variability in order to help frame appropriate scenarios for the spatial management scales of interest.

Introduction

As a response to declining fish stocks and threats to marine ecosystems, marine reserves (areas closed to fishing) have been widely advocated as conservation tools and means to achieving more sustainable use of marine resources [1]–[3]. The rationale behind their use lies in the dual opportunity they could offer to protect ecosystems and ecological processes while also enhancing fisheries via density-dependent spillover and larval dispersal of target species into fishing areas [1], [3]–[5]. However, while evidence has shown that marine reserves can meet conservation targets [6]–[8], the role they may have on fisheries is less understood. Previous studies have focused on benefits to adjacent fisheries via density-dependent spillover of adult fish from reserves [6], [9] or have been based primarily on larval transport models [3], [10]–[12], lacking validation through field monitoring and oceanographic data.

Models that inform effects of marine reserves via enhanced larval export rely on (a) assumptions about recruitment limitations in unprotected populations, and (b) connectivity between reserves and non-reserve sites [10]–[12]. If both assumptions are met, marine reserves could replenish adjacent, fished populations. However, effects could be localized or widespread depending on dispersal, which is in turn related to the complex interactions among local current patterns and larval duration and behavior [13]–[16]. Hence, actual effects are difficult to measure and understand. Furthermore, without explicit model predictions of patterns of enhanced recruitment, assumptions of reserve effects can neither be supported nor falsified by empirical results. These have been fundamental problems in investigations of marine reserves [17], and the basic question of if and over what temporal and spatial scales reserves can benefit fished populations via larval dispersal remains unanswered.

As a means to test the effects of reserves on adjacent fisheries via larval dispersal, we coupled predictions from a larval transport model with in situ field oceanography and monitoring of densities of individuals recruited since the establishment of a reserve network in Northwest Mexico. The Puerto Peñasco reserve network was established in summer 2002 primarily as a means to protect declining stocks of two commercial species of mollusks: rock scallop (Spondylus calcifer) and black murex snail (Hexaplex nigritus). The network includes an offshore reserve (San Jorge Island), and two coastal reserves (Las Conchas and Sandy) (Fig. 1). It covers approximately 18 km of coastline composed primarily of extended beach-rock (coquina) and granite reefs separated by beds of mussels and rhodoliths and shell/sandy patches.

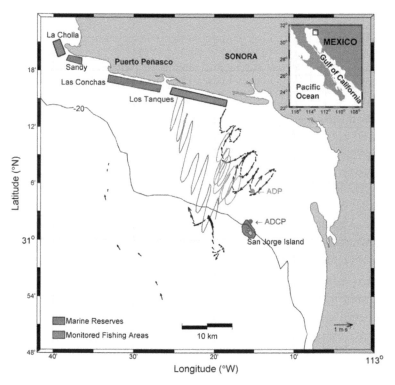

Figure 1. Location of reserve network, monitored fishing areas, and observations of regional currents.

The three diagrams in the center represent: track and velocity of a surface drifter (in black), progressive vector diagram (PVD) from velocity measured 15 m above the bottom at ADCP (Acoustic Doppler Current Profiler) site (in blue, bottom at 25 m), and the PVD from velocity measured 15 m above the bottom at ADP (Acoustic Doppler Profiler) site (in red, bottom at 18 m). Both PVDs have been shifted west for clarity. These three diagrams are made with hourly-mean data for the period 19:00:00 (UT) July 12 to 00:10:00 July 16 2006. The black arrows close to the 20 m isobath represent interpolated half-hourly data from a drifter drogued at 15 m, from 03:00 to 22:30, July 7 2006. The most offshore arrows are 6-hourly drifter data from the same drifter, from 18:00 June 24 to 12:00 June 25 2006.

Results

To assess the effects of the reserve network, we generated a larval export model and tested its predictions through observations of currents and bi-annual density counts of juvenile rock scallops and murex snails within reserves and fishing grounds prior and after reserve establishment (summer 2002-summer 2004). We developed a three-dimensional baroclinic numerical model that was based on the circulation pattern for the summer (the spawning season for both species), which is cyclonic overall [18], with northwestward flow in the area where the reserve network is located. We used the model to assess if the network could receive larvae from southern sources and to predict patterns of larval recruitment within the network. The model tracked passive particles for up to four weeks (exceeding the range of larval duration for both species) after being released (a) in the rocky reef south of the reserve network (~150 km south of San Jorge Island), (b) in San Jorge Island, the southern boundary of the network, and (c) every km from the Island to the northwestern portion of the network.

In case (a) released particles showed a median south-northwest travel distance of 148 km in four weeks (Fig. 2A). Larvae of both species, however, are competent to settle in less time [19], [20]. It is therefore highly unlikely that there could have been any substantial direct influence from southern reefs, particularly on coastal reserves which are 180–200 km north of these reefs. Furthermore, our model predictions are likely an overestimation of true dispersal distances, as larvae dispersal can be more constrained once behavior and habitat are accounted for in model predictions [15], [21]. Influence from western sources (Baja California peninsula) is highly unlikely, as previous studies indicate a clear cyclonic movement of the water during summer [22], when both of these species reproduce [23], [24]. On the eastern side of the Gulf of California, the water mass has a northbound movement whereas on the western side water moves towards the south and does not reach the eastern coastline [22].

Particles released in the area surrounding San Jorge Island (case b), the southern portion of the reserve network, showed a marked flow toward the coast and northwestern reserve sites (Fig. 2B). Direct evidence of this flow pattern is provided by the tracks of surface drifters released near the Island and progressive vector diagrams (PVDs) from concurrent acoustic current profilers (ADCPs, ADPs) (Fig. 1). Drifter tracks show the tidal ellipses plus a residual flow toward the coast (tidal ellipses refer to the trajectory that drifters followed with the ebb and flow of the tide; while residual flow refers to the net displacement of drifters over one or more tidal cycles, in this case, progressively moving north toward the coast). When modeling larval settlement as a function of distance from the Island to coastal reserves and monitored fishing areas (case c), for any period from 1–4 weeks the model predicted more settlement at northernmost sites (Sandy/La Cholla) (Fig. 3). Following this same modeling exercise, more settlement in southern reserve and fishing areas (Las Conchas/Los Tanques) compared to northern ones would only be evident if larvae were competent to settle no more than two days after release. However, larvae of both species are planctonic and competent to settle in >1 week [19], [20].

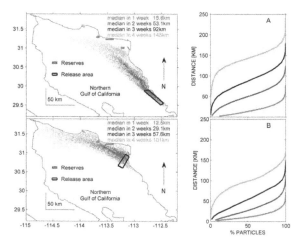

Figure 2. Final position of particles 1–4 weeks after having been released in (a) the nearest rocky reef located south of the marine reserve network, and (b) the network's southern boundary (San Jorge Island).

Panels on right show cumulative percentages of particles as a function of distance 1, 2, 3 and 4 weeks after release.

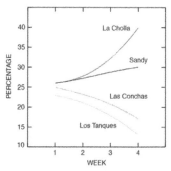

Figure 3. Modeled larval settlement (as relative percentages) within coastal reserves (Las Conchas, Sandy) and fishing areas (Los Tanques, La Cholla) as a function of the day larvae are competent to settle.

Model larvae were released every kilometer in the region of interest, from San Jorge Island to La Cholla. Earlier results suggested that there was no source of larvae to the south of the network. No sources were used to the north because models showed that larvae released to the north of the reserve network would be transported away from the network. The model assumed that larvae settled on the day of competency. If that assumption is relaxed, the difference in settlement between northern (Sandy/La Cholla) and southern (Las Conchas/Los Tanques) sites increases.

Observed spatial patterns of recruitment of juveniles of both species (individuals born and recruited since reserve establishment) were consistent with predictions of our larval transport model. Only two years after establishment of the reserves, both species showed evidence of changes in density as a function of time, protection from fishing (reserve effects) and site effects as a whole (repeated measures three-way MANOVA, time X protection X site; rock scallop: Pillai's Trace F4, 41 = 2.53, P = 0.05; black murex: Pillai's Trace F4, 41 = 3.02, P = 0.02). Density of juvenile rock scallop had increased by up to 40.7% within coastal reserves and by 20.6% in fished areas (repeated measures two-way MANOVA, time X protection from fishing, Pillai's Trace F4, 41 = 2.67, P<0.05). Changes were also evident for black murex, with more than a three-fold increase in density of juveniles within fished areas (repeated measures two-way MANOVA, time X protection from fishing, Pillai's Trace F4, 41 = 3.28, P<0.05). The pattern of increase in juveniles, however, was variable in space, evident only for the northwestern portion of the network (Fig. 4), as predicted by the larval transport model for any period between one and four weeks. Density of both species increased markedly in the reserve and fished northwestern sites (Sandy/La Cholla) and remained relatively constant in southeastern sites (Las Conchas/Los Tanques) (repeated measures two-way MANOVA, time X site; rock scallop: Pillai's Trace F4, 41 = 7.09, P<0.001; black murex: Pillai's Trace F4, 41 = 2.95, P<0.05).

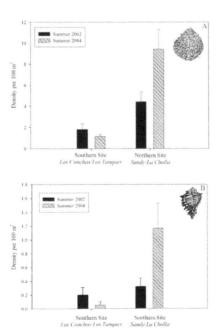

Figure 4. Differences in densities (S.E. bars included) of juvenile rock scallops (a) and black murex snails (b) in southern and northern sites before (summer 2002) and after (summer 2004) reserve establishment.

Discussion

Observed increase in recruitment was spatially-constricted to the northern portion of the reserve network and consistent with predictions of our larval transport model and field oceanographic observations. This recruitment pattern reflects effects of larval dispersal within the reserve network rather than other effects such as density-dependent adult spillover, a good year of high overall regional recruitment, or wider oceanographic processes. The restricted adult movement of both species (sessile in the case of rock scallop) constrains our inference to larval dispersal rather than density-dependent adult spillover. Furthermore, high overall recruitment and regional oceanographic processes would likely have resulted in changes throughout the study area, not only in the northern portion of the reserve network.

Surveys on San Jorge Island, as well as our modeling and current measurements (Figs. 1 and 2B) suggest that the Island could be acting as a key component of the network, providing a source for larval export to adjacent coastal reserves and fishing areas. Overall density of juvenile rock scallops on the Island actually decreased since reserve establishment (repeated measures 1-way ANOVA; $F_{4, 45}$ = 4.46, P<0.01) and those of black murex remained relatively constant (repeated measures 1-way ANOVA; $F_{4, 45}$ = 0.615, P = 0.65). However, although density of juveniles did not increase, even the lowest average numbers in five monitoring seasons were 80% higher than those of coastal reserves. Overall densities (adults and juveniles) were also six times higher than those of all coastal reserves and fishing sites combined, reaching up to $1.6/m2$ and exceeding any others reported for the Gulf of California [23], [25]. Given these high densities, a decrease in juveniles near the Island could be related to density-dependent processes.

Our findings provide needed insights for theory and empirical understanding of effects of marine reserves. First, we show evidence of rapid effects of reserve networks on adjacent fisheries via larval dispersal. Second, we also show that local retention of larvae within a network can take place with enhanced but spatially variable recruitment to local fisheries. Hence, effects should not be expected across an entire reserve network. Rather, they can be markedly variable within a local seascape.

These results have important implications for management. Reserves reduce the total area available for fishing, likely causing an initial economic cost to fishers. Therefore, in situations where there is local support for reserve establishment, evidence of rapid positive reserve effects, as we have here shown, could play a crucial role in reinforcing cooperation among fishers for further compliance [26]. Evidence of larval retention and enhanced recruitment to local fisheries also underscores the benefits of protecting reproductive larval sources and reconciles local management with social needs. Reserve networks with strong support from fishing communities are best designed if they enhance or maintain recruitment within the area of influence of these communities, not benefiting others at the expense of local management initiatives and, ultimately, initial costly decisions. In some situations, however, this may not be possible, as oceanographic processes could result in larval export outside the area of influence of the community or communities supporting the reserve network. Designs of reserve networks that cover broader spatial scales may be needed in these situations.

Finally, effects of marine reserves, positive or negative, may be overlooked when only focusing on overall responses and not considering finer spatially-explicit responses within a reserve network and its adjacent fishing grounds. Our results therefore call for future research on marine reserves that addresses this variability in order to help frame appropriate scenarios for the spatial management scales of interest. Not doing so could lead to false expectations among stakeholders.

Materials and Methods

Particle Tracking from a Three Dimensional Oceanographic Numerical Model

We released 2000 passive particles in two areas (between 0–60 m deep): San Jorge Island, and the nearest substantial rocky reef south of the marine reserve network. Particles were tracked for four weeks and the temporal scales resolved by the model (due to forcing) are tidal and seasonal. The model is described in detail for the Gulf of California by Marinone [27] and Mateos et al.. [28]. Briefly, the model domain has a mesh size of 2.5'×2.5' (~3.9×4.6 km) in the horizontal and 12 layers in the vertical with nominal lower levels at 10, 20, 30, 60, 100, 150, 200, 250, 350, 600, 1000 and 4000 m. Model equations are solved semi-implicitly with fully prognostic temperature and salinity fields. The model is forced with tides, climatological winds, climatological hydrography at the mouth of the Gulf of California, and climatological heat and fresh water fluxes at the air-sea interface. As shown by Marinone [27], the model adequately reproduces the main seasonal signals of surface temperature, heat balance, tidal elevation and surface circulation in the northern Gulf of California and also the tidal currents as shown by Marinone and Lavín [29].

Drifter Tracks and Current Profiles

We used two bottom-mounted acoustic current profilers moored at the sites marked ADCP and ADP in Fig. 1 and six PacificGyre (www.pacificgyre.com/

Lagrangian.aspx) Microstar surface drifters which drogues were centered at 1 m below the sea surface. These drifters provided GPS positions every 10 minutes. One PacificGyre ARGOS SVP (Surface Velocity Program) drifter was used to observe the current field offshore (west) of San Jorge Island; it was drogued with a 4.8 m- tall Holey Sock centered at 15 m depth.

Site ADCP (Fig. 1), equipped with a bottom-mounted 300 KHz Acoustic Doppler Current Profiler (by RDInstruments), was set just north of San Jorge Island where the mean bottom depth was 25 m. Site ADP, which contained a bottom-mounted 500 KHz Acoustic Doppler Profiler (by SonTek), was 8 km further north with the bottom at ~18 m. Both current profilers measured the mean velocity of every meter of the water column, every three minutes, for the periods June 2-July 4 and July 6-August 18 2006. For the purpose of this article, the best way to present the profiler current data are the Progressive Vector Diagrams (PVD), which are constructed by calculating the vector displacement that a water parcel would experience at the mooring position during each sampling interval, and drawing them sequentially, the tail of each vector on the head of the previous one. Note that they are not true tracks, but they can be plotted over maps to aid in the interpretation. In Fig. 1 we plotted the PVDs at 15 m above the bottom for both the ADCP and the ADP, from 19:00 UT on July 12 to 00:10 UT on July 16 2006. The surface drifters were deployed in groups of 4–6 units; several deployments were made between July 12 and July 23 2006, either over the ADP or ADCP sites or off the northern end of San Jorge Island. In Fig. 1, the track and velocity of one of the drifters is plotted together with the PVDs from the two current profilers, for the period covered by the drifter track. Fig. 1 also shows the tracks and velocities obtained outside San Jorge Island with the SVP drifter. Two tracks are shown, the most offshore comprises 6-hourly data from 18:00 (UT) June 24 to 12:00 (UT) June 25 2006, and the second interpolates half-hourly data from 02:57 (UT) to 22:23 (UT), July 7 2006.

Estimation of Population Parameters

We estimated changes in density of rock scallop (Spondylus calcifer) and black murex snail (Hexaplex nigritus) in reserve and fishing sites for two consecutive years beginning in May 2002, one month preceding reserve establishment. The region monitored encompassed the reefs of San Jorge Island and those found near the fishing town of Puerto Peñasco (within 3 km from highest tide line) in the eastern part of the northern Gulf of California, Mexico. This region extends from 31,22,18.1 N; 113,39,09.4 W to 31,15,03.8 N; 113,20,48.1 W.

We subdivided the region into 5 sampling areas: a) two coastal reserves (replicates), Las Conchas and Sandy; b) two coastal fishing areas ("controls"),

Los Tanques and La Cholla; c) one offshore island reserve, San Jorge Island (Fig. 1). We paired coastal reserves with appropriate coastal fishing areas (Sandy with La Cholla; Las Conchas with Los Tanques). We refer to each of these pairs as "sites." Given the lack of adequate comparison areas for San Jorge Island, we analyzed the response of this off-shore reserve independently. Assessments are based on bi-annual density counts in 58 100 m2 permanent plots before and after reserve establishment (repeated measures, five monitoring seasons).

To reduce heterogeneity associated with depth, we restricted all sampling to depths ranging from 40–65 ft. This also reduced health risks associated with diving and facilitated overall monitoring as we were able to remain underwater for longer periods of time. In all cases except San Jorge Island, this depth as well as the established constrained distance from the tide line covers the entire extension of the reefs. We restricted sampling in San Jorge Island to the reefs found on the eastern part of the island, as these are shallower and more similar to those found on the mainland coast.

Plot Design and Sample Unit Selection

We selected plots from within these five areas through simple random sampling. In the event that a specific plot selected happened to fall where at least 50% of sand was present, that plot was replaced by another one by swimming underwater in a straight line along the reef until reaching sufficient (>50%) rocky substrate. Once selected, all plots were permanently marked underwater.

Plots were 10×10 m subdivided into 16 quadrats of 2.5×2.5 m for ease of observation. Testing other sampling methods such as the use of 5×50 m or 5×30 m transects, distance sampling or others typically used for sessile organisms did not prove adequate for this region given the highly variable visibility of the region, the strong currents, and the overall patchiness of the reefs (i.e. patches of reef typically separated by patches of sand). We counted all individuals visible within each 2.5×2.5 m quadrat (subplot). For rock scallop, we estimated size of each individual to fall within one of three categories: small juveniles (up to 5 cm of height), medium-sized juveniles and young adults (>5 and <10 cm of height), and large adults (>10 cm of height). For black murex, sizes fell into two categories: juveniles and reproductively mature adults.

To reduce variation in detectability, the same person counted organisms on each sampling occasion and in the same designated plots while another diver assisted setting and maintaining the plot lines in place. To support this work, ten commercial divers with extensive experience searching for benthic mollusks (>5 years) were trained to participate in the monitoring process. We calculated variations in the detection of monitored species (s≤3 individuals/plot) and incorporated this variation to calculate statistical power of our sampling design (see below).

Sampling Frequency, Sample Sizes, and Allocation of Samples

We established a total of 58 sampling plots: San Jorge = 10, Las Conchas = 10, Los Tanques = 10, Sandy = 10, La Cholla = 18. Power analyses from baseline data on density of rock scallops found on these 58 plots gave us a high probability of detecting at least a 10% increase in their density (Power>95% for each reserve and fishing zone, α = 0.05 and s = 3 individuals/plot). Given that rock scallops are harder to detect than black murex (when closed they resemble rocks), we assume an even higher statistical power for detection of changes in population densities of black murex. We monitored each plot twice every year (Spring and Summer) for two consecutive years (Summer 2002, Spring 03, Summer 03, Spring 04, Summer 04). These months provide some of the best visibility underwater and are also usually devoid of algae beds covering the rocky reefs, which reduce detectability of species monitored.

Statistical Analysis

This is a longitudinal study with a repeated measures research design and various levels of analysis. We first generated profile plots of baseline data and applied square root transformations to improve homogeneity of variance. We then addressed "between subject" and "within subject" variability of baseline data graphically, determined the coefficient of variation, and tested for independence of plots and sampling sites in order to avoid pseudo-replication. We used multivariate analyses of variance (MANOVA) and relied on Multivariate Pillai's Trace P values to help assess time, protection from fishing, and treatment effects independently as well as combined factors. Univariate estimates were also obtained and analyzed to further understand observed.

Acknowledgements

We thank the Sociedad Cooperativa Buzos de Puerto Punta Peñasco, O. Morales, and J. Rupnow for their participation in subtidal field monitoring. C. Cabrera, A. Cinti, J. Duberstein, M. Figueroa, V. Godínez, R. Loaiza, A. Maldonado, M. Moreno, S. Pérez, E. Polanco, M. Rivera, R. Salazar, and G. Soria assisted in the field oceanography component of this research. We thank M. Carr, J. Donlan,

F. Michelli, R. Steidl, and J. Tewksbury for comments on earlier drafts of our paper. This is a scientific contribution of the PANGAS Project, www.pangas.arizona.edu.

Author Contributions

Conceived and designed the experiments: RCB WWS. Performed the experiments: RCB MFL SGM. Analyzed the data: RCB MFL SGM PTR. Contributed reagents/materials/analysis tools: RCB MFL SGM. Wrote the paper: RCB MFL SGM PTR WWS. Conducted subtidal monitoring: RCB.

References

1. Hastings A, Botsford LW (1999) Equivalence in yield from marine reserves and traditional fisheries management. Science 284: 1537–1538.

2. National Research Council (2001) Marine protected areas: tools for sustaining ocean ecosystems. Washington, DC: National Academy Press.

3. Gaylord B, Gaines SD, Siegel DA, Carr MH (2005) Marine reserves exploit population structure and life history in potentially improving fisheries yields. Ecological Applications 15: 2180–2191.

4. Roberts CM, Bohnsack JA, Gell F, Hawkins JP, Goodridge R (2001) Effects of marine reserves on adjacent fisheries. Science 294: 1920–1923.

5. Hastings A, Botsford LW (2003) Comparing designs of marine reserves for fisheries and biodiversity. Ecological Applications 13: S65–70.

6. Roberts CM, Hawkins JP (2000) Fully-protected marine reserves: a guide. WWF Endangered Seas Campaign, 1250 24th Street, NW, Washington D.C., 20037, USA and Environment Department, University of York, York, YO105-DD, UK.

7. Halpern BS, Warner RR (2002) Marine reserves have rapid and long lasting effects. Ecology Letters 5: 361–365.

8. Halpern BS (2003) The impact of marine reserves: do reserves work and does reserve size matter? Ecological Applications 13: S117–137.

9. Russ GR, Alcalá AC, Maypa AP, Calumpong HP, White AT (2004) Marine reserve benefits local fisheries. Ecological Applications 14: 597–606.

10. Botsford LW, Hastings A, Gaines SD (2001) Dependence of sustainability on the configuration of marine reserves and larval dispersal distances. Ecology Letters 4: 144–150.

11. Gaines SD, Gaylord B, Largier JL (2003) Avoiding current oversights in marine reserve design. Ecological Applications 13: S32–46.

12. Botsford LW, Micheli F, Hastings A (2003) Principles for the design of marine reserves. Ecological Applications 13: S25–31.

13. Cowen RK, Lwiza KMM, Sponaugle S, Paris CB, Olson DB (2000) Connectivity of marine populations: open or closed? Science 287: 857–859.

14. Sale PF (2004) Connectivity, recruitment variation, and the structure of reef fish communities. Integrative and Comparative Biology 44: 390–399.

15. Cowen RK, Paris CB, Srinivasan A (2006) Scaling of connectivity in marine populations. Science 311: 522–527.

16. Almany GR, Berumen ML, Thorrold SR, Planes S, Jones GP (2007) Local replenishment of coral reef fish populations in a marine reserve. Science 316: 742–744.

17. Sale PF, et al.. (2005) Critical science gaps impede use of no-take fishery reserves. Trends in Ecology and Evolution 20: 74–80.

18. Lavín MF, Durazo R, Palacios E, Argote ML, Carrillo L (1997) Lagrangian observations of the circulation in the Northern Gulf of California. Journal of Physical Oceanography 27: 2298–2305.

19. D'Asaro CN (1991) Gunnar Thorson's world-wide collection of prosobranch egg capsules: Muricidae. Ophelia 35: 1–101.

20. Parnell P (2002) Larval development, precompetent period, and a natural spawning event of the pectinacean bivalve Spondylus tenebrosus (Reeve, 1856). The Veliger 45: 58–64.

21. O' Connor MI, et al.. (2007) Temperature control of larval dispersal and the implications for marine ecology, evolution, and conservation. Proceedings of the National Academy of Sciences 104: 1266–1271.

22. Marinone SG, Gutiérrez OQ, Parés-Sierra A (2004) Numerical simulation of larval shrimp dispersion in the Northern Region of the Gulf of California. Estuarine, Coastal, Shelf Sciences 60: 611–617.

23. Villalejo-Fuerte M, Arellano-Martínez M, Ceballos-Vázquez BP, García-Domínguez F (2002) Reproductive cycle of Spondylus calcifer Carpenter, 1857 (Bivalvia: Spondylidae) in the "Bahía de Loreto" National Park, Gulf of California, Mexico. Journal of Shellfish Research 21: 103–108.

24. Cudney-Bueno R, Prescott R, Hinojosa-Huerta O (2008) The black murex snail, Hexaplex nigritus (Mollusca, Muricidae), in the Gulf of California, Mexico: I. reproductive ecology and breeding aggregations. Bulletin of Marine Science 83: 285–298.

25. Baqueiro E, Massó JA, Guajardo H (1988) Distribución y abundancia de moluscos de importancia comercial en Baja California Sur. Instituto Nacional de la Pesca, México, Serie de Divulgación 11: 1–32.

26. Ostrom E (1990) Governing the commons: the evolution of institutions for collective action. New York, NY: Cambridge University Press.

27. Marinone SG (2003) A three-dimensional model of the mean and seasonal circulation of the Gulf of California. Journal of Geophysical Research 108: 3325.

28. Mateos E, Marinone SG, Lavín MF (2006) Role of tides and mixing in the formation of an anticyclonic gyre in San Pedro Mártir Basin, Gulf of California. Deep Sea Research II 53: 60–76.

29. Marinone SG, Lavín MF (2005) Tidal current ellipses in a three-dimensional baroclinic numerical model of the Gulf of California. Estuarine, Coastal and Shelf Science 64: 519–530.

Chapter 11

Fishery-Independent Data Reveal Negative Effect of Human Population Density on Caribbean Predatory Fish Communities

Christopher D. Stallings

ABSTRACT

Background

Understanding the current status of predatory fish communities, and the effects fishing has on them, is vitally important information for management. However, data are often insufficient at region-wide scales to assess the effects of extraction in coral reef ecosystems of developing nations.

Originally published as Stallings CD (2009) Fishery-Independent Data Reveal Negative Effect of Human Population Density on Caribbean Predatory Fish Communities. PLoS ONE 4(5): e5333. https://doi.org/10.1371/journal.pone.0005333. © 2009 Stallings. https://creativecommons.org/licenses/by/4.0/

Methodology/Principal Findings

Here, I overcome this difficulty by using a publicly accessible, fisheries-independent database to provide a broad scale, comprehensive analysis of human impacts on predatory reef fish communities across the greater Caribbean region. Specifically, this study analyzed presence and diversity of predatory reef fishes over a gradient of human population density. Across the region, as human population density increases, presence of large-bodied fishes declines, and fish communities become dominated by a few smaller-bodied species.

Conclusions/Significance

Complete disappearance of several large-bodied fishes indicates ecological and local extinctions have occurred in some densely populated areas. These findings fill a fundamentally important gap in our knowledge of the ecosystem effects of artisanal fisheries in developing nations, and provide support for multiple approaches to data collection where they are commonly unavailable.

Introduction

It is well documented that humans have greatly altered predatory fish communities worldwide, especially through industrialized commercial and recreational fisheries [1]–[8]. These studies have based their conclusions on extensive databases of fisheries-dependent data (i.e., landings statistics), primarily from developed nations. However, fisheries statistics are commonly unavailable in developing nations where artisanal (subsistence or small-scale commercial) fisheries exist [9]–[11]. Despite the problem of insufficient data, it remains imperative to assess region-wide effects of extraction on predatory fish populations and to indicate whether indirect effects of human activities exist in the communities to which they belong (e.g., dominance shifts) in order to implement management and conservation strategies geared towards ecosystem-based approaches [12].

Artisanal fisheries supply food for millions of people in developing nations, and are the primary source of resource exploitation on coral reef systems [13]. Fishing on Caribbean reefs occurred long before the arrival of European settlers, but has returned increasingly diminished yields over the last 200 years as human populations have escalated in the region [14]–[16]. Similar to industrial and recreational counterparts in developed nations, artisanal fishing tends to target large-bodied, top trophic-level fishes, so greater numbers of fishermen per unit area should result in increased removal of larger species [17]–[20]. Indeed, populations of large-bodied fishes have become notoriously impoverished at some Caribbean locations with high densities of human populations (e.g., Jamaica)

[21], [22]. However, because fisheries data are generally unavailable or incomplete across the Caribbean, researchers have relied on either survey data from studies conducted on relatively small spatial scales or anecdotal and historical information. Therefore, the prevalence of these patterns and their potential indirect effects across the region remain unknown.

To address these issues on a larger scale, I used a publicly accessible, fisheries-independent database [23] to provide the first broad scale, quantitative analysis of the structure of predatory reef-fish communities across the greater Caribbean region (Fig. 1). The database consisted of over 38,000 presence/absence surveys conducted across 22 insular and continental nations (Table 1) by citizen scientists (i.e., trained volunteer SCUBA divers), a technique that has been used extensively by terrestrial ecologists (e.g., Breeding Bird Survey), but largely ignored by their marine colleagues. These community efforts can cover large geographic scales and produce sample sizes several order of magnitude greater than traditional efforts by either individual or small teams of scientists [24], effectively filling data gaps where fisheries-dependent data are currently unavailable. I also examined potential mechanisms, including factors that are both independent of and related to anthropogenic influences (Table 2), that may have affected the structure of these fish communities.

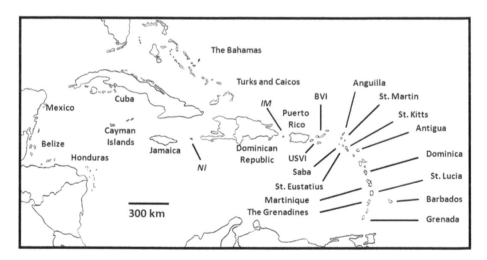

Figure 1. Map of Caribbean locations from which predator presence data were gathered.

The data were from all locations in which at least 10 volunteer diver surveys were conducted between 1994 and 2008. The locations of the two uninhabited islands are italicized: IM (Isla de Mona); NI (Navassa Island).

Table 1. Twenty-two nations from which REEF survey data were collected, including information of human population densities and sample sizes.

Country/region	HPD	Code	Survey locations	Total surveys
Belize	12	BZ	7	2304
Bahamas	21	BA	15	9457
Turks and Caicos	47	TC	10	3136
Mexican Caribbean	53	MC	5	5057
Honduras	62	HD	4	2124
Cuba	102	CU	3	567
Leeward Islands		LI	8	1819
--- Anguilla	129			13
--- Netherlands Antilles*	131			600
--- St. Kitts	149			285
--- Antigua	155			27
--- Dominica	91			894
British Virgin Islands	147	BV	3	2196
Cayman Islands	168	CI	4	4499
Dominican Republic	183	DR	4	515
Jamaica	248	JA	5	384
US Virgin Islands	308	UV	3	2347
Windward Islands		WI	8	2635
--- Martinique	359			163
--- St. Lucia	269			181
--- St. Vincent & The Grenadines†	302			1929
--- Barbados	647			173
--- Grenada	260			189
Puerto Rico	430	PR	7	1076
				TOTAL = 38116

*Netherlands Antilles (St Martin, Saba, St Eustatius).
†St Vincent & Grenadines (includes Bequia & Mustique).

Table 2. Pearson's correlations (r) between explanatory variables and the axes from the NMS ordination.

Variable	Axis 1	Axis 2
HPD (people/land km^2)	0.72	−0.01
HPReef (people/reef km^2)	0.09	0.05
GDP (PPP/capita)	−0.18	−0.08
Tourist (mean/year)	−0.23	0.11
Latitude	−0.64	−0.07

Results

A non-metric multidimensional scaling (NMS) ordination of 20 predatory taxa converged on a stable, 2-dimensional solution (final stress = 16.53, final instability = 0.00048, iterations = 74) (Fig. 2). The first axis accounted for the majority of variation in the NMS (r^2 = 0.67), was strongly correlated with human population density (r = 0.72) and slightly less so with latitude (r = −0.64; Table 2). The structure of the ordination was driven by strong associations of sharks

(Carcharhinidae), jacks (Carangidae), and large species of groupers (Serranidae) and snappers (Lutjanidae) with regions of low human population density (high latitude). The pattern was also driven by moderate associations of trumpetfish (Aulostomidae) and smaller species of groupers and snappers with regions of high human population density (low latitude; Fig. 2). The second axis accounted for less variation ($r^2 = 0.15$) and was driven by regional differences in which particular taxa of large or small predators predominated.

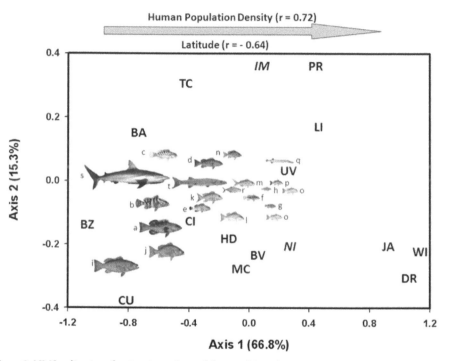

Figure 2. NMS ordination of regions in predatory fish space (20 taxa).

Regional centroids are displayed: BA (Bahamas); TC (Turks and Caicos); CU (Cuba); CI (Cayman Islands); JA (Jamaica); MC (Mexican Caribbean); BZ (Belize); HD (Honduras); DR (Dominican Republic), PR (Puerto Rico); UV (US Virgin Islands); BV (British Virgin Islands); LI (Leeward Islands); WI (Windward Islands). The axis 1 scores for the two uninhabited islands are italicized: IM (Isla de Mona); NI (Navassa Island). Along axis 1, latitude increases towards the left and human population density increases towards the right. Taxa locations are represented with coded fish displays: a (Mycteroperca bonaci); b (Epinephelus striatus); c (M. tigris); d (M. venenosa); e (E. guttatus); f (E. adscensionis); g (Cephalopholis cruentata); h (C. fulva); I (Lutjanus cyanopterus); j (L. jocu); k (L. analis); l (L. griseus); m (Ocyurus chrysurus); n (L. apodus); o (L. synagris); p (L. mahogoni); q (Aulostomus maculatus); r (Caranx spp.); s (Carcharhinus spp.); t (Sphyraena barracuda). Fish displays are scaled according to maximum attainable sizes of each taxa.

Because human population density and latitude were the primary factors related to the structure of the NMS ordination along the first axis, a multiple regression was used to investigate their independent effects. Although human population densities tend to decrease towards higher latitudes in the Caribbean region ($r = -0.57$), collinearity was low (variance inflation factor = 1.469); therefore the analysis was deemed robust. Both human population density ($p<0.00001$) and latitude ($p = 0.0121$) were related to the NMS scores after accounting for the effects of each. However, analysis of the standardized regression coefficients (1 standard deviation) revealed stronger evidence for a significant effect of human population density on NMS scores compared to latitude (i.e., lower p-values), and that the effect of the former (coefstandardized = 0.4583) was over twice as strong as the latter (coefstandardized = -0.2126).

Mean and median sighting frequencies of predators decreased 2.2–4.0% ($r2 = 0.19$, $p<0.0001$) and 4.1–7.1% ($r2 = 0.37$, $p<0.0001$), respectively, per incremental increase of 100 humans per km2. The predator communities exhibited lower richness ($r2 = 0.20$, $p<0.0001$) and Simpson's diversity ($r2 = 0.41$, $p<0.0001$) with increasing density of humans. At the taxon level, 15 of the 20 predators included in the analyses were sighted less frequently with increasing human population density (Table 3). The remaining five predatory taxa were sighted either evenly or at increasing frequencies with increasing human population density, and included the smallest species of grouper (graysby, Cephalopholis cruentata and coney, C. fulva) and snapper (mahogany snapper, Lutjanus mahogoni and lane snapper, L. synagris), as well as the relatively unfished trumpetfish (Aulostomus maculatus).

Table 3. Regression statistics of predatory reef-fish presence across a gradient of human population density.

Family	Taxa	Common name	TL_{max} (cm)	Intercept	SE	Coef	SE	t-Value	p-Value[c]
Aulostomidae	Aulostomus maculatus	trumpetfish	100	0.4827	0.0306	0.0005	0.0002	3.089	0.0027*
Carangidae	Caranx spp.	jacks[a]	69[b]	0.7690	0.0242	-0.0003	0.0001	-2.374	0.0199
Carcharhinidae	Carcharhinus spp.	requiem sharks[a]	300[b]	0.0887	0.0142	-0.0002	0.0001	-4.152	0.0001*
Lutjanidae	Lutjanus cyanopterus	cubera snapper[a]	160	0.0672	0.0095	-0.0002	0.0000	-5.572	<0.0001*
	L. jocu	dog snapper[a]	128	0.0975	0.0142	-0.0001	0.0001	-2.131	0.0361
	L. analis	mutton snapper[a]	94	0.1659	0.0198	-0.0002	0.0001	-1.770	0.0805
	L. griseus	gray snapper	89	0.1551	0.0165	-0.0002	0.0001	-2.568	0.0120
	Ocyurus chrysurus	yellowtail snapper	86	0.7602	0.0272	-0.0004	0.0001	-2.980	0.0038*
	L. apodus	schoolmaster	67	0.6091	0.0338	-0.0006	0.0002	-3.606	0.0005*
	L. synagris	lane snapper	60	0.0509	0.0126	0.0002	0.0001	3.015	0.0034*
	L. mahogoni	mahogany snapper	48	0.3445	0.0304	0.0003	0.0002	1.992	0.0497
Serranidae	Mycteroperca bonaci	black grouper[a]	148	0.1810	0.0190	-0.0006	0.0001	-6.858	<0.0001*
	Epinephelus striatus	Nassau grouper[a]	122	0.4607	0.0321	-0.0013	0.0002	-9.206	<0.0001*
	M. tigris	tiger grouper[a]	101	0.3112	0.0251	-0.0009	0.0001	-7.882	<0.0001*
	M. venenosa	yellowfin grouper[a]	100	0.0358	0.0042	-0.0001	0.0000	-4.753	<0.0001*
	E. guttatus	red hind[a]	76	0.0090	0.0015	-0.0001	0.0000	-3.778	0.0003*
	E. adscensionis	rock hind[a]	61	0.0873	0.0138	-0.0001	0.0001	-1.366	0.1756
	Cephalopholis cruentata	graysby	43	0.4705	0.0305	0.0004	0.0002	2.510	0.0140
	C. fulva	coney	41	0.4632	0.0385	0.0004	0.0002	1.873	0.0646
Sphyraenidae	Sphyraena barracuda	barracuda	200	0.4616	0.0278	-0.0006	0.0001	-4.447	<0.0001*

[a]Regression coefficient and intercept values computed from untransformed data; test statistics computed from arcsine($\bar{x}$0.5) transformed data (Zar 1999).
[b]Size data for sharks and jacks are from Caribbean reef shark (Carcharhinus perezii) and bar jack (Caranx ruber), respectively, which were the most common family representatives.
[c]Significant test after correction for multiple comparisons using sequential Bonferroni noted (*).
Note: Barbados was removed from the regressions since its high HPD (642people/km2) was approximately 50% greater than the second highest nation (i.e., outlier), and therefore quantitatively exaggerated the effect of HPD; trends were qualitatively unaffected.

NMS ordinations within both the grouper (final stress = 11.18, final instability = 0.00045, iterations = 59) and snapper (final stress = 11.21, final instability = 0.00045, iterations = 59) families each converged on stable, 3-dimensional solutions. The first axes of both ordinations accounted for the majority of variation (grouper r2 = 0.55; snapper r2 = 0.59) and were strongly correlated with human population density (grouper r = 0.75; snapper r = 0.57). Linear regressions within both families indicated strong decreases in maximum sizes of the species associations with regions along an index from low to high human population densities (Fig. 3).

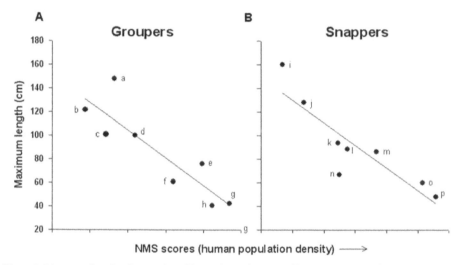

Figure 3. Maximum lengths of serranids and lutjanids as a function of human population density.

Taxon codes are in caption to Figure 2. Regression statistics (n = 8 species each): (A) serranid maximum published lengths (r^2 = 0.78, p = 0.004); (B) lutjanid maximum published lengths (r^2 = 0.78, p = 0.003). NMS scores are from the axis that accounted for the most variation in the data. Axis variation explained and correlation with human population density: (A) serranid ordination (axis r^2 = 0.55, r with axis = 0.75); (B) lutjanid ordination (axis r^2 = 0.46, r with axis = 0.50).

Discussion

The analyses presented here suggest human population density is strongly, negatively related to both richness and total presence (a surrogate of abundance) of predatory reef fishes in the Caribbean. Large predatory species were rare or absent in locations of high human population density, where smaller predators have become dominant, indicating the potential of indirect effects through competitive release. Although correlatives with both human activities and latitude may have had an influence on the structure of predatory communities, fishing was likely the most important mechanism driving the documented patterns.

Human population density and latitude were both correlated with the ordination of all taxa, but close examination of the data allow the relative effects of each predictor to be disentangled. In addition to compelling results from multiple regression analyses (see Results), further evidence reinforces that human population density was the dominant signal. First, although all taxa included in the analysis are naturally distributed across all locations in the study region, most fishes, particularly the larger-bodied ones, were rare or completely absent in surveys conducted in areas of high human population density. These patterns were evident in both the compressed, multivariate space (i.e., all large-bodied predators grouped on the left/negative side of axis 1, Fig. 2) and in the presence data of individual taxa (Table 3). In addition, historical data further illustrate that large groupers, snappers, and sharks were once abundant throughout the Caribbean, including reefs located in the Greater and Lesser Antilles where several of the species examined here are now ecologically or locally extinct [15], [25], [26].

Second, comparisons between inhabited and uninhabited islands within otherwise densely populated regions highlight potential human induced effects [27], [28]. For example, Isla de Mona and Navassa Island are uninhabited, relatively isolated nature reserves near the densely populated islands of Puerto Rico and Jamaica, respectively (Fig. 1). Although both islands have historically been fished and have experienced other anthropogenic effects, the intensity of such effects on these relatively remote locations is likely lower compared to nearby inhabited islands. Indeed, the similarities between the predator communities at these locales and other locations of low human density can be detected in both the ordinated space (i.e., italicized locations IM and NI further to the left on axis 1 than centroids of neighboring PR and JA, respectively, Fig. 2) and the presence/absence data for each taxon. Sighting frequencies of large-bodied predators, such as sharks, jacks, barracuda, and large groupers and snappers, were two to three times higher on reefs adjacent to the uninhabited islands relative to nearby inhabited ones. The more extensive presence of these predators within regions where they are otherwise rare or completely absent indicates that anthropogenic effects, not latitudinal gradients, limit the presence of these large-bodied fishes.

The relationship between human population density and ecological communities has been investigated far more extensively in terrestrial systems than marine ones [29]. However, several recent studies from the Line Islands [20], [30], [31] and the Hawaiian Islands [27], [32] have found higher abundances and biomass of large predatory fishes in locations of low human population densities compared to those that are densely populated. Similar results were found in the current study, with large predators becoming increasingly rare or locally extinct with increasing human population densities. Human activities can negatively affect populations and communities of coral reef fishes directly through harvesting and indirectly through habitat loss [32]. Worldwide degradation of coral reefs has been well documented [33]–[35], and although the effects of global climate change (and associated effects of bleaching, acidification, and disease) are thought to be the major drivers, local effects related to human population density (e.g., destructive fishing, pollution) exacerbate the destruction to coral habitats [36]–[41]. Decreased coral cover can result in declines to the abundance, biomass, and diversity of coral reef fishes [42]–[46], but most evidence is for small fishes occupying lower trophic levels, while that for predatory fishes is less clear. For example, Wormald [47] found varying relationships (positive and negative) of coral volume on two snappers (schoolmaster and lane snapper, respectively) while Graham et al.. [43] was unable to detect a relationship between coral loss and fishes larger than 20 cm. Using meta-analysis, Paddack et al.. [45] suggested declines in Caribbean fishes from several trophic groups were due to loss of coral, but were unable to detect a significant effect of habitat degradation on piscivores. Separating the effects of habitat loss from those of fishing have proven difficult since they commonly co-occur [48], but Williams et al.. [32] was able to do so and concluded fishing to be the dominant factor affecting Hawaiian fish communities. The effects of fishing generally precede other stressors [49] and typically have the strongest human induced consequences on predatory marine fishes [18], [40], [50]. Although multiple and interactive local effects related to increasing human population density cannot be ignored, fishing is the most parsimonious mechanism driving the loss of predatory fishes in the Caribbean.

Artisanal fishing is the predominant source of resource extraction on coral reefs in the Caribbean [51]. Although commonly considered to be relatively benign compared to industrialized fisheries, increasing evidence from around the world suggests otherwise. Even at relatively low fishing intensities, artisanal fishing has been shown to strongly reduce populations and biomass of targeted species on coral reefs in the Indo-Pacific [52]–[54], eastern Pacific [55], and the Caribbean [18]. Fishermen tend to target and directly reduce populations of large-bodied fishes that are typically longer lived, mature more slowly than smaller ones, and often form spawning aggregations, all of which increase their vulnerability to overfishing [56]–[60]. Fishing can also have indirect effects on predatory fish communities. For example, removal of large-bodied predators may have allowed smaller ones to increase in abundance due to release from competition or predation [30], [61], [62]. Indeed, the relatively unfished trumpetfish, and the two smallest species of both grouper (i.e., graysby and coney) and snapper (i.e., lane and mahogany snappers) were found to increase in presence with decreasing presence of large predators (Table 3, Figs. 2 and 3). Although the temporal trends were not significant, it is notable that only graysby, lane snapper, and mahogany snapper exhibited increasing presence across the 15-year period of surveys.

Latitude was the second strongest correlative with the structure of predatory fish communities (Table 2). Most studies that have addressed latitudinal patterns of fish communities in the western Atlantic have done so across biogeographic provinces [63], [64], while few have been confined to the greater Caribbean and none have focused solely on predators in the region. Temperature and productivity can each vary greatly over large spatial scales and both have been linked to species richness gradients in the Atlantic [65] and Indo-Pacific [66]. However, neither annual temperature [67] nor productivity [68] varies greatly across the relatively warm, oligotrophic waters of the current study; their roles in affecting the structure of reef fish communities in the Caribbean, including that of the predatory fishes examined here, has therefore remained elusive. In a study that included various habitats including coral reefs, Bouchon-Navaro et al.. [69] found latitude to explain a small but significant amount of the variance (8.4%) on the structure of fish assemblages across the Antilles, with increasing species richness towards lower latitudes. The authors suggested the patterns may have been attributable to the types and area of available habitat, but also acknowledged that it is difficult to attribute mechanism to latitudinal gradients of fishes in the Caribbean given our current knowledge. Following island biogeography theory [70], Sandin et al.. [71] found fish richness on Caribbean reefs from insular nations to increase with both island area and decreasing isolation. Although distance between islands in the Caribbean tends to increase towards lower latitudes (r = between 0.40 and 0.65, depending on metric of isolation), richness was not correlated with latitude per se (r = -0.08; S.A. Sandin, unpublished data). Therefore the mechanisms behind the latitude signal in the current study are not very clear, but may have been due to a combination of gradients in both isolation and area of reefs confounded by the effects of human population density in a general north-south orientation.

The remaining three factors explained far less variance in the structure of predatory fish communities. The lack of a strong signal from the tourism data (i.e., the number of visitors) was somewhat surprising, since increased number of tourists should theoretically have had effects similar to those of increased number of residents. However, a recent study from the Bahamas indicated that residents account for the vast majority of seafood consumed (88%) compared to tourists, with the former preferring fishes (especially grouper and snapper) and the latter preferring conch and lobster (unpublished data, L. Talaue-McManus). Chronic demand for seafood from residents (particularly fishes) may supersede the effects from visitors.

Predicting the ecological consequences of changes to the structure of predator communities is difficult [72], [73]. Different sized predatory fishes may perform various functional roles and can have drastically different effects on the diversity and abundance of prey species [74], [75]. Furthermore, loss of functional roles can lead to decreased ecological stability [76] and ecosystems can become both less resilient to catastrophic phenomena such as cyclones [39] and less resistant to invasions by exotic species [77]. The recent invasion of Indo-Pacific lionfishes (Pterois volitans and P. miles) in the Caribbean may have been facilitated by overfishing large predators capable controlling their rapid spread and population explosion [78] and is alarming considering the strong predatory effects lionfish can have on native fishes [79]. Management of human impacts on entire functional groups may therefore be more important than targeting specific taxa [80], but tests of functional redundancy among predatory marine fishes is sorely needed [81]. In addition, incorporating the effects of environmental variation [82], multiple human stressors [83], and linkages in interaction webs [84], [85] with socioeconomic factors that lead to overfishing [86] may improve management and conservation in coral reef systems.

On a global scale, 37% of human populations are within 100 km of a coastline [87]. As human populations continue to increase, the associated negative effects on coastal ecosystems are not likely to be easily resolved. Continued efforts at broad spatial scales are necessary to better understand individual and interactive effects of anthropogenic activities on marine ecosystems [19], [39], [88], [89]. If we are to overcome the challenges of collecting data in developing nations and on a region-wide scale, these studies will require multiple disciplinary approaches [90] including publicly available survey data collected by citizen scientists and other community volunteers.

Materials and Methods

Survey Data

Predator presence/absence data from locations across the greater Caribbean region (Fig. 1) were queried using the Reef Environmental Education Foundation's (REEF) online database (World Wide Web electronic publication; www.reef.org, date of download: 20 August 2008). The data included coral reef habitats located in 22 continental and insular nations and consisted of 38,116 surveys conducted between 1994 and 2008. Within each of the 22 nations, I chose survey locations with a minimum of 10 surveys (Table 1; 86 total locations). The data were collected by trained volunteer SCUBA divers using the Roving Diver Technique (RDT) where divers swim freely around a survey site and record all species that can be positively identified [91]. The RDT was specifically designed for volunteer data and is effective at rapid assessment of both fish distribution and abundance [92].

The analysis included all predators (trophic level≥4) [93] that met two fundamental criteria: 1) previously documented natural distributions for each of the 22 nations [93]–[96], and 2) only data for conspicuous species because the data were collected by volunteer divers. Although cryptic species (e.g., moray eels, Muraenidae; lizardfishes, Synodontidae) were recorded by the divers, the accuracy of the RDT at estimating their presence was unclear, so those data were not included. Twenty taxa of predatory fishes met the above criteria and included eight species of grouper (Family Serranidae), eight species of snapper (Lutjanidae), one species each of trumpetfish (Aulostomidae) and barracuda (Sphyraenidae), and both jacks (Carangidae) and requiem sharks (Carcharhinidae) summarized at the family levels (Table 3). The 20 taxa ranged in maximum attainable total lengths from 40 cm to over 300 cm. The average depth of each survey was recorded by REEF participants in 10 feet (3.05 meter) increments. Across all surveys included in the analyses here, the majority of dives (82%) were made at depths between 10–30 m, with decreasing proportions made at shallower (<10 m; 12%) and deeper (30–45 m; 7%) depths. Importantly, all surveys were conducted within the natural depth ranges of the 20 predatory taxa [93]–[96].

Data Analysis

The predator presence/absence data had extremely low Whitaker's beta diversity ($\beta = 0.1$) and low values of the coefficient of variation for both taxa (CV = 87.6) and sample locations (CV = 22.8); therefore data transformation was not required. To investigate spatial patterns in the data, a matrix of sample locations by taxa presence was ordinated using non-metric multidimensional scaling (NMS) [97], [98]. NMS can investigate potential drivers influencing the final structure of the ordination by examining correlations between the main dataset (i.e., predator presence) and variables in a second matrix. Therefore a second matrix was constructed that included four variables related to human influences as well as latitude to account for biogeographic patterns that may have naturally existed across the 22 nations (Table 2). The four variables related to human influences included: 1) the size of human populations corrected for land area (the standard measure of human population density) [99], 2) human population size corrected for reef area [99], [100], 3) per capita gross domestic product [101], and 4) average tourist arrivals per year [102].

The ordinations of sample locations in species space were presented graphically, with overlays of the environmental data from the second matrix. The presentation was simplified by displaying national centroids and by grouping nations from the Lesser Antilles into 'Windward' (i.e., Barbados, Grenada, Martinique, St. Lucia, St. Vincent and the Grenadines) and 'Leeward' (i.e., Anguilla, Antigua,

Dominica, Netherlands Antilles, St. Kitts) islands. The resulting ordination displayed 14 regions across the greater Caribbean region. All NMS ordinations were conducted in PC-ORD 5.14 using the 'Autopilot Mode' with Sorensen distance measure and random starting configurations [103].

In addition to the ordination, linear regressions were conducted between human population densities and several metrics of the predator presence data per sample location: 1) mean and median presence across all taxa, 2) richness (S, the total number of species), and 3) Simpson's diversity (D = 1–Σ (pi2).

Groupers and snappers are among the most speciose families of predatory reef fishes in the Caribbean, with a range of maximum total lengths for the species included here from <0.5 m to >1.5 m. Therefore, additional NMS ordinations were conducted on both families to investigate their within family associations with the survey locations relative to the maximum sizes of each species. The first axes of both ordinations were strongly correlated with human population densities. The NMS scores therefore served as an index of human population density in multivariate space for both ordinations. The relationship between how sizes of the associated species changed across the index of human population densities was analyzed using linear regression of the NMS scores versus the maximum attainable lengths of each species.

Acknowledgements

I thank N. Baron, F. Coleman, S. Heppell, M. Hixon, D. Johnson, B. McCune, B. McLeod, K. McLeod, B. Menge, P. Murtaugh, L. Petes, J. Samhouri, S. Sandin, M. Stallings, A. Stoner, G. Von Glavenvich, W. White, E. Wood-Charlson, and two anonymous reviewers for helpful comments and discussion, the REEF volunteer divers for collecting data, and C. Pattengill-Semmens for assistance with REEF data acquisition and organization.

Author Contributions

Conceived and designed the experiments: CDS. Performed the experiments: CDS. Analyzed the data: CDS. Wrote the paper: CDS.

References

1. Christensen V, Guenette S, Heymans JJ, Walters CJ, Watson R, et al.. (2003) Hundred-year decline of North Atlantic predatory fishes. Fish and Fisheries 4(1): 1–24.

2. Coleman FC, Figueira WF, Ueland JS, Crowder LB (2004) The impact of United States recreational fisheries on marine fish populations. Science 305(5692): 1958–1960.

3. Essington TE, Beaudreau AH, Wiedenmann J (2006) Fishing through marine food webs. Proceedings of the National Academy of Sciences of the United States of America 103(9): 3171–3175.

4. Myers RA, Worm B (2003) Rapid worldwide depletion of predatory fish communities. Nature 423(6937): 280–283.

5. Myers RA, Worm B (2005) Extinction, survival or recovery of large predatory fishes. Philosophical Transactions of the Royal Society B-Biological Sciences 360(1453): 13–20.

6. Pauly D, Christensen V, Dalsgaard J, Froese R, Torres F (1998) Fishing down marine food webs. Science 279(5352): 860–863.

7. Worm B, Barbier EB, Beaumont N, Duffy JE, Folke C, et al.. (2006) Impacts of biodiversity loss on ocean ecosystem services. Science 314(5800): 787–790.

8. Worm B, Sandow M, Oschlies A, Lotze HK, Myers RA (2005) Global patterns of predator diversity in the open oceans. Science 309(5739): 1365–1369.

9. Polunin NVC, Roberts CM, Pauly D (1996) Developments in tropical reef fisheries science and management. In: Polunin NVC, Roberts CM, editors. Reef Fisheries. Chapman and Hall.

10. Russ GR (1991) Coral reef fisheries: effects and yields. In: Sale PF, editor. The Ecology of Fishes on Coral Reefs. San Diego: Academic Press. pp. 601–635.

11. Sadovy Y, Domeier M (2005) Are aggregation-fisheries sustainable? Reef fish fisheries as a case study. Coral Reefs 24(2): 254–262.

12. Francis RC, Hixon MA, Clarke ME, Murawski SA, Ralston S (2007) Fisheries management - Ten commandments for ecosystem-based fisheries scientists. Fisheries 32(5): 217–233.

13. Munro JL (1996) The scope of tropical reef fisheries and their management. In: Polunin NVC, Roberts CM, editors. Reef Fisheries. Chapman and Hall. pp. 1–14.

14. Jackson JBC (1997) Reefs since Columbus. Coral Reefs 16: S23–S32.

15. Jackson JBC (2001) What was natural in the coastal oceans? Proceedings of the National Academy of Sciences of the United States of America 98(10): 5411–5418.

16. Wing SR, Wing ES (2001) Prehistoric fisheries in the Caribbean. Coral Reefs 20(1): 1–8.

17. Abesamis RA, Russ GR (2005) Density-dependent spillover from a marine reserve: Long-term evidence. Ecological Applications 15(5): 1798–1812.

18. Hawkins JP, Roberts CM (2004) Effects of artisanal fishing on Caribbean coral reefs. Conservation Biology 18(1): 215–226.

19. Newton K, Cote IM, Pilling GM, Jennings S, Dulvy NK (2007) Current and future sustainability of island coral reef fisheries. Current Biology 17: 655–658.

20. Stevenson C, Katz LS, Micheli F, Block B, Heiman KW, et al.. (2007) High apex predator biomass on remote Pacific islands. Coral Reefs 26(1): 47–51.

21. Hughes TP (1994) Catastrophes, phase shifts, and large-scale degradation of a Caribbean coral reef. Science 265(5178): 1547–1551.

22. Munro JL (1983) Coral reef fish and fisheries of the Caribbean Sea. ICLARM Stud Rev 7: 1–9.

23. REEF (2008) Reef Environmental Education Foundation. World Wide Web electronic publication. www.reef.org, date of download (20 August 2008).

24. Cohn JP (2008) Citizen science: Can volunteers do real research? Bioscience 58(3): 192–197.

25. Dampier W (1729) A new voyage around the world. Dover, New York.

26. Levin PS, Grimes CB (2002) Reef fish ecology and grouper conservation and management. In: Sale P, editor. Coral Reef Fishes: Dynamics and Diversity in a Complex Ecosystem. San Diego: Academic Press.

27. Friedlander AM, DeMartini EE (2002) Contrasts in density, size, and biomass of reef fishes between the northwestern and the main Hawaiian islands: the effects of fishing down apex predators. Marine Ecology Progress Series 230: 253–264.

28. Miller MW, Gerstner CL (2002) Reefs of an uninhabited Caribbean island: fishes, benthic habitat, and opportunities to discern reef fishery impact. Biological Conservation 106(1): 37–44.

29. Luck GW (2007) A review of the relationships between human population density and biodiversity. Biological Reviews 82(4): 607–645.

30. DeMartini EE, Friedlander AM, Sandin SA, Sala E (2008) Differences in fish-assemblage structure between fished and unfished atolls in the northern Line Islands, central Pacific. Marine Ecology Progress Series 365: 199–215.

31. Sandin SA, Smith JE, DeMartini EE, Dinsdale EA, Donner SD, et al.. (2008) Baselines and degradation of coral reefs in the northern Line Islands. PLoS ONE 3(2): e1548.

32. Williams ID, Walsh WJ, Schroeder RE, Friedlander AM, Richards BL, et al.. (2008) Assessing the importance of fishing impacts on Hawaiian coral reef fish assemblages along regional-scale human population gradients. Environmental Conservation 35(3): 261–272.

33. Bellwood DR, Hughes TP, Folke C, Nystrom M (2004) Confronting the coral reef crisis. Nature 429(6994): 827–833.

34. Bruno JF, Selig ER (2007) Regional decline of coral cover in the Indo-Pacific: timing, extent, and subregional comparisons. PLoS ONE 2(8): e711.

35. Gardner TA, Cote IM, Gill JA, Grant A, Watkinson AR (2003) Long-term region-wide declines in Caribbean corals. Science 301(5635): 958–960.

36. Aronson RB, Bruno JF, Precht WF, Glynn PW, Harvell CD, et al.. (2003) Causes of coral reef degradation. Science 302(5650): 1502–1502.

37. Carpenter KE, Abrar M, Aeby G, Aronson RB, Banks S, et al.. (2008) One-third of reef-building corals face elevated extinction risk from climate change and local impacts. Science 321(5888): 560–563.

38. Hoegh-Guldberg O, Mumby PJ, Hooten AJ, Steneck RS, Greenfield P, et al.. (2007) Coral reefs under rapid climate change and ocean acidification. Science 318(5857): 1737–1742.

39. Hughes TP, Baird AH, Bellwood DR, Card M, Connolly SR, et al.. (2003) Climate change, human impacts, and the resilience of coral reefs. Science 301(5635): 929–933.

40. Mora C (2008) A clear human footprint in the coral reefs of the Caribbean. Proceedings of the Royal Society B-Biological Sciences 275(1636): 767–773.

41. Rogers C (2009) Coral bleaching and disease should not be underestimated as causes of Caribbean coral reef decline. Proceedings of the Royal Society B-Biological Sciences 276(1655): 197–198.

42. Cheal AJ, Wilson SK, Emslie MJ, Dolman AM, Sweatman H (2008) Responses of reef fish communities to coral declines on the Great Barrier Reef. Marine Ecology Progress Series 372: 211–223.

43. Graham NAJ, McClanahan TR, MacNeil MA, Wilson SK, Polunin NVC, et al.. (2008) Climate warming, marine protected areas and the ocean-scale integrity of coral reef ecosystems. PLoS ONE 3(8): e3039.

44. Jones GP, McCormick MI, Srinivasan M, Eagle JV (2004) Coral decline threatens fish biodiversity in marine reserves. Proceedings of the National Academy of Sciences of the United States of America 101(21): 8251–8253.

45. Paddack MJ, Reynolds JD, Aguilar C, Appeldoorn RS, Beets J, et al.. (in press) Recent region-wide declines in Caribbean reef fish abundance. Current Biology.

46. Syms C, Jones GP (2000) Disturbance, habitat structure, and the dynamics of a coral-reef fish community. Ecology 81(10): 2714–2729.

47. Wormald CL (2007) Effects of density and habitat structure on growth and survival of harvested coral reef fishes [PhD Dissertation]: University of Rhode Island.

48. Sadovy Y (2005) Trouble on the reef: the imperative for managing vulnerable and valuable fisheries. Fish and Fisheries 6(3): 167–185.

49. Jackson JBC, Kirby MX, Berger WH, Bjorndal KA, Botsford LW, et al.. (2001) Historical overfishing and the recent collapse of coastal ecosystems. Science 293(5530): 629–638.

50. Jenkins M (2003) Prospects for biodiversity. Science 302(5648): 1175–1177.

51. Breton Y, Brown DN, Haughton M, Ovares L (2006) Social sciences and the diversity of Caribbean communities. In: Breton Y, Brown DN, Davy B, Haughton M, Ovares L, editors. Coastal Resource Management in the Wider Caribbean: Resilience, Adaptation, and Community Diversity. Kingston, Jamaica: Ian Randle Publishers.

52. Jennings S, Polunin NVC (1996) Effects of fishing effort and catch rate upon the structure and biomass of Fijian reef fish communities. Journal of Applied Ecology 33(2): 400–412.

53. McClanahan TR, Hicks CC, Darling ES (2008) Malthusian overfishing and efforts to overcome it on Kenyan coral reefs. Ecological Applications 18(6): 1516–1529.

54. Russ GR, Alcala AC (1996) Marine reserves: Rates and patterns of recovery and decline of large predatory fish. Ecological Applications 6(3): 947–961.

55. Ruttenberg BI (2001) Effects of artisanal fishing on marine communities in the Galapagos Islands. Conservation Biology 15(6): 1691–1699.

56. Huntsman GR, Potts J, Mays RW, Vaughan D (1999) Groupers (Serranidae, Epinephelinae): Endangered Apex Predators of Reef Communities; In: Musick JA, editor. Bethesda, MD: American Fisheries Society.

57. Jennings S, Reynolds JD, Polunin NVC (1999) Predicting the vulnerability of tropical reef fishes to exploitation with phylogenies and life histories. Conservation Biology 13(6): 1466–1475.

58. Levin PS, Grimes CB (2002) Reef fish ecology and grouper conservation and management. In: Sale P, editor. Coral Reef Fishes: Dynamics and Diversity in a Complex Ecosystem. San Diego: Academic Press. pp. 377–389.

59. Roberts CM (1997) Ecological advice for the global fisheries crisis. Trends in Ecology & Evolution 12(1): 35–38.

60. Sala E, Ballesteros E, Starr RM (2001) Rapid decline of Nassau grouper spawning aggregations in Belize: Fishery management and conservation needs. Fisheries 26(10): 23–30.

61. Fogarty MJ, Murawski SA (1998) Large-scale disturbance and the structure of marine system: Fishery impacts on Georges Bank. Ecological Applications 8(1): S6–S22.

62. Watson M, Ormond RFG (1994) Effect of an artisanal fishery on the fish and urchin populations of a Kenyan coral reef. Marine Ecology Progress Series 109(2–3): 115–129.

63. Briggs JC (1974) Marine Zoogeography. New York: McGraw-Hill.

64. Floeter SR, Rocha LA, Robertson DR, Joyeux JC, Smith-Vaniz WF, et al.. (2008) Atlantic reef fish biogeography and evolution. Journal of Biogeography 35(1): 22–47.

65. Macpherson E (2002) Large-scale species-richness gradients in the Atlantic Ocean. Proceedings of the Royal Society of London Series B-Biological Sciences 269(1501): 1715–1720.

66. Bellwood DR, Hughes TP, Connolly SR, Tanner J (2005) Environmental and geometric constraints on Indo-Pacific coral reef biodiversity. Ecology Letters 8(6): 643–651.

67. Leichter JJ, Helmuth B, Fischer AM (2006) Variation beneath the surface: Quantifying complex thermal environments on coral reefs in the Caribbean, Bahamas and Florida. Journal of Marine Research 64(4): 563–588.

68. Dandonneau Y, Deschamps PY, Nicolas JM, Loisel H, Blanchot J, et al.. (2004) Seasonal and interannual variability of ocean color and composition of phytoplankton communities in the North Atlantic, equatorial Pacific and South Pacific. Deep-Sea Research Part II 51(1–3): 303–318.

69. Bouchon-Navaro Y, Bouchon C, Louis M, Legendre P (2005) Biogeographic patterns of coastal fish assemblages in the West Indies. Journal of Experimental Marine Biology and Ecology 315(1): 31–47.

70. MacArthur RH, Wilson EO (1967) The theory of island biogeography. Princeton, NJ: Princeton University Press.

71. Sandin SA, Vermeij MJA, Hurlbert AH (2008) Island biogeography of Caribbean coral reef fish. Global Ecology and Biogeography 17(6): 770–777.

72. Bruno JF, Cardinale BJ (2008) Cascading effects of predator richness. Frontiers In Ecology And The Environment 6(10): 539–546.

73. Heithaus MR, Frid A, Wirsing AJ, Worm B (2008) Predicting Ecological Consequences of Marine Top Predator Declines. Trends in Ecology and Evolution 23(4): 202–210.

74. Hixon MA, Carr MH (1997) Synergistic predation, density dependence, and population regulation in marine fish. Science 277(5328): 946–949.

75. Stallings CD (2008) Indirect effects of an exploited predator on recruitment of coral-reef fishes. Ecology 89(8): 2090–2095.

76. McCann KS (2000) The diversity-stability debate. Nature 405(6783): 228–233.

77. Elton CS (1958) The ecology of invasions by plants and animals. London, UK: Methuen.

78. Whitfield PE, Hare JA, David AW, Harter SL, Munoz RC, et al.. (2007) Abundance estimates of the Indo-Pacific lionfish Pterois volitans/miles complex in the Western North Atlantic. Biological Invasions 9(1): 53–64.

79. Albins MA, Hixon MA (2008) Invasive Indo-Pacific lionfish Pterois volitans reduce recruitment of Atlantic coral-reef fishes. Marine Ecology Progress Series 367: 233–238.

80. Hughes TP, Bellwood DR, Folke C, Steneck RS, Wilson J (2005) New paradigms for supporting the resilience of marine ecosystems. Trends in Ecology & Evolution 20(7): 380–386.

81. Stallings CD (in press) Predatory identity and recruitment of coral-reef fishes: indirect effects of fishing. Marine Ecology Progress Series.

82. Pikitch EK, Santora C, Babcock EA, Bakun A, Bonfil R, et al.. (2004) Ecosystem-based fishery management. Science 305(5682): 346–347.

83. McLeod KL, Lubchenco J, Palumbi SR, Rosenberg AA (2005) Scientific Consensus Statement on Marine Ecosystem-Based Management. Signed by 219 academic scientists and policy experts with relevant expertise and published by the Communication Partnership for Science and the Sea at http://compassonline.org/?q=EBM. COMPASS.

84. Crowder LB, Hazen EL, Avissar N, Bjorkland R, Latanich C, et al.. (2008) The Impacts of Fisheries on Marine Ecosystems and the Transition to Ecosystem-Based Management. Annual Review of Ecology Evolution and Systematics 39: 259–278.

85. Walters C, Christensen V, Pauly D (1997) Structuring dynamic models of exploited ecosystems from trophic mass-balance assessments. Reviews in Fish Biology and Fisheries 7(2): 139–172.

86. Cinner JE, McClanahan TR, Daw TM, Graham NAJ, Maina J, et al.. (2009) Linking social and ecological systems to sustain coral reef fisheries. Current Biology 19: 206–212.

87. Cohen JE, Small C, Mellinger A, Gallup J, Sachs J (1997) Estimates of coastal populations. Science 278(5341): 1211–1212.

88. Mumby PJ, Dahlgren CP, Harborne AR, Kappel CV, Micheli F, et al.. (2006) Fishing, trophic cascades, and the process of grazing on coral reefs. Science 311(5757): 98–101.

89. Pandolfi JM, Jackson JBC, Baron N, Bradbury RH, Guzman HM, et al.. (2005) Are US coral reefs on the slippery slope to slime? Science 307(5716): 1725–1726.

90. Pinnegar JK, Engelhard GH (2008) The 'shifting baseline' phenomenon: a global perspective. Reviews in Fish Biology and Fisheries 18(1): 1–16.

91. Schmitt EF, Sullivan KM (1996) Analysis of a volunteer method for collecting fish presence and abundance data in the Florida Keys. Bulletin of Marine Science 59(2): 404–416.

92. Schmitt EF, Sluka RD, Sullivan-Sealey KM (2002) Evaluating the use of roving diver and transect surveys to assess the coral reef fish assemblages off southeastern Hispaniola. Coral Reefs 21: 216–223.

93. Froese R, Pauly D (2005) FishBase (http://www.fishbase.org).

94. Allen GR (1985) FAO species catalogue. Snappers of the world. Rome: Food and Agriculture Organization of the United Nations.

95. Heemstra PC, Randall JE (1993) FAO species catalogue. Vol. 16. Groupers of the world (family Serranidae, subfamily Epinephelinae). Rome: FAO.

96. Humann P, DeLoach N (2002) Reef fish identification. Jacksonville, FL: New World Publications.

97. Kruskal JB (1964) Nonmetric multidimensional scaling: a numerical method. Psychometrika 29: 115–129.

98. Mather PM (1976) Computational methods of multivariate analysis in physical geography. London: J. Wiley and Sons.

99. United Nations (2005) World Populations Prospects Report. New York, NY:

100. Spalding MD, Ravilous C, Green EP (2001) World atlas of coral reefs. Berkeley: University of California Press.

101. Central Intelligence Agency (2008) The world factbook. Washington, DC: Central Intelligence Agency.

102. Caribbean Tourism Organization (2008) Tourist stop over arrivals

103. McCune B, Mefford MJ (1999) PC-ORD. Multivariate Analysis of Ecological Data. 5.14 beta ed. Gleneden, Beach, Oregon, U.S.A: MjM Software.

Chapter 12

Fishers' Knowledge and Seahorse Conservation in Brazil

Ierecê M. L. Rosa, Rômulo R. N. Alves,
Kallyne M. Bonifácio, José S. Mourão, Frederico M. Osório,
Tacyana P. R. Oliveira and Mara C. Nottingham

ABSTRACT

From a conservationist perspective, seahorses are threatened fishes. Concomitantly, from a socioeconomic perspective, they represent a source of income to many fishing communities in developing countries. An integration between these two views requires, among other things, the recognition that seahorse fishers have knowledge and abilities that can assist the implementation of conservation strategies and of management plans for seahorses and their habitats. This paper documents the knowledge held by Brazilian fishers on the biology and ecology of the longsnout seahorse Hippocampus reidi. Its aims were to explore collaborative approaches to seahorse conservation and management

Originally published as Rosa, I.M., Alves, R.R., Bonifácio, K.M. et al. Fishers' knowledge and seahorse conservation in Brazil. J Ethnobiology Ethnomedicine 1, 12 (2005). https://doi.org/10.1186/1746-4269-1-12. © 2021 BioMed Central Ltd. https://creativecommons.org/licenses/by/2.0

in Brazil; to assess fishers' perception of seahorse biology and ecology, in the context evaluating potential management options; to increase fishers' involvement with seahorse conservation in Brazil. Data were obtained through questionnaires and interviews made during field surveys conducted in fishing villages located in the States of Piauí, Ceará, Paraíba, Maranhão, Pernambuco and Pará. We consider the following aspects as positive for the conservation of seahorses and their habitats in Brazil: fishers were willing to dialogue with researchers; although captures and/or trade of brooding seahorses occurred, most interviewees recognized the importance of reproduction to the maintenance of seahorses in the wild (and therefore of their source of income), and expressed concern over population declines; fishers associated the presence of a ventral pouch with reproduction in seahorses (regardless of them knowing which sex bears the pouch), and this may facilitate the construction of collaborative management options designed to eliminate captures of brooding specimens; fishers recognized microhabitats of importance to the maintenance of seahorse wild populations; fishers who kept seahorses in captivity tended to recognize the conditions as poor, and as being a cause of seahorse mortality.

Introduction

Over a decade ago, Ruddle [1] pointed out the great potential value of local knowledge as an information base for local management of marine environments and resources, especially in the tropics, where conventionally-used data were usually scarce to non-existent. A number of subsequent studies have documented and recognized the value of local knowledge to conservation and management of fisheries [2-12].

A pragmatic view of the relevance of fishers knowledge to fisheries management has been expressed by Ames [8]: "fishermen and their subjective, anecdotal descriptions have a pivotal role to play in the development and function of sustainable fisheries (.......) fishermen are, in fact, the only available source of local, historical, place-based information." Nevertheless, lack of sound management practices have led to the collapse of particular types of fisheries in some parts of the world, and, as pointed out by Meewig et al.. [13], interest in participatory approaches in resource management in part reflects the failure of top-down, centralized approaches to manage natural resources.

Definitions of artisanal, subsistence fisheries have traditionally focused on the capture and trade of food fish. However, a growing number of examples of fish species being traded worldwide for purposes other than alimentary (e.g., as pets, remedies, souvenirs) has revealed the existence of an international and multi-faceted commerce, supported by a diffuse (and generally poorly quantified) harvesting of a number of species.

Those forms of exploitation have received little attention when compared with the trade of animals for alimentary purposes [14]. In the marine realm, the scarcity of data has rendered the identification of key elements for conservation and management, and the assessment of impacts difficult.

Seahorses (Hippocampus spp.) are among the few non-food marine fishes whose trade has been documented, initially in Asia [15], where the demand for those fishes was primarily for use in the Traditional Chinese Medicine (TCM) and its derivatives.

Brazil has been involved in the dried seahorse trade, and has been a major exporter of live seahorses at least since 1999. However, only recently the magnitude and impacts of the seahorse fishery in the country began to be assessed and translated into regulatory measures [16]. The seahorse fishery involves many fishing communities in Brazil, to whom seahorses represent an important source of income, particularly in the Northeast of the country.

This paper represents the first attempt to use an ethnoecological approach to examine issues relevant to seahorse conservation and management in Brazil. Its aims were to explore collaborative approaches to seahorse conservation and management in Brazil; to assess fishers' perception on seahorse biology and ecology, in the context evaluating potential management options; to increase fishers' involvement with seahorse conservation in Brazil.

Background

Seahorses (genus Hippocampus) are traded worldwide for use in traditional medicines, as aquarium pets and as curios. Initial surveys of the seahorse trade conducted in the nineties showed that the market for seahorses was economically important, threatened wild seahorse populations, and involved 32 countries [15]. Recent surveys have shown that the number of countries known to be involved in the trade has risen to at least 77 [17]. Additionally, seahorses' coastal habitats, such as reefs and mangroves, are among the most threatened in the world.

The combination of those two factors has resulted in the listing of 33 seahorse species on the IUCN Red List [18], and in the inclusion of the entire genus Hippocampus in the Appendix II of the Convention on International Trade in Endangered Species of Wild Fauna and Flora (CITES). Besides requiring that source countries demonstrate that exports are non-detrimental to the long-term persistence of wild seahorse populations, the listing highlighted the need for monitoring seahorse wild populations, so that the international seahorse trade can be effectively managed.

Hippocampus reidi, commonly known as the longsnout seahorse, is one of the most sought after seahorse species in the aquarium trade. The species figures in the IUCN Red List as Data Deficient [18].

The recognition that seahorses constitute an important source of income, and the need to ensure that their trade is non-detrimental to the long-term persistence of wild populations create a need to supplement the existing regulatory measures with other initiatives, such as cooperative resource management.

In recent years, researchers have emphasized the importance of the knowledge produced and orally transmitted by traditional fishermen and the potential role of traditional fishing and related environmental knowledge can play for the development and implementation of fisheries management [1,10,19-21,4-8]. With regards to seahorses, Pajaro et al..[2] and Meewig et al..[13] have highlighted the importance of fishers' knowledge to the conservation of those fishes in the Philippines.

Lessons learned through examination of the multi-faceted fishery and trade of seahorses can perhaps be used to increase our understanding of interesting questions encompassing social, economic, environmental and cultural aspects of other artisanal fisheries.

Materials and Methods

The study was based on field surveys conducted between February/2002 and October/2005 in fishing villages at the Northeastern Brazilian States of Piauí, Ceará, Paraíba, Maranhão and Pernambuco, and at the Northern State of Pará (Figure 1, Table 1), and focused on the species Hippocampus reidi Ginsburg, 1933.

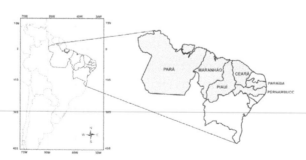

Figure 1. Map showing the surveyed States.

Table 1. Localities where fishers were interviewed in Brazil.

REGION / STATE / MUNICIPALITY	COORDINATES	LOCALITIES WHERE INTERVIEWS WERE MADE
NORTHEASTERN		
Ceará		
Acaraú	02°53'08" S 40°26'57" W	Rio Acaraú, Arpoeiras
Aquiraz	03°54'05" S 38°23'28" W	Prainha, Iguape, Batoque
Beberibe	04°10'47" S 38°07'50" W	Parajuru, Rio Choro, Uruau, Sucatinga
Camocim	02°54'08" S 40°50'28" W	Rio Coreaú, Tatajuba
Cascavel	04°07'59" S 38°14'31" W	Rio Mal Cozinhado, Caponga, Balbino, Águas Belas
Caucaia	03°44'10" S 38°39'11" W	Rio São Gonçalo, Barra do Cauípe
Cruz	02°55'04" S 40°10'18" W	Preá
Eusébio	03°53'24" S 38°27'02" W	Rio Pacoti
Fortim	04°27'07" S 37°47'50" W	Rio Jaguaribe, Rio Piranji
Icapuí	04°23'15" S 37°21'48" W	Manibu, Requenguela, Retiro Grande, Redonda, Peixe Gordo, Braço de mar da Barra Grande
Itarema	02°55'13" S 39°54'54" W	Braço de mar de Porto dos Barcos
Jijoca	02°53'42" S 40°26'57" W	Rio Guriu
Paracuru	03°24'36" S 39°01'50" W	Rio Curu
São Gonçalo do Amarante	03°36'26" S 38°58'06" W	Pecém, Taíba
Trairi	03°16'40" S 39°16'08" W	Rio Trairi, Rio Mundau, Guajiru, Flexeiras, Mundau
Maranhão		
Raposa	02°25'23" S 44°06'12" W	Raposa
Paraíba		
Marcação	06°48'11" S 35°04'50" W	Tramataia
Rio Tinto	06°48'11" S 35°04'50" W	Barra de Mamanguape
Pernambuco		
Goiana	07°33'38" S 35°00'09" W	Estuário de Itapessoca
Piauí		
Cajueiro da Praia	02°55'40" S 41°20'10" W	Barra Grande, Cajueiro de Cima
NORTH		
Pará		
Bragança	01°07'30" S 46°37'30" W	Rio Caeté, Rio Maguari, Praia de Ajuruteua
Augusto Corrêa	01°07'30" S 46°37'30" W	Rio Urumajó

Pilot surveys were conducted to delimit the areas where a seahorse fishery existed along the Brazilian coast. During that period (January–July/2002), observations of the application of ecological knowledge, and acquisition of baseline information on fishing communities were done in praxis.

Ethnoecological data were gathered through semi-structured questionnaires and semi-directive interviews, with some questions left open-ended [22-24]. Interviews were conducted on a one-to-one basis. Based on information provided by community leaders, we initially sought out intentional seahorse fishers, who exclusively collect or have collected seahorses over a period of time. A total of 36 intentional seahorse fishers was found. Subsequently, we interviewed fishers who occasionally come in contact with seahorses, either through their own fishing gear while harvesting for other resources, or as bycatch from the commercial shrimp, fish and lobster nets. Additional interviewees were chosen by using the snowball technique, based on information initially provided by the intentional fishers [25].

Throughout the study, 47 localities (22 municipalities) were visited, in which 181 fishers were interviewed (42 in Ceará State, 29 in Pernambuco, 29 in

Maranhão, 19 in Paraíba, 32 in Piauí and 30 in Pará). Questionnaires and interviews encompassed questions on seahorse ethnotaxonomy, behavior, feeding ecology, reproduction and habitats. No information provided by repondents was excluded from the analysis, following Marques [26]. Interviewees sometimes provided more than one answer to the same question (i.e., seahorses inhabit estuaries and reefs), therefore in some cases the sum of sample sizes provided for a given answer may be higher than the total number of people interviewed.

Surveyed Communities

Our surveys focused on the North and Northeastern regions of Brazil, and encompassed six States (Figure 1). In four of the States seahorses were or are targetted by local fishers, while in the remaining States there was no fisheries directed to seahorses.

Pará State, N Brazil – Two localities were surveyed: Bragança and Augusto Corrêa, both portuary cities. Interviewees at the two localities did not target seahorses in their fisheries, nevertheless, seahorses were caught as bycatch in commercial shrimp, food-fish or lobster nets. Interviewed fishers came in contact with seahorses either through the incidental captures, or by direct observation in the wild while fishing for food. Seahorses obtained as bycatch in the two localities generally enter the dried trade, and occasionally were taken home by fishers to be used as medicine. Fishers' age ranged from 23 to 66 years. With regards to schooling, 43.3% (n = 13) of the fishers interviewed were illiterate, 6.7% (n = 2) attended school for eight years (completing what is known in Brazil as "ensino fundamental"), while 50% (n = 19) attended school for less than eight years.

Maranhão State, NE Brazil – The surveyed locality (Raposa municipality) encompasses the largest and most important fishing community in the State of Maranhão [27]. Seahorses were not targetted by the local fishery, nevertheless they were caught as bycatch in commercial shrimp, food-fish or lobster nets. Interviewed fishers came in contact with seahorses either through the incidental captures, or by direct observation in the wild while fishing for food. Seahorses obtained as bycatch generally enter the dried trade, and occasionally are taken home by fishers to be used as medicine.

Fishers' age ranged from 24 to 67 years. With regards to schooling, 31% (n = 9) of the fishers interviewed were illiterate, 6.9% (n = 2) attended school for eight years (completing what is known in Brazil as "ensino fundamental"), 58.6% (n = 17) attended school for less than eight years and 3.4% (n = 1) attended the three years of high school (completing what is known in Brazil as "ensino médio").

Piauí State, NE Brazil – Two localities were surveyed: Cajueiro da Praia and Barra Grande, both within the limits of an Environmental Protected Area. The main economic activities in that area are commercial fisheries and shrimp farms adjacent to the Timonha-Ubatuba estuarine system. Seahorses were targetted in the area, being sold as ornamental fish and/or dried; occasionally seahorses were taken home by fishers to be used as medicine. Interviewed fishers came in contact with seahorses through the seahorse fishery, local incidental captures, or by direct observation in the wild while fishing for food. Seahorses obtained as bycatch generally enter the dried trade, and occasionally are taken home by fishers to be used as medicine.

Fishers' age ranged from 19 to 76 years. With regards to schooling, 71.9% (n = 23) of the fishers interviewed were illiterate, and, 28.1% (n = 9) attended school for less than eight years.

Ceará State, NE Brazil – At Ceará 15 coastal municipalities were surveyed (Table 1). Fisheries conducted in Ceará's coastal areas are of great social importance [28]. Interviewed fishers came in contact with seahorses through the seahorse fishery, incidental captures (local or external to the community), or by direct observation in the wild while fishing for food.

With regards to schooling, 72.7% of the fishers interviewed attended school for less than eight years, 18.2% were illiterate and 9.1% attended school for 11 years (completing what is known in Brazil as "2° grau").

Paraíba State – The surveyed locality (Mamanguape Estuary, 06° 43' e 06°51' S e 35°07' e 34° 54' W), is part of an Environmental Protected Area. Interviewees belong to two communities located along opposite margins of that estuary, and have been traditionally involved with artisanal fisheries. Seahorses have been amply harvested in that area (the activity peaked from 1999 to 2003) and were mostly sold as ornamental fishes. Occasionally seahorses were taken home by fishers to be used as medicine. Interviewees came in contact with seahorses through the seahorse fishery, incidental captures, or by direct observation in the wild while fishing for food.

Age of interviewed fishers ranged from 30 to 65 years. With regards to schooling, 47.4% (n = 9) of the fishers interviewed were illiterate, 10.5% (n = 8) attended school for eight years (completing what is known in Brazil as "ensino fundamental"), 42.1% (n = 8) attended school for less than eight years.

Pernambuco State – At the surveyed locality (Itapessoca estuary, 07° 37', 07°41' S and 34° 50', 34° 55' W), all fishers interviewed were male, their age ranging from 16 to 69 years old. Artisanal fisheries constitute a main source of income in that area. Seahorses have been traditionally harvested at Itapessoca and sold as ornamental fishes (for at least 10 years); occasionally seahorses were taken home by fishers to be used as medicine. With regards to schooling, 10% (n = 3) of them were illiterate, 3% (n = 1) attended the four years of elementary school, 67% (n = 19) attended elementary for less than four years; 10% (n = 3) attended the three years of high school, while 10% (n = 3) attended high school for less than three years.

Results and Discussion

Answers provided by fishers are summarized in Tables 2 and 3. Interviewed fishers said that they acquired their knowledge on the biology and ecology of seahorses by direct contact with the seahorse fishery (n = 36, 20%), through handling specimens caught as bycatch (n = 59, 32.5%), or by in direct contact with specimens the wild, while harvesting for other resources, such as molluscs, crustaceans or food-fish (n = 86, 47,5%).

Table 2. Summary of information provided by fishers in the surveyed communities in Brazil.

Regions / Brazilian States						
			NORTHEASTERN			NORTH
	Ceará (n = 42)	Maranhão (n = 29)	Paraíba (n = 19)	Pernambuco (n = 29)	Piauí (n = 32)	Pará (n = 30)
Intentional	12		02	12	10	
Occasional	30	29	17	17	22	30
Age groups of fishers	10 a 20 (25%)	20 a 30 (10.3%)	30 a 40 (26.3%)	10 a 30 (10.3%)	10 a 20 (3.1%)	20 a 30 (23.3%)
	20 a 30 (50%)	30 a 40 (17.2%)	40 a 50 (26.3%)	20 a 30 (10.3%)	20 a 30 (15.6%)	30 a 40 (30%)
	30 a 40 (25%)	40 a 50 (41.4%)	50 a 60 (26.1%)	30 a 40 (31.5%)	30 a 40 (15.6%)	40 a 50 (33.3%)
		50 a 60 (27.6%)	60 a 70 (21.1%)	40 a 50 (20.7%)	40 a 50 (25%)	50 a 60 (6.7%)
		60 a 70 (3.5%)		50 a 60 (20.7%)	50 a 60 (18.8%)	60 a 70 (6.7%)
				60 a 70 (7%)	60 a 70 (12.5%)	
					70 a 80 (9.4%)	
Ethnoclassification	Colour only (n = 20), colour and skin appendages (n = 12), no answers (n = 10)	Colour only	Colour only	Colour only (n = 10), colour and skin appendages (n = 19)	Colour only (n = 30), colour and skin appendages (n = 2)	Colour only
Population decline	Yes (n = 8), no (n = 10), no answers (n = 24)	Yes (n = 15), no (n = 14)	Yes (n = 19)	Yes (n = 25), no (n = 4)	Yes (n = 13), no (n = 19)	Yes (n = 8), no (n = 22)
Cited habitats	Only estuarine areas (camboas) (n = 26), only sea (n = 13), estuarine areas (camboas) and sea (n = 3)	Only estuarine areas (camboas) (n = 24), only sea (n = 1), estuarine areas (camboas) and sea (n = 4)	Only estuarine areas (camboas) (n = 3), estuarine areas (camboas) and sea (n = 16)	Only estuarine areas (camboas) (n = 29)	Only estuarine areas (camboas) (n = 30), estuarine areas (camboas) and sea (n = 2)	Only estuarine areas (camboas) (n = 12), only sea (n = 10), estuarine areas (camboas) and sea (n = 8)
Seasonal migration	Yes (n = 19), no (n = 2), no answers (n = 21)	No (n = 29)	Yes (n = 19)	Yes (n = 15), no (n = 1), no answers (n = 13)	No (n = 32)	No (n = 30)
Cited preys	No answers (n = 31), "dirt in the water" (n = 3), fish larvae (n = 1), shrimp larvae (n = 3), crabs (n = 2), algae (n = 2), "lodo" (n = 1)	No answers (n = 14), "dirt in the water" (n = 5), algae (n = 1), "lodo" (n = 11)	No answers (n = 4), fish and shrimp larvae (n = 6), algae (n = 4), "lodo" (n = 5)	No answers (n = 12), adult shrimp (n = 3), shrimp larvae (n = 11), algae (n = 1), mud and "lodo" (n = 2)	No answers (n = 7), "dirt in the water" (n = 5), shrimp larvae (n = 1), algae (n = 4), mud and "lodo" (n = 16)	No answers (n = 6), "dirt in the water" (n = 4), fish larvae (n = 1), shrimp larvae (n = 1), algae (n = 7), mud and "lodo" (n = 19)
Cited Predators	No answers (n = 33), "baiacu" (n = 4), "moréias" (n = 3), "ciobas" (n = 1), "cavalas" (n = 1), "beijupirás" (n = 1), other fishes (n = 1)	No answers (21), none (n = 1), large fish ("mero") (n = 1), other fishes (n = 3), crabs (n = 3)	None (all answers)	No answers (n = 21), none (n = 6), crabs (n = 1), large fish ("mero") (n = 1)	No answers (n = 20), none (n = 3), large fish ("mero") (n = 1), "baiacu" (n = 4), crabs (n = 2), "pacamão" (n = 2), "dourado velho" (n = 1), "bagre" (n = 3), "curapeang" (n = 1), "anacó" ("cioba") (n = 1)	No answers (n = 11), none (n = 1), "pescada amarela" (n = 3), "garoupa" (n = 4), "gorjuba" (n = 3), "caçpó" (n = 6), "camurupim" (n = 1), "pirauna" (n = 1), "pargo" (n = 2), "xingoló" (n = 1), large fish ("mero") (n = 1), "baiacu" (n = 2), "bagre" (n = 1), "anacó" ("cioba") (n = 3), other fishes (n = 5), crabs (n = 1)
Feeding behavior	Not mentioned	Not mentioned	One respondent	Not mentioned	Not mentioned	Not mentioned
Social structure (adults)	Solitary (generally) (n = 12), groups of 2 or 3 individuals (n = 5), solitary and groups of 2 or 3 individuals (n = 1), no answers (n = 24)	Solitary (n = 14), groups of 2 a 3 individuals (n = 11), solitary and groups of 2 or 3 individuals (n = 4)	Not mentioned	Solitary (n = 21), groups of 2 a 3 individuals (n = 6), solitary and groups (n = 1), no answers (n = 1)	Solitary (n = 12), groups of 2 a 3 individuals (n = 16), solitary and groups of 2 a 3 individuals (n = 4)	Solitary (n = 24), groups of 2 a 3 individuals (n = 5), solitary and groups of 2 or 3 individuals (n = 1)
Pregnancy	No answers (n = 28), male (n = 11), female (n = 3)	Female (n = 29)	Male (n = 2), female (n = 17)	No answers (n = 6), male (n = 11), female (n = 12)	Male (n = 3), female (n = 29)	Male (n = 0), female (n = 30)
Courtship behavior	Not mentioned	Not mentioned	Mentioned by one respondents	Not mentioned	Mentioned by three respondents	Not mentioned
Reproductive period	No answers (n = 31), winter (n = 7), summer (n = 1), throughout the year (n = 3)	No answers (n = 21), winter (n = 5), summer (n = 3)	No answers (n = 2), summer (n = 16), throughout the year (n = 1)	No answers (n = 22), winter (n = 5), summer (n = 1), throughout the year (n = 1)	No answers (n = 23), winter (n = 6), summer (n = 2), throughout the year (n = 1)	No answers (n = 23), winter (n = 3), summer (n = 3), throughout the year (n = 1)

Table 3. Summary of information provided by fishers (intentional and occasional) in the surveyed communities in Brazil.

	Brazilian States							
	Paraíba (n = 19)		Pernambuco (n = 29)		Ceará (n = 42)		Piauí (n = 32)	
	Intentional fishers (n= 2)	Occasional fishers (n = 17)	Intentional fishers (n= 12)	Occasional fishers (n = 17)	Intentional fishers (n= 12)	Occasional fishers (n = 30)	Intentional fishers (n = 10)	Occasional fishers (n = 22)
Ethnoclassification	Colour (n = 2)	Colour (n = 17)	Colour (n = 3), colour and skin appendages (n = 9)	Colour (n = 7), colour and skin appendages (n = 10)	Colour (n = 2), colour and skin appendages (n = 10)	Colour (n = 19), colour and skin appendages (n = 2), No answers (n = 9)	Colour (n = 8), colour and skin appendages (n = 2)	Colour (n = 22)
Population decline	Yes (n = 2)	Yes (n = 17)	Yes (n = 9), no (n = 3)	Yes (n = 16), no (n = 1)	Yes (n = 5), no (n = 6) no answers (n = 1)	Yes (n = 3), no (n = 4) no answers (n = 23)	Yes (n = 8), no (n = 2)	Yes (n = 5), no (n = 17)
Habitats	Estuarine areas only (camboas) (n = 2)	Estuarine areas (camboas) and sea (n = 17)	Estuarine areas only (camboas) (n = 12)	Estuarine areas only (camboas) (n = 17)	Estuarine areas only (camboas) (n = 12)	Estuarine areas only (camboas) (n = 12), sea only (n = 12), estuarine areas (camboas) and sea (n = 6)	Estuarine areas only (camboas) (n = 9), estuarine areas (camboas) and sea (n = 1)	Estuarine areas only (camboas) (n = 21), estuarine areas (camboas) and sea (n = 1)
Seasonal migration	Yes (n = 2)	Yes (n = 17)	Yes (n = 9), no answers (n = 3)	Yes (n = 5), no (n = 1), no answers (n = 11)	Yes (n = 9), no (n = 1), no answers (n = 2)	Yes (n = 10), no (n = 1), no answers (n = 19)	No (n = 10)	No (n = 22)
Quoted prey types	Shrimp larvae (n = 2)	Shrimp and larvae (n = 4), algae (n = 4), "lodo" (n = 5), no answers (n = 4)	Adult shrimp (n = 2), shrimp larvae (n = 7), algae (n = 1), no answers (n = 2)	Adult shrimp (n = 1), shrimp larvae (n = 4), mud and "lodo" (n = 2), no answers (n = 10)	Fish and algae (n = 1), shrimp larvae (n = 3), crabs (n = 2), "dirt in the water" (n = 3), "lodo" (n = 1), no answers (n = 2)	"Dirt in the water" (n = 1), no answers (n = 29)	Egg of fish (1), algae (n = 1), "dirt in the water" (n = 1), mud and "lodo" (n = 6), no answers (n = 1)	"Dirt in the water" (n = 4), lama and "lodo" (n = 10), shrymp larvae (1), algae (n = 3), no answers (6)
Quoted predators	None (all answers)	None (all answers)	None (n = 5), large fish ("mero") (n = 1), no answers (n = 6)	None (n = 1), crabs (n = 1), no answers (n = 15)	"Baiacu" (n = 4), "moréias" (n = 3), no answers (n = 5)	Other fishes (n = 1); "cioba", "cavala" e "beijupirá" (n = 1), no answers (n = 28)	"Baiacu" (n = 1), "dourado velho" (n = 1), "bagre" (n = 1), "carapitanga" (n = 1), "siri" (n = 2), no answers (n = 7)	"Baiacu" (n = 3), "pacamão" (n = 2), "bagre" (n = 2), "cioba" (n = 1), no answers (n = 16)
Feeding behavior	One respondent	Not mentioned	Not mentioned	Not mentioned	Not mentioned	Not mentioned	Not mentioned	Not mentioned
Social structure (adults)	Not mentioned	Not mentioned	Solitary (generally) (n = 7), groups of 2 or 3 individuals (n = 5)	Solitary (generally) (n = 14), groups of 2 or 3 individuals (n = 1), solitary and groups of 2 or 3 individuals (n = 1), no answers (n = 1)	Solitary (generally) (n = 4), groups of 2 or 3 individuals (n = 5), solitary or in groups of 2 or 3 individuals (n = 1), no answers (n = 2)	Solitary (generally) (n = 8), no answers (n = 22)	Solitary (generally) (n = 3), groups of 2 or 3 individuals (n = 7)	Solitary (generally) (n = 9), groups of 2 or 3 individuals (n = 9), solitary or in groups of 2 or 3 individuals (n = 4)
Presence of brooding pouch	Female (n = 2)	Male (n = 1), female (n = 16)	Male (n = 6), female (n = 4), no answers (n = 2)	Male (n = 5), female (n = 8), no answers (n = 4)	Male (n = 10), female (n = 2)	Male (n = 1), female (n = 1), no answers (n = 28)	Male (n = 3), female (n = 7)	Female (n = 22)
Courtship behavior	One respondent	Not mentioned	Not mentioned	Not mentioned	Not mentioned	Not mentioned	Three respondents	Not mentioned
Reproductive period	Summer (n = 2)	Summer (n = 14), throughout the year (n = 1), no answers (n = 2)	Winter (n = 1), throughout the year (n = 1), no answers (n = 10)	Summer (n = 1), Winter (n = 4), no answers (n = 12)	Summer (n = 1), Winter (n = 6), throughout the year (n = 3), no answers (n = 2)	Winter (n = 1), no answers (n = 29)	Summer (n = 1), Winter (n = 1), throughout the year (n = 2), no answers (n = 6)	Winter (n = 5), no answers (n = 17)

Ethnotaxonomy

The first frame of reference for gathering and organizing traditional environmental knowledge is taxonomic [29]. The classification of plants and animals by traditional societies is viewed as a reflex of cognitive and intellectual principles to understand the world, being mainly moved by "interest" [30]. In the present study most interviewees (N = 171, 94.5%) said that they recognized different types of seahorses; ten fishers (5.5%) from Ceará State did not answer the question on ethnoclassification. All interviewed fishers (n = 181) used color to differentiate morphotypes; among them, 33 (19.3%) also used the presence of skin appendages to differentiate morphotypes. Fishers who targeted seahorses for the live trade recognized colorful seahorses as more valuable than black seahorses; 72.2% of them (n = 26), said that seahorse wild populations had declined over the years, affecting their capacity to choose seahorses by color.

The number of color morphs recognized by fishers ranged from five to 10, and closely corresponded to the color patterns recorded to H. reidi by [31]. Generally, categories were consistently employed by fishers, irrespective of fishers being intentional or occasional, of differences in age or circumstances of use. At Pará nine color morphs were recognized (black, grey, yellow, brown, orange, red, green, white and burgundy), while at Maranhão fishers recognized seven color patterns (black, grey, yellow, yellow with darker dots, red, white and white with darker dots). Fishers from Piauí mentioned eight color morphs (black, black with white dots, yellow, brown, brown with lighter dots, red, greenish and whitish), while fishers from Ceará recognized five color morphs (black, yellow, red, black and white, and red and yellow); fishers from Paraíba mentioned five color morphs (black, yellow, yellow with black marks, red, and green), while fishers from Pernambuco recognized 10: black, grey, yellow, brown, red, red with black dots, white, orange and white, orange and green and burgundy. Seahorses are cryptic species which use camouflage as a defense mechanism Their color patterns can be highly variable intraspecifically [32].

Knowledge about the presence of skin appendages was more pronounced among intentional fishers. Skin appendages are known to occur in a number of seahorses species, and can also vary intraspecifically. Their presence adds up to seahorses' ability to blend in with their surroundings (see Figure 2), therefore it would be expected that fishers who targetted seahorses would be more knowledgeable about that characteristic than occasional fishers – a reflex of their greater ability to find seahorse specimens in the wild.

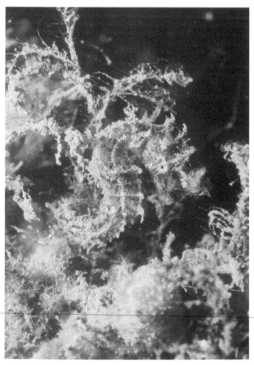

Figure 2. Camouflaged Hippocampus reidi, showing skin appendages. Photo: Bertran M. Feitoza.

A field study conducted at Rio Grande do Norte, NE Brazil [33], found that most specimens with skin appendages were juveniles, a view only expressed by one fisher from Ceará. One fisher from Pernambuco associated skin appendages with older seahorses; at Ceará, seven fishers indicated that skin appendages were found in black specimens, an information that generally concides with the results of underwater surveys conducted in that State, where most specimens with skin appendages were either black or brown. Fishers from Pernambuco mentioned that skin appendages were very common, and that "nearly all seahorses have them." That information agrees with data obtained through underwater surveys, and should be further investigated from a taxonomic viewpoint.

Habitat

Seahorses are cryptic species, which are known to have a patchy distribution, site fidelity, small home ranges and low mobility [34]. Therefore, we anticipated that intentional fishers have precise knowledge about seahorses' habitat use.

All fishers interviewed mentioned that seahorses were sedentary animals which used their tail to grab holdfasts (e.g., roots or leaves of mangrove trees, algae, corals and rocks), or that are seen floating or leaning their bodies against the muddy substrate. A comparison of fishers' perception with data available in the scientific literature [35-38] reveals that the two sources of information were generally consistent with each other.

Most respondents said that seahorses exclusively inhabit shallow estuarine areas (Figure 3), particularly along their margins or in places where currents are not strong – locally known as "camboas" (n = 124, 68.5%). 13.2% of the fishers (n = 24) said that seahorses lived exclusively in the sea, while 33 (18.3%) mentioned that seahorses inhabit both estuaries and the sea. Among the fishers who mentioned that seahorses live in the sea, 14 from Pará, five from Maranhão, 13 from Ceará and 12 from Paraíba mentioned that those fishes inhabit deep waters (up to 80 m), being incidentally captured in nets in areas where algae are found. One fisher from Pará and 10 from Ceará added that seahorses sometimes are found in traps designed to capture other fish species or lobster, locally known as "manzuá;" three fishers (two from Pernambuco and one from Pará) mentioned that seahorses sometimes use the screens of fish corrals as anchor points. The incidental capture of seahorses in shrimp and occasionally in lobster nets has been reported in Brazil [16], however, neither accidental captures in traps nor the occurrence of seahorses in fish corrals had been previously reported in the country.

Figure 3. Seahorse fishers, Pernambuco State. Photo: Bertran M. Feitoza.

On the other hand, four fishers from Paraíba, four from Pará, two from Piauí and three from Ceará quoted shallow areas (including the intertidal zone) as seahorses' habitat. Some fishers (16 from Ceará and one from Piauí) added that seahorses can also be be sighted from the boats, near the sea surface. Most habitats quoted by fishers coincide with the scientific literature [35-38], except for their occurrence in the intertidal zone.

H. reidi is known to occur in a relative wide range of depths (10 cm to 60 m) and salinities (up to 45‰) in Brazilian waters [16,36]. These variations were encompassed by the answers provided by fishers, and possibly reflect the types of habitats used by them as fishing areas (i.e., intentional fishers only harvesting specimens in estuaries, and occasional fishers using both the estuary and the sea as fishing grounds).

Locations of rare or endangered species are more likely to be identified by local resource users involved in such mapping exercises than by outside researchers doing site inventories [29]. The knowledge held by fishers on the distribution and characteristics of different microhabitats used by H. reidi can provide effective shortcuts for researchers investigating the local resource base, and be instrumental to the mapping of seahorses' geographic and spatial distribution. This is of particular relevance in a country with one of the longest coastlines in the world, and with limited resources for conducting surveying and population monitoring.

All respondents reported that seahorses cannot survive in freshwater. However, intentional fishers could provide more information on seasonal migrations. Intentional fishers who said that seahorses migrated (n= 20, 55.5%) provided three diferent answers: migrations occurred a) during the winter, when seahorses moved to the river's mouth or the sea "looking for higher salinity" b) during the summer or c) throughout the year (see Table 3). Vertical migration to deeper, more saline portions of the water column has been suggested to occur in H. reidi [31]; other seahorse species, such as H.comes, H.erectus and H. whitei are known to have made some seasonal migration in the winter [34].

The information provided by interviewees on H. reidi's possible migratory patterns should be further investigated, as it has direct implications for the interpretation of local fisheries data, to the development of management options encompassing temporal closures, and to increase our understanding of H. reidi's population dynamics. As pointed out by Johanes [29], animal migration pathways and aggregation sites known to local people will not always coincide with areas judged to be important based on common criteria for identifying sensitive areas such as aesthetic qualities or species diversity. However, in these areas the value of the resources which are known to local people is sometimes very great.

Interviewed fishers either walked to their fishing areas or used small boats, which they could manouver in shallow waters. Seahorses were generally hand-picked or caught with the aid of throw-nets dragged along the margins of the estuaries, which were recognized by fishers as areas where seahorses occur. Fishers' knowledge on H. reidi's preferential habitats was used to maximize captures, while minimizing the time spent in the captures.

Water visibility and tides played an important role in the organization of the seahorse fishery. The influence of the tides on the activities carried out by fishers has been reported in the literature [6,39-41], and in the present study most (n = 176, 97.2%) of the interviewees mentioned that the tide directly influenced in the organization of their fishing activity.

Feeding Ecology

Seahorses are voracious carnivores, preying upon crustaceans, larval fishes and plankton. The few studies on their feeding ecology suggest that they may play a substantial role in structuring at least some benthic faunal communities [34].

Interviewees provided relevant information on H. reidi's' diet, a still unstudied aspect of that species' ecology. The food items quoted by them generally agreed with the items reported for other species of the family Syngnathidae. Wilson and Vincent [42] showed that amphipods and copepods were the main food items consumed by H. erectus, and Kendrik and Hyndes [43] showed that small crustaceans dominated the diets of 12 syngnathid species. Two items quoted (fish eggs and larvae) had not been previoulsly recorded in the scientific literature.

Knowledge of how food was ingested by H. reidi, on the other hand, was virtually non-existent among interviewees. Only one respondent at Paraíba said that during feeding seahorses sucked prey into the "long thing" (meaning tubular snout), produced a snapping sound ("estralo"), and then swallowed the prey, a description that can be also found in Lourie et al..[32].

Captivity Care

All 36 intentional fishers provided information on captivity care. According to them, seahorses were kept in plastic or styrofoam containers, in glass aquaria equipped with an air pump (in most cases supplied by the buyer with whom the fisher has a non-written exclusivity "contract"), in PVC tanks or in containers ("gaiolas") left in the natural habitat; four intentional fishers from Piauí added that when the activity started, seahorses were kept in shrimp ponds of a local aquaculture farm owned by the main marine ornamental fish exporters from Ceará.

With regards to food items provided by seahorses, one fisher from Pernambuco said that seahorses were not fed in captivity; five reported that seahorses were fed in captivity, but did not specify which type of food they provided to the specimens; two fishers from Cajueiro da Praia mentioned that seahorses were fed with Artemia (supplied by the buyer), while eight did not know if seahorses were fed or not, because they handed the specimens they collected to other fishers who had tanks at home. Artemia nauplii has been widely used in seahorse aquaculture [42,44-46]. An Artemia only diet, however, is considered as nutritionally deficient [44,47]. Five fishers (four from Pernambuco and one from Paraíba) mentioned that seahorses were fed with "newly hatched" shrimp. One fisher from Pernambuco mentioned that seahorses were fed with "small fish." The other intentional fishers did not provide information on this topic.

According to two respondents from Piauí State, mortality rate in the tanks could reach 13%, inadequate feeding and poor sanitary conditions possibly playing a role in the mortality rate reported by the respondents. They also added that on one occasion all 700 seahorses they had kept for 15 days died, possibly due to the poor quality of the water in the tank.

Fishers said that although they contacted buyers when specimens were available, in some cases they had to keep the seahorses for up to 15 days. Only 13.8 % of the intentional fishers (three from Pernambuco and two from Piauí) mentioned that they kept brooding seahorses until release of the offspring, and then released the young in the wild. At first glance, release of young seahorses in the estuary where the adults were captured may look like a good conservationist practice. However, unplanned release of organisms in the wild can cause a number of problems, particularly in cases where capitivity conditions are poor (see figure 4, as an example).

Figure 4. Type of container commonly used to keep the specimens of Hippocampus reidi harvested for the aquarium trade. Photo: Bertran M. Feitoza.

Maintenance of seahorses in their natural habitat reported by some interviewees may be more related to survival of the specimens (readily available food), and to limitations imposed by a few buyers with regards to purchase of brooding seahorses, rather than to conservationist reasons. Nevertheless, the initiative to keep the specimens in their natural habitat is positive, particularly given the present poor conditions and equipment available to fishers to maintain seahorse specimens in captivity, and should be further discussed with fishers.

Predation

Most fishers (n = 106, 58.5%) did not provide information on predators of adult seahorses, and 16.6% (n = 30) of them said that the adult seahorse had no predators. The respondents who mentioned the existence of predators (n = 45, 24.8%) said that "seahorses were often found with cuts on the tail, body or belly, the wounds being inflicted by crabs" (Callinectes sp.), or mentioned that crabs and fishes fed on seahorses, the fishes being cobias (Rachycentron canadum), pufferfishes (Sphoeroides sp, Colomesus psittacus), morey eels (Muraenidae), mackerels (Scomberomorus sp.), groupers (Epinephelus), snappers (Lutjanus sp.), "gurijuba" (a type of catfish) "piraúna" (Cephalopholis fulva), cação (small shark), pacamão (Amphichthys cryptocentrus) and pescada amarela (Sciaenidae). Five fishers (two from Pará, one from Maranhão, and one from Piauí) said that they had found seahorse specimens in the stomach of food-fish. Zavala-Camin [48] recorded the presence of seahorse in the stomach of C. hippurus, and two adult specimens of H. reidi have been found in the stomach of that same species (I. L. Rosa, unpublished data), thus corroborating the information provided by the fishers.

Crabs and seahorses share the same microhabitats in the surveyed estuaries, and therefore the wounds observed by fishers could be the result of agonistic interactions. However, although adult seahorses are recognized as having few predators [32], crabs and large pelagic fishes are considered as such in the literature [49]. Direct observations of seahorses with shortened tails also suggest that partial predation by crabs may be a threat to seahorses [50].

With regards to young seahorses, only three respondents could provide information on predation. According to them, soon after birth many young seahorses are rapidly eaten by other fish species, such as bagres (Ariidae), carapitanga (Lutjanus sp.) and dourado velho (Coryphaena sp.). Those remarks agree with Lourie et al..[32], who mention that after birth seahorses receive no parental care, being very vulnerable to predation at that life stage.

Although cannibalism had not been previously recorded to H. reidi in the wild, two intentional fishers from Piauí said that young seahorses may be eaten by the adults: "the large seahorse eats the young ones. A young seahorse might be eaten if it stays in front of the adult." Interestingly, the two fishers observed cannibalism while seahorses were in captivity, and perhaps cannibalism occurred because the adult seahorses were underfed.

It is known that seahorses mostly rely on camouflage to avoid predators. A single mechanism of defense has been suggested in which, when threatened, seahorses react by bending their head and withdrawing their tail [32]. This agrees with the view expressed by fishers in this study, who said that seahorses were extremely "tame," not possessing any type of defense.

Reproduction

Most respondents (n = 144) said that seahorses are born in the estuary; nine said that seahorses are born in the sea, and 28 did not answer the question. The dominant perception that seahorses are born in the estuary possibly reflects the fact the most interviewees use the estuary as fishing ground, where they encounter the young seahorses.

Only two fishers (from Piauí) could provide information on brood size, and that knowledge was acquired when keeping "pregnant" seahorses in tanks (and posteriorly releasing the offspring in the wild). According to the two fishers, the number of newborn young ranged from 800 to 1,000 per "pregnancy." The same respondents said that although "many seahorses are born, more than 100 at each contraction," after birth, "less than 1/4 survive" in the wild.

Seven respondents (three from Pará, two from Piauí and two from Pernambuco) said that the offspring looks like the adult specimens – "seahorses are born perfect, they are born complete." Two other respondents (from Pernambuco) mentioned that a young seahorse resembles a "mosquito larva." The information provided by fishers agrees with the literature. Seahorses are born as miniature adults, and during their initial phases mortality rates are high [32].

Egg-bearing H. reidi are known to occur in Brazilian estuaries (I. L. Rosa, unpublished data), and gives birth to approximately 1000 offspring at each "pregnancy" [51]. Teixeira and Musick [52] recorded 97 to 1552 eggs/embryos in H. erectus. Foster and Vincent [34] mentioned that, after brooding, male seahorses released from c. 5 to 2000 young, depending on species and adult size, and that newborn young measured from 2 to 20 mm in length. Fishers' limited knowledge on H. reidi's initial phases evidentiated the need to develop educational activities in the surveyed communities to raise awareness with regards to seahorses' relatively low reproductive rates and high mortality during the initial phases, and the links between such life-history, fisheries and conservation.

Sexual Behavior

Some species of seahorse have elaborate courtship behavior, which includes daily greetings [53,54]. Only four of the intentional fishers (three from Piauí and one from Paraíba) knew about seahorses' courtship behavior. One described the behavior as "seahorses are always in the company of each other, sometimes males and females are together, I have seen them with their tails entangled." This description agrees with the observations made by Dias [31] who mentioned that males and females of H. reidi sometimes move as a pair, with their tails entangled. No occasional fisher could provide information on that topic.

It is known that H. reidi's reproductive period lasts over eight months, and varies with temperature [51]. In Brazil, the species reproduces year-round (I. L. Rosa, unpublished data). Of the total number of interviewees, only 59 (22.6%) could provide information on seahorses' reproductive period. From these, 44.1%, (n = 26) mentioned that seahorses reproduced in the winter, while 42.4% (n = 25) said that seahorses reproduced during the summer. Eight fishers said that seahorses reproduced throughout the year. Among intentional fishers, 50% (n = 18) could provide information on seahorses' reproductive period.

Sex Differentiation

Seahorses exhibit a unique reproductive system in which males have a ventral pouch (Figure 5) where the eggs are nourished [32,42,51]. Out of the 181 fishers interviewed, 147 (81,21%) said that they could differentiate males from females by the presence of a ventral pouch in the former. 120 (81.6%) associated the presence of the ventral pouch with the female seahorse – "females have the pouch, they are the ones that caries the embryos – the natural way." Among the intentional fishers (n = 36), 15 (41.6%) associate the presence of the ventral pouch with the female seahorse, 19 (52.7%) associated it with the male seahorse. Two fishers did not answer the question. Among occasional fishers (n = 145), 105 (72.4%) associated the presence of the pouch with the females, while only eight (5.5%) associated it with the male seahorse. A total of 32 occasional fishers (22.1%) did not provide information on the topic.

Figure 5. "Pregnant" male of Hippocampus reidi exhibiting ventral pouch. Photo: Bertran M. Feitoza.

The prevailing view among interviewd fishers was that the female seahorse has the pouch (as expressed by some fishers, the "natural way"). Although that type of perception can sometimes be related to "machismo" [20], some fishers admitted that they were changing their opinion, because "a television program said that the male seahorses are the ones that have the pouch." At the Raposa locality (where seahorses were not targeted), all of the respondents said that they could differentiate a male seahorse from a female, only one respondent of them said that they learned how to recognize males and females from a TV show. At Ceará, out of the 11 fishers who said that the male seahorse has the pouch, nine said that they learned it from the buyers (previously they believed that the females had the pouch), the same being the case with the three fishers from Piauí who mentioned that the male seahorse had the pouch. The changing perception on the seahorse pouch (female versus male) agrees with what has been pointed out by Johannes [55]- that traditional knowledge may be biased due to cultural influences, consumers'market, fishing gear, individual preferences and ability to observe and memorize information. Two other aspects that may influence traditional knowledge are ease of observation and perceived importance of a problem [56,57].

Martin-Smith et al..[58] have shown that in the Philippines, seahorse fishers understand that the "pregnant" males are important to the sustainability of the wild populations, and relate population declines to habitat damage.

Although only a small percentage of the fishers interviewed could provide information on seahorses' reproductive period, all of them recognized the presence of the brooding pouch as an indicator of reproduction; additionally, all intentional fishers understood that brooding specimens were important to the sustainability of the wild populations (nevertheles, 50% (n = 18) of them said that they collected "pregnant" specimens). Intentional captures of "pregnant" seahorses have also been recorded in other parts of Brazil [16]. As an exporter country, Brazil has to comply to the listing of seahorses in appendix II of CITES, and demonstrate that captures of seahorses are non-detrimental to wild populations. Therefore, feasible management options addressing the capture of brooding seahorses should be sought after in the country.

Social Structure

With regards to social structure, 137 (75.7%) of the interviewees provided information on social structure, while 24.3% (n = 44) did not provide information. Among the fishers who provided information, 60.6% (n = 83) respondents said that adult seahorses are solitary, 31.4% (n = 43) said that seahorses formed groups, generally with two or three individuals, and a maximum of six individuals, and 8% (n = 11) said that seahorses could either be found solitary or in groups. That perception is corroborated by scientific studies conducted in Brazil [31,36], in which H. reidi was found solitary or in small groups of up to 4 individuals.

Population Declines

Despite the importance of seahorses for some of the interviewees, 48.6% of the respondents (n = 88) mentioned that seahorse populations had declined over the years.

Intentional seahorse fishers 66.6% (n = 24) said that populations have declined over the years, their present catch levels being much smaller than they used to be, and their capacity to choose seahorses by color reduced. The following causes for the declines were mentioned by fishers: habitat degradation (aquaculture ponds, siltation, mangrove destruction, use of the natural ichthyocide known in the Amazon region as "timbó"), exploitation (too many seahorses captured for the live trade or bycatch).

In practical terms, the population declines mentioned by fishers may result in the development of a less selective fishery in terms of color and size among fishers who harvest seahorses for the aquarium trade. Colorful seahorses attain higher monetary value than black ones, therefore a reduction in their availability could lead to the capture of a higher number of specimens to minimize monetary losses. This aspect, when coupled with the non-selectivity of the dried trade of seahorses captured as bycatch in Brazil, could further impact wild populations of seahorses in the country. Nevertheless, unlike many food fish species targetted by artisanal fishers, seahorses require almost no gear, are comparatively easier to catch, and have a comparatively high economic value. Recognition of population declines by fishers (which are also recognized by researchers and decision-makers) perhaps could be used as a point of consensus upon which dialogue can be initiated. Examples of potential uses for the information provided by fishers to the manangement of seahorses and their habitats in Brazil are shown in Table 4.

Table 4. Examples of how information provided by fishers can be used to manage seahorses and their habitats in Brazil

Objective	Information provided by fishers
Identify threats to seahorses populations Determine those populations targetted by fisheries, the incidental capture in fisheries, and other sources of mortality	Population declines through overharvesting and bycatch; habitat damage.
Identify and document useful practices for maintaining seahorses in captivity (live trade)	Keeping seahorses in the wild until selling the specimens
Identify economic incentives that threaten seahorse wild populations	Brooding seahorses are captured because they are accepted by buyers; No control of seahorses caught as bycatch in commercial nets; specimens enter the dried trade
Reduce to the greatest extent practicable the incidental capture and mortality of seahorses in nets through the development of spatial and seasonal closures	Fishers detain a broad knowledge of seahorses' habitats and main areas of occurrence.
Establish necessary measures to protect and conserve seahorse habitats, through the identification of areas of critical habitat.	Fishers detain a broad knowledge of seahorses' habitats and main areas of occurrence, and of possible migrations
Gather information on seahorse populations and their habitats Initiate and/or continue long-term monitoring of priority seahorse populations	Fishers' knowledge on seahorses' habitat use, colour patterns and skin filaments can be used to monitor seahorse populations, and to better delimit seahorse populations from a taxonomic viewpoint.

Conclusion

All interviewed fishers considered seahorses important either for their economic value or for their value as a medicinal resource for the community. Fishers collectively demonstrated a broad knowledge of the ecology of Hippocampus reidi, and comparison of ethnoecological data with scientific publications showed them to be compatible in most cases. Limitations of the ethnoecological data set included a shortage of useful information on feeding behaviour and courtship behaviour in all surveyed localities.

Most of the ethnoecological information provided by fishers appeared to be acquired through personal experience; in the only case where information was acquired through means other than personal experience, fishers were open and specific about the other sources of information (television, dealers).

Similarly to what has been found by Donovan and Puri [59] in a study on non-timber forest products in Indonesian Borneo, intentional seahorse fishers' traditional knowledge was primarily associated with exploitation, and encompassed aspects relevant to the location and collection of seahorses (e.g., preferential habitats, color patterns, companion species). In other words, they knew how, where and when to fish for seahorses.

With regards to the capture of seahorses for ornamental purposes, although interviewed fishers apparently understood the links between uncontrolled harvesting and the present availability of seahorses, they still harvested specimens they themselves perceive as important to the persistence of seahorses' wild populations. Guidelines, mainly through education, must be designed to reduce the impacts of local resource users, and, more important, strategies to empower the local communities must be developed.

The decline of the artisanal fisheries focused on food fish has opened new "exploitation niches" among fishing communities in Brazil, the marine ornamental trade being one of them. However, unlike other categories of artisanal fishers who have gained some (but still limited) access to consultation processes, marine ornamental fishers, which often are not even registered as fishers, so far have had no participation in decisions that directly affect their activity. As an example, none took part in the consultation process which led to the creation of the first regulatory measure for the marine fish ornamental trade in Brazil, the industry being solely represented by exporters of marine aquarium fish.

Controls and management of different forms of marine artisanal fisheries have traditionally been attempted (and in many cases not achieved) in Brazil through measures which often found little local social resonance. With regards to the marine ornamental fisheries, no attempt to translate local knowledge into management has been made, even when fishers harvested seahorses within the boundaries of environmental protected areas. In that context, avenues for true participatory approaches dwindled, and words such as mistrust, misuse, misreporting, mismanagement and misundertandings thrived.

Nevertheless, in the last few years the management of marine artisanal fisheries in Brazil has produced some interesting examples of species participatory management (e.g., land crabs), and concomitantly, more recent categories of protected areas in Brazil ("Reservas Extrativistas" and "Reservas de Desenvolvimento Sustentável") allow for grater community participation in resource management [60]. A similar approach can be sought after in conjunction with seahorse fishers, by coupling their knowledge with strategies such as temporal and/or spatial closure of fishing areas, and maintenance of harvested seahorses in the wild until selling the specimens.

Information provided by fishers clearly indicate the need to also address the issue of seahorse bycatch in the North and Northeastern regions of Brazil. Previously, focus has been directed to the SE-S regions of the country, where a second species of seahorse is caught in shrimp trawls, and sold as part of the dried trade. Also: recognition that seahorse bycatch is an important issue to seahorse conservation in Brazil will require new strategies on the part of the environmental agencies. So far, the focus of seahorse conservation in Brazil has been the live trade, which has been tentatively controlled through export quotas. Nevertheless, the information provided by fishers interviewed in this study indicate the need to urgently address the incidental capture of seahorses.

In summary, we consider the following aspects as positive for the conservation of seahorses and their habitats in Brazil: 1) fishers were willing to dialogue with researchers; 2) although capture and/or trade of brooding seahorses occurred, most interviewees recognized the importance of reproduction to the maintenance of seahorses in the wild (and therefore of their source of income), and expressed concern over population declines; 3) fishers associated the presence of a ventral pouch with reproduction in seahorses (regardless of them knowing which sex bears the pouch), and this may facilitate the construction of collaborative management options designed to eliminate captures of brooding specimens; 4) fishers recognized microhabitats of importance to the maintenance of seahorse wild populations; 5) fishers who kept seahorses in captivity tended to recognize the conditions as poor, and as being a cause of seahorse mortality.

Acknowledgements

To PROBIO/MMA/IBRD/GEF/CNPq/ and PADI FOUNDATION for the financial support which made the field surveys possible. To all colleagues who participated in the broader project on seahorse conservation, for their enthusiasm and assistance. To IBAMA (Instituto Brasileiro do Meio Ambiente e dos Recursos Naturais Renováveis) for providing research permits, and for being a partner on seahorse conservation in Brazil. To André Castro, for his assistance with the illustrations. Special thanks are due to Rita and Danilo (PROBIO/MMA) for taking such good care of PROBIO's seahorse conservation project, and to all fishers who warmly received us in their homes and shared their knowledge with us.

References

1. Ruddle K: A Guide to the Literature on Traditional Community-Based Fishery Management in the Asia-Pacific Tropics. In Rome: FAO Fisheries Circular Number 869,FIPP/C869;1994.1. Vincent ACJ: The International Trade in Seahorses. Cambridge: TRAFFIC International; 1996.

2. Pajaro MG, Vincent ACJ, Buhat DY, Perante NC: The role of seahorse fishers in conservation and management. Proceedings of the 1st International Symposium in Marine Conservation Hong Kong 1997, 118–126.

3. Duffield C, Gardner JS, Berkes F, Singh RB: Local knowledge the assessment or resources sustainability: case studies in Himachal Pradesh, India, and Britsh Columbia, Canada. Mountain research and development 1998, 18(1):35–49.

4. Kurien J: Traditional ecological knowledge and ecosystem sustainability: new meaning to asian Coastal proverbs. Ecological Application Special Issue 1998, 52–55.

5. Berkes F: Sacred Ecology – Traditional Ecological Knowledge and Resource Management. Taylor & Francis, Philadelphia; 1999.

6. Alves RRN, Nishida AK: A ecdise do caranguejo-uçá, Ucides cordatus (Crustacea, Decapoda, Brachyura) na visão dos caranguejeiros. Interciencia 2002, 27(3):110–117.

7. Ruddle K: Systems of knowledge: Dialogue, relationships and process. Environment, Development and Sustainability 2000, 2:277–304.

8. Ames T: Putting fishermen's knowledge to work: the promise and pitfalls. In Putting Fishers' Knowledge to Work: Conference Proceedings: 27–30 August 2001; Vancouver. Edited by: Haggan N, Brignall C, Wood L. Vancouver: FCRR; 2003:228–204.

9. Faulkner A, Silvano RAM: Status of research on traditional fisher's knowledge in Australia and Brazil. [http://www.fisheries.ubc.ca/publications/reports/report11_1.php] In Putting Fisher's Knowledge to Work: Conference proceedings: 27–30 August 2001; Vancouver Edited by: Haggan N, Brignall C, Wood L. Vancouver: FCRR; 2001, 110–116.

10. Power MD, Chuenpagdee R: Fisher and fishery scientist: nolonger foe, but not yet friend. [http://www.fisheries.ubc.ca/publications/reports/report11_1.php]

In Putting Fisher's Knowledge to Work: Conference proceedings: 27–30 August 2001; Vancouver Edited by: Haggan N, Brignall C, Wood L. Vancouver: FCRR; 2001, 110–116.

11. Diegues AC: Sea tenure, traditional knowledge and management among brazilian artisanal fishermen. NUPAUB. Universidade de São Paulo; 2002.

12. Alves RRN, Nishida AK, Hernandez : Environmental perception of gatherers of the crab 'caranguejo-uca' (Ucides cordatus, Decapoda, Brachyura) affecting their collection attitudes. [http://www.ethnobiomed.com/content/1/1/10] Journal of Ethnobiology and Ethnomedicine 2005, 1:10.

13. Meeuwig J, Samoilys MA, Erediano J, Hall H: Fishers' Perceptions on the Seahorse Fishery in Central Philippines: Interactive approaches and an evaluation of results. In Putting Fishers' Knowledge to Work: Conference Proceedings: 27–30 August 2001; Vancouver. Edited by: Haggan N, Brignall C, Wood L. Vancouver: FCRR; 2003:228–204.

14. Wood E, Wells S: The marine curio Trade. Conservation Issues. A report for the Marine Conservation Society 1988.

15. Vincent ACJ: The International Trade in Seahorses. Cambridge: TRAFFIC International; 1996.

16. Rosa IL: National Report – Brazil. In The Proceedings of the International Workshop on CITES Implementation for Seahorse Conservation and Trade: 3–5 February 2004; Mazatlan. Edited by: Bruckner AW, Fields JD, Daves N. Silver Spring: NOAA Techinical Memorandum NMFS-OPR-36; 2005:46–53.

17. Project seahorse.

18. IUCN: IUCN Red List of Threatened Species. [http://www.redlist.org/] 2004.

19. Silvano RAM: Ecologia de três comunidades de pescadores do rio Piracicaba (SP). In MSc thesis. Universidade Estadual de Campinas; 1997.

20. Marques JGW: Pescando pescadores: Etnoecologia abrangente no baixo São Francisco Alagoano. São Paulo: NUPAUB/USP; 1995.

21. Costa Neto EM: A cultura pesqueira do litoral Norte da Bahia: etnoictiologia, desenvolvimento e sustentabilidade. Salvador, Maceió: EDUFBA/EDUFAL; 2001.

22. Mello LG: Antropologia cultural. Rio de Janeiro: Editora Vozes; 1995.

23. Chizzoti A: Pesquisa em ciências humanas e sociais. São Paulo: Cortez Editora; 2000.

24. Huntigton HP: Using Traditional ecological knowledge in science: Methods and applications. Ecological Applications 2000, 10(5):1270–1274.

25. Bailey K: Methods of social research. 4th edition. New York: The Free press; 1994.

26. Marques JGW: Aspectos ecológicos na etnoictiologia dos pescadores do complexo estuarino-lagunar de Mundaú-Manguaba, Alagoas. PhD thesis. Universidade Estadual de Campinas; 1991.

27. Stride RK: Dignóstico da pesca artesanal no Estado doMaranhão, Brasil. ODA, FINEP, LABOHIDRO. UFMA: São Luís; 1998.

28. Aquasis: A Zona Costeira do Ceará: Diagnóstico para Gestão Integrada. Coordenadores Alberto Alves Campos Fortaleza 2003, 248.

29. Johannes RE: Integrating traditional ecological knowledge and management with environmental impact assessment. In Traditional ecological knowledge: concepts and cases. Edited by: Inglis JT. Ottawa : International Program on Traditional Ecological Knowledge and International Development Research Centre; 1993:33–39.

30. Berlin B: Ethnobiological Classification. Principles of Categorization of plants and animals in traditional societies. New Jersey: Princeton University Press; 1992.

31. Dias TLP: Ecologia populacional de Hippocampus reidi Ginsburg, 1933 (Teleostei: Syngnathidae) no Estado do Rio Grande do Norte, Brasil. In MSc thesis. Universidade Federal da Paraíba; 2002.

32. Lourie SA, Vincent ACJ, Hall HJ: Seahorses: an identification guide to the world's species and their conservation. London: Project Seahorse; 1999.

33. Dias TLP, Rosa IL: Habitat preferences of a seahorse species, Hippocampus reidi (Teleostei: Syngnathidae). Aqua, Journal of Ichthyology and Aquatic Biology 2003, 6(4):165–176.

34. Foster SJ, Vincent ACJ: Life history and ecology of seahorses: implications for conservation and management. Journal of Fish Biology 2004, 65:1–61.

35. Dias TLP, Rosa IL, Baum JK: Threatened fishes of the world: Hippocampus reidi Ginsburg, 1933 (Syngnathidae). Environmental Biology of Fishes 2002, 65:326.

36. Rosa IL, Dias TLP, Baum JK: Threatened fishes of the world: Hippocampus reidi Ginsburg, 1933 (Syngnathidae). Environmental Biology of Fishes 2002, 64:378.

37. Humann P: Reef Fish Identification – Florida, Caribbean, Bahamas. 2nd edition. Jacksonville: New World Publications; 1994.

38. Lieske E, Myers R: Coral Reef Fishes: Caribbean, Indian Ocean and Pacific Ocean including the Red Sea. London:Harper Collins; 1994.

39. Maneschy MC: Pescadores nos manguezais: estratégias técnicas e relações sociais de produção na captura de caranguejo. In Povos das Águas: Realidade e Perspectivas na Amazônia. Edited by: Furtado LG, Leitão W, Fiúza A. Belém: MCT/CNPq; 1993:19–62.

40. Nordi N: A captura do caranguejo-uçá (Ucides cordatus) durante o evento reprodutivo da espécie: o ponto de vista dos caranguejeiros. Revista Nordestina de Biologia 1994, 9(1):41–47.

41. Mourão JS: Classificação e ecologia de peixes estuarinos por pescadores do Estuário do Rio Mamanguape – PB. PhD thesis. Universidade Federal de São Carlos; 2000.

42. Wilson MJ, Vincent ACJ: Preliminary success in closing the life cycle of exploited seahorse species, Hippocampus spp., in captivity. Aquarium Sciences and Conservation 2000, 2:179–196.

43. Kendrick AJ, Hyndes GA: Patterns in the abundance and size-distribution of syngnathid fishes among habitats in a seagrass-dominated marine environment. Estuary Coast Shelf Sci 2003, 57:631–640.

44. Payne MF, Rippingale RJ: Rearing West Australian seahorse, Hippocampus subelongatus, juveniles on copepod nauplii and enriched Artemia. Aquaculture 2000, 188:353–361.

45. Woods CMC: Improving initial survival in cultured seahorses, Hippocampus abdominalis Leeson, 1827 (Teleostei: Syngnathidae). Aquaculture 2000, 190:377–388.

46. Job SD, Do HH, Meeuwig JJ, Hall HJ: Culturing the oceanic seahorse, Hippocampus kuda. Aquaculture 2002, 214:333–341.

47. Hoff FH, Snell TW: Plankton Culture Manual. Dade City: Florida Aqua Farms; 1987.

48. Zavala-Camin LA: Conteúdo estomacal e distribuição do dourado Coryphaena hippurus e ocorrência de C. equiselis no Brasil (24°S–33°S). B Inst Pesca 1986, 12(3):5–14.

49. Whitley G, Allan J: The seahorse and its relatives. Melbourne: Griffin Press; 1958.

50. Baum JK, Meeuwig JJ, Vincent ACJ: Bycatch of lined seahorses (Hippocampus erectus) in a Gulf of Mexico shrimp trawl fishery. Fish Bull 2003, 101:721–731.

51. Vincent ACJ: Reproductive ecology of seahorses. PhD thesis. University of Cambridge; 1990.

52. Teixeira RL, Musick JA: Reproduction and food habits of the lined seahorse, Hippocampus erectus (Teleostei: Syngnathidae) of Chesapeake Bay, Virginia. Rev Bras Biol 2001, 61(1):79–90.

53. Vincent ACJ: A role for daily greetings in maintaining seahorse pair bonds. Animal Behaviour 1995, 49:258–260.

54. Vincent ACJ, Sadler LM: Faithful pair bonds in wild seahorses, Hippocampus whitei. Animal Behavior 1995, 50:1557–1569.

55. Johannes RE: Fishing and traditional knowledge. In Traditional Ecological Knowledge: a Collection of Essays. Edited by: Johannes RE. Gland: IUCN; 1989:39–42.

56. Bentley J: Alternatives to pesticides in Central America applied studies of local knowledge. Culture and Agriculture 1992, 41:10–13.

57. Bentley JW: The epistemology of plant protection: Honduran campesino knowledge of pests and natural enemies. In Proceedings of the CTA/NRI Seminar on Crop Protection for Resource-Poor Farmers, University of Sussex, 4–8 November 1991. Chatham, UK: Natural Resources Institute; 1992:107–117.

58. Martin-Smith KM, Samoilys MA, Meeuwig JJ, Amanda CJV: Collaborative development of management options for an artisanal fishery for seahorses in the central Philippines. Ocean & Coastal Management 2004, 47(3–4):165–193.

59. Donovan D, Puri R: Learning from traditional knowledge of non-timber forest products: Penan Benalui and the autecology of Aquilaria in Indonesian Borneo. [http://www.ecologyandsociety.org/vol9/iss3/art3/] Ecology and Society 2004, 9(3):3.

60. Glaser M, Oliveira RS: The prospects for co-management of mangrove ecosystems on the North Brazilian coast – Whose rights, whose duties and whose priorities? In Natural Resources Forum. Volume 28. Blackwell Publishing; 2004:224–233.

Chapter 13

Fish Invasions in the World's River Systems: When Natural Processes are Blurred by Human Activities

Fabien Leprieur, Olivier Beauchard, Simon Blanchet, Thierry Oberdorff and Sébastien Brosse

ABSTRACT

Because species invasions are a principal driver of the human-induced biodiversity crisis, the identification of the major determinants of global invasions is a prerequisite for adopting sound conservation policies. Three major hypotheses, which are not necessarily mutually exclusive, have been proposed to explain the establishment of non-native species: the "human activity" hypothesis, which argues that human activities facilitate the establishment of non-native species by disturbing natural landscapes and by increasing propagule pressure; the "biotic resistance" hypothesis, predicting that species-rich communities will readily impede the establishment of non-native species; and

Originally published as Leprieur F, Beauchard O, Blanchet S, Oberdorff T, Brosse S (2008) Correction: Fish Invasions in the World's River Systems: When Natural Processes Are Blurred by Human Activities. PLOS Biology 6(12): e322. © 2008 Leprieur et al. https://creativecommons.org/licenses/by/4.0/

the "biotic acceptance" hypothesis, predicting that environmentally suitable habitats for native species are also suitable for non-native species. We tested these hypotheses and report here a global map of fish invasions (i.e., the number of non-native fish species established per river basin) using an original worldwide dataset of freshwater fish occurrences, environmental variables, and human activity indicators for 1,055 river basins covering more than 80% of Earth's surface. First, we identified six major invasion hotspots where non-native species represent more than a quarter of the total number of species. According to the World Conservation Union, these areas are also characterised by the highest proportion of threatened fish species. Second, we show that the human activity indicators account for most of the global variation in non-native species richness, which is highly consistent with the "human activity" hypothesis. In contrast, our results do not provide support for either the "biotic acceptance" or the "biotic resistance" hypothesis. We show that the biogeography of fish invasions matches the geography of human impact at the global scale, which means that natural processes are blurred by human activities in driving fish invasions in the world's river systems. In view of our findings, we fear massive invasions in developing countries with a growing economy as already experienced in developed countries. Anticipating such potential biodiversity threats should therefore be a priority.

Introduction

The deliberate or accidental introduction of species outside their native range is a key component of the human-induced biodiversity crisis, harming native species and disturbing ecosystems processes [1–3]. The greater the introduction of non-natives in a region, the higher the probability that some of them become invasive and will hence cause ecological or economic damage [4,5]. Patterns of non-native species richness are therefore relevant in forecasting the overall impact of invasions on a global scale [5] and should help management authorities to adopt sound, effective conservation policies [5–7].

The process of species invasion consists of three successive stages: initial dispersal, establishment of self-sustaining populations, and spread into the recipient habitat. The last two stages are contingent upon the first one, i.e., if initial dispersal is interrupted, establishment and spread do not occur [8]. Three major hypotheses, which are not necessarily mutually exclusive, have been proposed to explain invasion patterns: the "human activity" [9], "biotic acceptance" [10], and "biotic resistance" [11] hypotheses. The "human activity" hypothesis refers to the three stages of the invasion process (initial dispersal, establishment, and spread), whereas the "biotic resistance" and "biotic acceptance" hypotheses address only the establishment and spread stages [12]. With regards to the establishment stage, the "human activity" hypothesis predicts that, by disturbing natural landscapes and increasing propagule pressure (i.e., the number of individuals released and the frequency of introductions in a given habitat), human activities facilitate the establishment of non-native species [9,13,14]. Everything else being equal, a positive relationship is therefore expected between non-native species richness and quantitative surrogates of propagule pressure and habitat disturbance (e.g., gross domestic product [GDP], percentage of urban area, and human population density [5]). Then, the "biotic acceptance" hypothesis predicts that the establishment of non-native species would be greatest in areas that are rich in native species and with optimal environmental conditions for growth (i.e., "what is good for natives is good for non-natives too" [10]). Everything else being equal, native and non-native species richness should co-vary positively with environmental factors such as energy availability and habitat heterogeneity, which are already recognised as the primary global determinants of native species richness [15,16]. In contrast, the "biotic resistance" hypothesis predicts that species-poor communities will host more non-native species than species-rich communities, the latter being highly competitive and hence readily impede the establishment of non-native species [11,17]. Therefore, a negative relationship is expected between native and non-native species richness. To date, the relative importance of these hypotheses in explaining the variation in non-native species richness had never been tested at the global scale.

We tested these hypotheses and report a global map of fish invasions (i.e., the number of non-native fish established per river basin) by using an extensive worldwide dataset of freshwater fish occurrences (i.e., more than 40,000 occurrences of 9,968 fish species) on the river basin scale (1,055 basins covering more than 80% of Earth's surface). Freshwater fish offer a unique opportunity to identify factors that are responsible for large-scale gradients in non-native species richness for at least two main reasons. First, among vertebrate groups, freshwater fish have been widely introduced over the world [18], which often had subsequent negative consequences on native species and ecosystems integrity [19–23]. Second, as rivers are separated from one another by barriers insurmountable for freshwater fish (land or ocean), they form kind of "biogeographical islands," whose space is delimited [15]. This implies that the natural and human factors shaping global patterns of non-native species richness can be easily separated.

Results

Our results revealed six global invasion hotspots where non-native species represent more than a quarter of the total number of species per basin: the Pacific coast of North and Central America, southern South America, western and southern Europe, Central Eurasia, South Africa and Madagascar, and southern Australia and New Zealand (Figure 1A). According to The World Conservation Union (IUCN) Red List [25], these areas were also characterised by the highest proportion of fish species having a high risk of extinction in the wild (Figure 2).

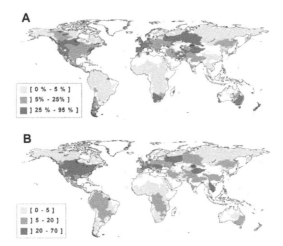

Figure 1. Worldwide Distribution of Non-Native Freshwater Fish

(A) The percentage of non-native species per basin (i.e., the ratio of non-native species richness/total species richness) and (B) the non-native species richness per basin. Each basin was delimited by a GIS using 0.5° × 0.5° unit grid. The maps were drawn using species occurrence data for 9,968 species in 1,055 river basins covering more than 80% of continental areas worldwide. Invasion hotspots are defined as areas where more than a quarter of the species are non-native (red areas on map (A)), leading to define six invasion hotspots: the Pacific coast of North and Central America, southern South America, western and southern Europe, central Eurasia, South Africa and Madagascar, southern Australia, and New Zealand.

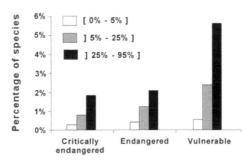

Figure 2. Percentage of Threatened Species for the Three Invasion Levels

Threatened species were identified from the IUCN Red List (vulnerable, endangered, critically endangered). We calculated the percentage of threatened species, listed in the IUCN Red List, for the three invasion levels considered in Figure 1A. Each invasion level expessed as the percentage of non-native species. ([0%–5%],]5%–25%],]25%–95%]) account for 8,363, 2,257, 1,241 native species and 544, 240, 271 river basins, respectively.

Analysing the absolute number of species, we found that river basins of the Northern Hemisphere host the highest number of non-native fish species (Figure 1B). The human factors considered here to test the "human activity" hypothesis (GDP, population density, percentage of urban area) were found to be positively related to non-native species richness (Table 1), after controlling for the effects of environmental conditions and native species richness. In contrast, the positive correlation between native and non-native richness that was expected by the "biotic acceptance" hypothesis was not significant after controlling for the effects of propagule pressure and habitat disturbance (Table 1). Indeed, the environmental factors displayed either no (net primary productivity) or a weak positive correlation (altitudinal range and basin area) with non-native species richness, after controlling for the effects of propagule pressure and habitat disturbance (Table 1). The negative correlation between native and non-native richness, expected by the "biotic resistance" hypothesis, was not significant after controlling for the effects of environmental conditions, propagule pressure, and habitat disturbance (Table 1).

Table 1. Spearman Rank Correlation (rs) between the Number of Non-Native Fish Species (Residuals) and Each Explanatory Variable Related to the "Human Activity," "Biotic Acceptance," and "Biotic Resistance" Hypotheses (n = 597)

Hypothesis	Variable	r_s	p
Human activity hypothesis	Gross domestic product	0.550*	<0.0001
	Percentage of urban area	0.556*	<0.0001
	Population density	0.306*	<0.0001
Biotic acceptance hypothesis	Number of native species	0.093	0.062
	Altitudinal range	0.264*	<0.0001
	Basin area	0.175*	<0.0001
	Net primary productivity	−0.008	0.842
Biotic resistance hypothesis	Number of native species	−0.034	0.400

For each hypothesis, the relationship between the number of non-native fish species and the explanatory variables considered was quantified by controlling for the effects of the explanatory variables relevant to the other hypotheses (see Materials and Methods for more details).
* $p < 0.006$ (Bonferroni correction, $\alpha = 0.006$).

Then, we applied hierarchical partitioning [26–28] that aims to quantify the independent explanatory power of each variable by considering all possible submodels. The deviance explained by the 128 submodels computed in hierarchical partitioning accounted in average for 52% of the total deviance (±7% standard deviation [SD], min = 37%, max = 67%). The human factors had together the greatest independent effect on non-native species richness (70%, Table 2). Among the human factors, the GDP (an economical index of human activities [9]) had the greatest independent explanatory power (43%; Table 2). To a lesser extent, the habitat heterogeneity (i.e., basin area and altitudinal range) and the number of native species also contribute to the variation in non-native species between river basins (Table 2).

Table 2. Independent Effect of Each Environmental and Human Activity–Related Variable on the Number of Non-Native Species per Basin

Variable	Independent Effect (%) (n = 597)	95% Boostrap Confidence Interval (n = 100)
Gross domestic product	43.06	[36.68; 45.08]
Percentage of urban area	13.94	[10.93; 16.63]
Population density	13.36	[11.63; 14.91]
Number of native species	5.16	[3.57; 7.46]
Altitudinal range	7.11	[4.24; 10.23]
Basin area	15.08	[9.96; 19.06]
Net primary productivity	2.26	[1.35; 4.63]

Hierarchical partitioning was applied to the 597 basins for which the seven variables selected to test the "human activity," "biotic acceptance," and "biotic resistance" hypotheses were available. The independent effect of a variable was expressed as a percentage of the total independent contribution associated with the seven variables. To test potential bias due to sample size, hierarchical partitioning was run on 1,000 random subsets of 100 basins among the total of 597 basins. For each variable, the independent effect based on 597 basins did not differ from the 95% bootstrap percentile confidence interval, testifying that sample size hardly affected the results. Both analyses underline the predominant role of the three human variables that together represent more than 70% of the independent effect.

To test for potential bias in our results due to differences in sampling effort between continents, bootstrap analysis was performed by applying hierarchical partitioning to 1,000 random subsets of 100 basins. For each variable, the independent effect observed did not differ from the 95% bootstrap percentile confidence interval (Table 2), testifying that potential differences in sampling effort between continents hardly affected the results.

Discussion

By using an explanatory modelling approach, we showed that the human activity indicators of the world's river basins were positively related to the number of established non-native fish species. In addition, they account for most of the global variation in non-native species richness, giving support for the "human activity" hypothesis. More particularly, we highlight that the level of economic activity of a given river basin (expressed by the GDP) strongly determines its invasibility. Three non-exclusive mechanisms may account for this pattern. First, economically rich areas are more prone to habitat disturbances (e.g., dams and reservoirs modifying river flows) that are known to facilitate the establishment of non-native species [7,23,29]. Second, high rates of economic exchanges increase the propagule fluxes of non-native species [6,9] via ornamental trade, sport fishing, and aquaculture [18]. Third, the increased demand for imported products associated with economic development increase the likelihood of unintentional introductions through the import process [6].

The "biotic resistance" hypothesis cannot explain the pattern of fish invasions observed, because no negative relationship between native and non-native species richness was found after controlling for the effects of environmental conditions, propagule pressure, and habitat disturbance. This means that regional species-rich communities are not necessarily a barrier against the establishment of non-native species [17]. Our results are consistent with several studies showing that species-rich fish communities can support higher species richness if the pool of potential colonisers is increased by species introductions [24,30,31]. More generally, our results agree with studies on various taxa that do not report biotic resistance at broad spatial scales [10,11]. Then, we provide no real support for the alternative "biotic acceptance" hypothesis [10] even if native and non-native species richness do respond similarly to some of the environmental gradients tested (i.e., altitudinal range and basin area). Actually, the absence of a significant positive relationship between native and non-native species richness implies that species-rich river basins do not support more non-native species than basins with a low native species richness (i.e., "the rich do not get richer"). This contrasts with numerous continental and regional-scale studies on plants and animals that report a strong matching between native and non-native species richness [10,32–35]. More generally, our results do not agree with the expectation that native and non-native species richness covary positively at macroecological scales [36].

The interpretation of the exact role of human activities (i.e., propagule pressure and habitat disturbance) in driving broad-scale patterns of non-native species richness faced major difficulties in previous continental and regional-scale studies due to covariations between human and natural factors [9,13,34,35]. Indeed, because humans may have preferred to settle in areas providing diverse natural resources, human population was found to be largest in regions with high levels of habitat heterogeneity and energy availability that favour species-rich native fauna and flora [34,37]. This therefore makes it difficult to determine whether the often-reported positive relationship between native and non-native species richness is driven by (i) common responses to habitat heterogeneity and energy availability or (ii) increased propagule pressure and habitat disturbance. Such difficulties were probably related to the spatial extent considered (i.e., a continental or regional extent). Indeed, we found a weak covariation between environmental and human descriptors of the world's river basins at the global scale (Pearson's correlation coefficients: $r < 0.35$). This allowed us to clearly disentangle the relative roles of human activities and environmental conditions in shaping the global pattern of fish invasions. We show that the biogeography of fish invasions at the global scale matches the geography of human impact but not the biogeography of native species.

Because increasing the number of non-native species increases the risk of biodiversity loss [4,5], our results have two major implications for future conservation strategies. First, the six global invasion hotspots identified here account for the highest proportion of threatened fish species listed on the IUCN Red List [25]. These areas are also recognised as being biodiversity hotspots (particularly southern Europe, South Africa and Madagascar, southern Australia, and New Zealand [38,39]). Although species classified on the IUCN Red List are threatened by various sources of disturbance (e.g., habitat loss, pollution, species invasion, and overexploitation [25]), non-native species are recognised as a major threat to biodiversity after habitat loss [25,40]. For example, 20% of the 680 species extinctions listed by the IUCN were directly caused by species invasions [2]. Freshwater fish follow the same tendency, as 20% of the species listed by the IUCN are threatened by non-native species [41]. In that context, we recommend that non-native species importations in the six invasion hotspots be prohibited without detailed risk and long term cost-benefits assessments [42]. Special attention should also be given to these areas to design efficient control programs of already-established non-native species.

Second, as we provide strong evidence for the "human activity" hypothesis (with a special emphasis on economic activity), we expect that river basins of developing countries will host an increasing number of non-native fish species as a direct result of economic development. This constitutes a serious threat to global biodiversity, because rivers of most developing areas (e.g., southern Asia, western and central Africa) are characterised by high levels of endemism [38]. Anticipating potential biodiversity threats should therefore be a priority, because once they are established, the eradication of a non-native species is extremely difficult and result in high economic costs [43].

Despite the increasing literature on non-native species, this study is, to our knowledge, the first to provide a global map of species invasions for a given taxonomic group and should stimulate others to test the generality of these findings for other taxa at this spatial scale. Such broad-scale analyses would help local researches to focus on non-native species control in the most sensitive areas (e.g., the six invasion hotspots we identified here for freshwater fish). This study should also stimulate researches on freshwater ecosystems by combining the existing global scale databases of physical disturbances [44,45] and the global pattern of fish invasions given here. This would permit to quantify river basins threats by considering simultaneously different sources of disturbance. Such an approach is urgently needed as rivers are among the most threatened ecosystems of the world [46] and as freshwater fish constitute a major source of protein for a large part of the world population [46].

Materials and Methods

Databases

We conducted an extensive literature survey of native and non-native freshwater fish species check lists. Only complete species lists at the river basin scale were considered, and we discarded incomplete check lists such as local inventories of a stream reach or based only on a given family. The resulting database was gathered from more than 400 bibliographic sources including published papers, books, and grey literature databases (references available upon request). Our species database contains species occurrence data for the world's freshwater fish fauna at the river basin scale (i.e., 80% of all freshwater species described [47] and 1,055 river basins covering more than 80% of Earth's surface). It constitutes the most comprehensive global database for freshwater fish occurrences at the river basin scale and, to our knowledge, the largest database for a group of invaders. We considered as non-native a species (i) that did not historically occur in a given basin and (ii) that was successfully established, i.e., self-reproducing populations. Estuarine species with no freshwater life stage were not considered in our analyses.

The environmental and human databases contain seven variables selected to test (i) the "human activity" hypothesis: human population density (number of people km–2), percentage of urban area and purchase power parity GDP (in US$); (ii) the "biotic acceptance" hypothesis: number of native fish species, basin area (km2), altitudinal range (m), net primary productivity (NPP in kg-carbon m–2 year–1), and (iii) the "biotic resistance hypothesis:" number of native fish species. The area of each river basin was taken from published and unpublished data. The altitudinal range for each river basin was determined from a geographical atlas. We calculated the mean value of NPP, human population density, GDP, and percentage of urban area over the surface area of each basin from 0.5° × 0.5° grid data available in the Center for International Earth Science Information Network (CIESIN) and the Atlas of Biosphere [48,49]. The surface area and altitudinal range at the river basin scale are used as quantitative surrogates for habitat heterogeneity [16], which is known to influence native freshwater fish species richness [15,16]. Net primary productivity is used as a quantitative surrogate to river basin energy availability [16] and strongly correlates to native freshwater fish species richness [15,16]. This is verified in our data, as we found that both basin area and NPP are positively correlated to native species richness (partial Pearson's correlation coefficient: $r = 0.592$ and $p < 0.0001$ for basin area while controlling for the effect of NPP; $r = 0.514$ and $p < 0.0001$ for NPP while controlling for the effect of the basin area). Then, the human population density, percentage of urban area, and GDP were used as quantitative surrogates for propagule pressure and habitat disturbance [5,9,33]. The GDP measures the size of the economy and is defined as the market value of all final goods and services produced within a region in a given period of time.

Fish Invasions Mapping

We first mapped the worldwide distribution of (i) the non-native species richness per basin and (ii) the percentage of non-native species per basin (i.e., the ratio of non-native species richness/total species richness). To do that, each basin was delimited by a geographic information system (GIS) using a grid reference of 0.5° latitude and 0.5° longitude and then reported on a world map. We used three classes of percentage (Figure 1A) and richness (Figure 1B) of non-native species to draw color maps. Other maps with more classes were tried and provided similar results. We selected the one that minimised differences in sample size (i.e., number of river basins) between classes. The percentage of non-native species per basin was used to define invasion hotspots where more than a quarter of the species are non-native (i.e., the third class of percentage of non-native species; red areas in Figure 1A). It was preferred to the richness in non-natives due to its independence from native richness and basin area. For each of the three levels of fish invasion ([0%–5%], [5%–25%], [25%–95%]), we determined the percentage of species facing a high to extremely high risk of extinction in the wild, i.e., the vulnerable, endangered, and critically endangered fish species according to the IUCN Red List [25]. The percentage of threatened species should be regarded with caution, because the IUCN Red List for freshwater fish is still incomplete. The percentages of threatened species for the three levels of fish invasion are therefore probably underestimated. Although we recognise the potential biases and limitations of the IUCN listing procedure, the IUCN Red List of threatened species remains the most objective and authoritative system for classifying species in terms of the risk of extinction at the global scale [41,50].

Modeling Method

In this study, to test the three hypotheses (i.e., "human activity," "biotic acceptance," and "biotic resistance"), we did not build the best single and parsimonious model by using stepwise selection of a subset of independent variables having a significant effect on the number of non-native species per basin (i.e., predictive approach). Indeed, a single best model is not necessarily the best explanatory model, because minimizing the overall difference between the observed and predicted values does not necessarily equate to determining probable influence in a multivariate setting [26–28,51,52]. In addition, a simple regression model cannot identify situations in which potentially important independent variables are suppressed by other variables due to their high colinearity. When there is colinearity between independent variables, the direct response of the dependent variable to a independent variable may in fact only be an indirect effect owing to high dependence of the considered variable with one or many others [27].

In our dataset, the seven environmental and human variables are not independent (Pearson's correlation coefficient values ranging from –0.25 to 0.79). We therefore evaluated the independent explanatory power of each environmental and human variable by using hierarchical partitioning [26–28,51,52], a method based on the theorem of hierarchies in which all possible models in a multiple regression setting are considered jointly to attempt to identify the most likely causal factors (explanatory approach).

If we consider k, the number of explanatory variables (X1, Xi,…, Xk), there are 2k possible models (i.e., 128 submodels by considering the seven explanatory variables), including the null model (M0). The Ri is a measure of fit between one independent variable Xi and the dependent variable Y. The fit between each of the seven explanatory variables and the dependent variable Y (number of non-native fish species per basin) was measured by the reduction of deviance generated by introducing a given variable into all of the possible models built with the six other variables within the considered hierarchies. We used a generalised linear model (GLM) with a Poisson error to treat our count data (i.e., the number of non-native fish species per basin). Each explanatory variable was log-transformed to meet the assumptions of normality and homoscedasticity.

We consider k! hierarchical orderings of models that always begin with M0 and end with $M_{x_{1,2,3\ldots k}}$. For any given initial variable Xi, there are (k − 1)! possible hierarchies containing k(k − 1)! models in which Xi appears. For each hierarchy, we evaluate the influence of Xi on each of the k models including Xi (increase in model fit generated by including the variable Xi within each model). The independent influence (Ii) of Xi on Y was obtained by averaging all of the k(k − 1)! increases of fit. This averaging alleviates multicolinearity problems that are ignored by using a simple regression model [26–28]. The joint component Ji (effect caused jointly with the k − 1 other variables) is obtained by subtracting Ii from Ri, with Ri = Ii + Ji. If all explanatory variables were completely independent of one another, there would be no joint contributions [26]. For each variable, the independent and joint contributions are expressed as the percentage of the total explained deviance (R)

$$R = I + J = \sum_{i=1}^{k} R_i = \sum_{i=1}^{k} I_i + \sum_{i=1}^{k} J_i.$$

In our models, the total independent contribution accounts for 75% of the total explained deviance, which means that the joint contribution of each explanatory variable was weak in explaining the global variation in non-native species richness. We therefore quantified the independent effect (IEi) of each variable on the dependent variable Y as the percentage of the total independent contribution, i.e $IE_i = I_i / \sum_{i=1}^{k} I_i$. The significance of the independent effect (IEi) of each variable was determined by a randomization approach (n = 100) which yielded

Z-scores [52]. Statistical significance was based on an upper confidence limit of 0.95. Each variable display a significant independent effect.

We applied hierarchical partitioning to a subsample of 597 basins (Afrotropical: 72; Australian: 94; Nearctic: 127; Neotropical: 68; Oriental: 29; Palearctic: 207) for which all seven environmental and human variables used were available. To test potential bias due to differences in sampling effort between continents, hierarchical partitioning was run on 1,000 random subsets of 100 basins among the total of 597 basins. For each variable, we calculated the 95% bootstrap percentile confidence interval of the independent effect (IEi). Hierarchical partitioning was conducted using the 'hier.part' package [52] version 1.0–1 implemented on the open source R software [53]. Hierarchical partitioning implemented for linear relationships was relevant to our data, because preliminary analyses did not detected any significant effect of polynomial terms. The hierarchical partitioning results were compared with those obtained with another method (i.e., variation partitioning, [54]). Overall, the results of the two methods were similar, and the variables highlighted as significant by the two approaches were the same.

Hierarchical partitioning does not provide information on the form of the relationship (positive or negative) between the number of non-native species and each explanatory variable. To test the "human activity" hypothesis, we analysed the form and the significance of the relationship between each variable related to the "human activity" hypothesis (GDP, percentage of urban area, and population density) and the residuals from a GLM with a Poisson error. This model explains the number of non-native species by using independent variables related to the "biotic resistance" and "biotic acceptance" hypotheses (number of native species, altitudinal range, basin area, and net primary productivity). This allowed us to control for the effects of environmental conditions and native species richness. Then, to test the "biotic acceptance" hypothesis, we analysed the form and the significance of the relationship between each variable related to the "biotic acceptance" hypothesis (i.e., number of native species, altitudinal range, basin area, and net primary productivity) and the residuals from a GLM explaining the number of non-native species by using the human activity–related variables (i.e., GDP, percentage of urban area, and population density). This allowed us to control for the effects of propagule pressure and habitat disturbance. Lastly, to test the "biotic resistance" hypothesis, we analysed the form and the significance of the relationship between the number of native species and the residuals from a GLM explaining the number of non-native species by using independent variables related to the "biotic acceptance" and "human activity" hypotheses (i.e., altitudinal range, basin area, net primary productivity, GDP, percentage of urban area, and population density). This allowed us to control for the effects of environmental conditions, propagule pressure and habitat disturbance. To test the relationship between the model residuals and each explanatory variable, we performed a Spearman rank correlation test, because the model residuals were not normally distributed.

Acknowledgements

We thank J. Chave, E. Danchin, C.R. Townsend, P. Winterton, and three anonymous referees for their insightful comments, which have improved the manuscript.

Author Contributions

F. Leprieur, O. Beauchard, S. Blanchet, and S. Brosse conceived and designed the experiments, performed the experiments, and analyzed the data. F. Leprieur, O. Beauchard, S. Blanchet, T. Oberdorff, and S. Brosse contributed reagents/materials/analysis tools and wrote the paper.

References

1. Vitousek PM, Mooney HA, Lubchenco J, Melillo JM (1997) Human domination of Earth's ecosystems. Science 278: 494–499.

2. Clavero M, García-Berthou E (2005) Invasive species are a leading cause of animal extinctions. Trends Ecol Evol 20: 110.

3. Byrnes JE, Reynolds PL, Stachowicz JJ (2007) Invasions and extinctions reshape coastal marine food webs. PLoS ONE 2(3).e295 doi:10.1371/journal.pone.0000295.

4. Jeschke JM, Strayer DL (2005) Invasion success of vertebrates in Europe and North America. Proc Natl Acad Sci USA 102: 7198–7202.

5. Pyšek P, Richardson DM (2006) The biogeography of naturalization in alien plants. J Biogeogr 33: 2040–2050.

6. Levine JM, D'Antonio CM (2003) Forecasting biological invasions with increasing international trade. Conserv Biol 17: 322–326.

7. Williamson M (1996) Biological invasions. London: Chapman & Hall. 244 p.

8. Puth M, Post DM (2005) Studying invasion: have we missed the boat? Ecol Lett 8: 715–721.

9. Taylor BW, Irwin RE (2004) Linking economic activities to the distribution of exotic plants. Proc Natl Acad Sci U S A 101: 17725–17730.

10. Fridley JD, Stachowicz JJ, Naeem S, Sax DF, Seabloom EW, et al.. (2007) The invasion paradox: reconciling pattern and process in species invasions. Ecology 88: 3–17.

11. Levine JM (2000) Species diversity and biological invasions: relating local process to community pattern. Science 288: 852–854.

12. Lockwood JL, Hoopes MF, Marchetti MP (2007) Invasion Ecology. Oxford: Blackwell Publishing. 304 p.

13. Chown SL, Gremmen NJM, Gaston KJ (1998) Ecological biogeography of southern ocean islands: species–area relationships, human impacts, and conservation. Am Nat 152: 562–575.

14. Meyerson LA, Mooney HA (2007) Invasive alien species in an era of globalization. Front Ecol Environ 5: 199–208.

15. Oberdorff T, Guégan JF, Hugueny B (1995) Global scale patterns of fish species richness in rivers. Ecography 18: 345–352.

16. Guéguan JF, Lek S, Oberdorff T (1998) Energy availability and habitat heterogeneity predict global riverine fish diversity. Nature 391: 382–384.

17. Kennedy TA, Naeem S, Howe KM, Knops JMH, Tilman D, et al.. (2002) Biodiversity as a barrier to ecological invasion. Nature 417: 636–638.

18. Lever C (1996) Naturalized Fishes of the World. London: Academic Press. 408 p.

19. Vander Zanden MJ, Casselman JM, Rasmussen JB (1999) Stable isotope evidence for the food web consequences of species invasions in lakes. Nature 401: 464–467.

20. Kolar CS, Lodge DM (2002) Ecological predictions and risk assessment for alien fishes in North America. Science 298: 1233.

21. McDowall RM (2006) Crying wolf, crying foul, or crying shame: alien salmonids and a biodiversity crisis in the southern cool-temperate galaxioid fishes? Rev Fish Biol Fish 16: 233–422.

22. Leprieur F, Hickey M, Arbuckle CJ, Closs G, Brosse S, et al.. (2006) Hydrological disturbance benefits a native fish at the expense of an exotic fish. J Appl Ecol 43: 930–939.

23. Light T, Marchetti M (2007) Distinguishing between invasions and habitat changes as drivers of diversity loss among California's freshwater fishes. Conserv Biol 21: 434–446.

24. Gido KB, Brown JH (1999) Invasion of North American drainages by alien fish species. Freshw Biol 42: 387–399.

25. IUCN (2006) 2006 IUCN Red List of Threatened Species. Available: http://www.iucnredlist.org. Accessed 3 January 2008.

26. Chevan A, Sutherland M (1991) Hierarchical partitioning. Am Stat 45: 90–96.

27. Mac Nally R (2002) Multiple regression and inference in conservation biology and ecology: further comments on identifying important predictor variables. Biodivers Conserv 11: 1397–1401.

28. Heikkinen RK, Luoto M, Kuussaari M, Pöyry J (2005) New insights to butter-fly–environment relationships with partitioning methods. Proc R Soc Lond B 272: 2203–2210.

29. Havel JE, Lee CE, Vander Zanden MJ (2005) Do reservoirs facilitate invasions into landscapes? BioScience 55: 518–252.

30. Smith SA, Bell G, Bermimgham E (2004) Cross-Cordillera exchange mediated by the Panama Canal increased the species richness of local freshwater fish assemblages. Proc R Soc Lond B 271: 1889–1896.

31. Leprieur F, Beauchard O, Hugueny B, Grenouillet G, Brosse S (2008) Null model of biotic homogenization: a test with the European freshwater fish fauna. Divers Distrib. In press. doi:10.1111/j.1472-4642.2007.00409.x.

32. Stohlgren TJ, Barnett DT, Kartesz JT (2003) The rich get richer: patterns of plant invasions in the United States. Front Ecol Environ 1: 11–14.

33. Marchetti MP, Light TS, Moyle PB, Viers J (2004) Invasion and extinction in California fish assemblages: testing hypotheses using landscape patterns. Ecol Appl 14: 1507–1525.

34. Evans KL, Warren PH, Gaston KJ (2005) Does energy availability influence classical patterns of spatial variation in exotic species richness? Glob Ecol Biogeogr 14: 57–65.

35. Chown SL, Hull B, Gaston KJ (2005) Human impacts, energy availability and invasion across Southern Ocean Islands. Glob Ecol Biogeogr 14: 521–528.

36. Melbourne BA, Cornell HV, Davies KF, Dugaw CJ, Elmendorf S, et al.. (2007) Invasion in a heterogeneous world: resistance, coexistence or hostile takeover? Ecol Lett 10: 77–94.

37. Rejmánek M (2003) The rich get richer – responses. Front Ecol Environ 1: 122–123.

38. Moyle PB, Cech JJ (2004) Fishes: An introduction to ichthyology. New Jersey: Prentice-Hall. 726 p.

39. Wilcove DS, Rothstein D, Bubow J, Phillips A, Losos E (1998) Quantifying threats to imperiled species in the United States. BioScience. 48: 607–615.

40. Myers N, Mittermeier RA, Mittermeier CG, Da Fonsega GAB, Kent J (2000) Biodiversity hotspots for conservation priorities. Nature 403: 853–858.

41. Olden JD, Hogan ZS, Vander Zanden JV (2007) Small fish, big fish, red fish, blue fish: size-biased extinction risk of the world's freshwater and marine fishes. Global Ecol Biogeo 16: 694–701.

42. Lodge DM, Williams S, MacIaac HJ, Hayes KR, Leung B, et al.. (2006) Biological invasions: recommendations for U.S. policy and management. Ecol Appl 16: 2035–2054.

43. Pimentel D, Zuniga R, Morrison D (2005) Update on the environmental and economic costs associated with alien–invasive species in the United States. Ecol Econ 52: 273–88.

44. Vörösmarty CJ, Green P, Salisbury J, Lammers RB (2000) Global water resources: vulnerability from climate change and population growth. Science 289: 284–288.

45. Nilsson C, Reidy CA, Dynesius M, Revenga C (2005) Fragmentation and flow regulation of the World's large river systems. Science 308: 405–408.

46. Millennium Ecosystem Assessment (2005) Ecosystems and human well being: current state and trends. Freshwater (Vol 1, Chapter 7). In: Rijsberman F, Costanza R, Jacobi P, editors. Washington (D.C.): World Resources Institute. pp. 165–207.

47. Lêvêque C, Oberdorff T, Paugy D, Stiassny M, Tedsco PA (2008) Global diversity of fish (Pisces) in freshwater. Hydrobiologia.

48. Center for International Earth Science Information Network (CIESIN) (2005) Gridded Gross Domestic Product (GDP). Available: http://islscp2.sesda.com/ISLSCP2_1/html_pages/groups/soc/gdp_xdeg.html. Accessed 3 January 2008.

49. Centre for Sustainability and the Global Environment (SAGE) (2002) Atlas of the Biosphere. Available: http://www.sage.wisc.edu. Accessed 3 January 2008.

50. Rodrigues ASL, Pilgrim JD, Lamoreux JF, Hoffmann M, Brooks TM (2006) The value of the IUCN Red List for conservation. Trends Ecol Evol 21: 71–76.

51. Pont D, Hugueny B, Oberdorff T (2005) Modelling habitat requirement of European fishes: do species have similar responses to local and regional environmental constraints? Can J Fish Aquat Sci 62: 163–173.

52. MacNally R, Walsh CJ (2004) Hierarchical partitioning public-domain software. Biodivers Conserv 13: 659–660.

53. Ihaka R, Gentleman RJ (1996) R: a language for data analysis and graphics. J Comput Graph Stat 5: 299–314.

54. Borcard D, Legendre P, Drapeau P (1992) Partialling out the spatial component of ecological variation. Ecology 73: 1045–1055.

Chapter 14

Integrated Ecosystem Assessment: Lake Ontario Water Management

Mark B. Bain, Nuanchan Singkran and Katherine E. Mills

ABSTRACT

Background

Ecosystem management requires organizing, synthesizing, and projecting information at a large scale while simultaneously addressing public interests, dynamic ecological properties, and a continuum of physicochemical conditions. We compared the impacts of seven water level management plans for Lake Ontario on a set of environmental attributes of public relevance.

Methodology and Findings

Our assessment method was developed with a set of established impact assessment tools (checklists, classifications, matrices, simulations, representative taxa, and performance relations) and the concept of archetypal geomorphic shoreline classes. We considered each environmental attribute and shoreline

Originally published as Bain MB, Singkran N, Mills KE (2008) Integrated Ecosystem Assessment: Lake Ontario Water Management. PLoS ONE 3(11): e3806. https://doi.org/10.1371/journal. pone.0003806 © 2008 Bain et al. https://creativecommons.org/licenses/by/4.0/

class in its typical and essential form and predicted how water level change would interact with defining properties. The analysis indicated that about half the shoreline of Lake Ontario is potentially sensitive to water level change with a small portion being highly sensitive. The current water management plan may be best for maintaining the environmental resources. In contrast, a natural water regime plan designed for greatest environmental benefits most often had adverse impacts, impacted most shoreline classes, and the largest portion of the lake coast. Plans that balanced multiple objectives and avoided hydrologic extremes were found to be similar relative to the environment, low on adverse impacts, and had many minor impacts across many shoreline classes.

Significance

The Lake Ontario ecosystem assessment provided information that can inform decisions about water management and the environment. No approach and set of methods will perfectly and unarguably accomplish integrated ecosystem assessment. For managing water levels in Lake Ontario, we found that there are no uniformly good and bad options for environmental conservation. The scientific challenge was selecting a set of tools and practices to present broad, relevant, unbiased, and accessible information to guide decision-making on a set of management options.

Introduction

Ecosystem-scale management is increasingly being initiated around the world to cope with complex problems spanning diverse environmental attributes over large areas. Methods to assess ecosystem management impacts and benefits are slowly developing through practice. Some notable US examples of ecosystem management are the landscape habitat modeling used for restoration of the Florida's Everglades [1], [2]; the indicator set used to track Chesapeake Bay management progress [3]; a key environmental tradeoffs comparison among scenarios for the Sacramento-San Joaquin Delta in California [4]; and long-term empirical monitoring of the Mississippi River [5]. In these and others cases, managed changes are expected to have numerous and widespread effects across many attributes of an ecosystem. Methods for anticipating and predicting magnitudes of change are needed to assess management options and identify a preferred alternative. Governments and decision-makers need concise and comparative information on their policy options, and the ecological science community should provide methods for forecasting ecosystem change [6], [7].

Assessing impacts of ecosystem scale change will commonly require a broad scope in space, ecosystem features of public interest, and different kinds of information. Nevertheless, the fundamental needs of impact assessment remain:

quantitative estimates of effects on priority environmental resources under each proposed alternative [8]. We provide a method for comparing the environmental impacts of different policies for managing water levels in Lake Ontario. The governments of the United States and Canada determined [9] that the water management plan for this ecosystem needed re-evaluation and the potential effects on coastal environments were expected to be important and diverse [10]. We present a set of common assessment tools that can be used in concert to forecast diversified environmental impacts of different water level management plans. The tools are applied to the actual policy options under consideration to demonstrate our ecosystem impact assessment approach. The methods and results are sufficiently described to allow application to other ecosystem management programs.

Methods

Case Study and Area

Lake Ontario is the most downstream of North America's Great Lakes and it is positioned between Canada's Province of Ontario and the USA state of New York. Among the Great Lakes, it is the smallest (surface area of 18,960 km^2) and second deepest (average depth of 86 m, maximum depth of 244 km) with the largest drainage area for its size (1:3.4; watershed area of 64,030 km^2). Nevertheless, a large majority (80%) of water input comes from Lake Erie through the Niagara River, and almost all water (93%) leaves by way of the St. Lawrence River [11]. The flow of water out of Lake Ontario is constrained by dams on the St. Lawrence River although lake level is affected by inflows, evapotranspiration, diversions, precipitation, and other hydrologic factors. Regulation of Lake Ontario water levels began in 1960 with a subsequent mean annual variation of 0.5 m (74.49–75.01 m, International Great Lakes Datum of 1985; IGLD 1985) [12]. However, seasonal variation in water level ranges from 0.3 to 1.1 m [13]. Previous to 1960, the lake had a greater range of water levels: 73.76–75.77 m.

The United States and Canadian governments adopted a treaty in 1909 establishing the International Joint Commission to manage binational waterways. In 1952 a set of water management rules was adopted by the International Joint Commission, and in 2000 the International Lake Ontario and St. Lawrence River Study Board was formed to re-evaluate options for St. Lawrence River discharge regulation. The Study Board [9] was assigned to consider environmental resources that were poorly assessed in the original water management plan and other important factors: economic costs, coastal erosion, commercial navigation, water supply, hydrology, hydroelectric power, tourism, and recreational boating. Any change in lake regulation would be the first substantial modification of Lake Ontario water management in a half century, and a new plan would likely remain in place for decades.

The coastal zone of Lake Ontario was the focus of water management impacts because the anticipated range of water level change was not expected to be important in the open waters of the lake [10]. The coastal zone includes a variety of water and shoreline formations that support a high diversity of species and a substantial portion of the flora and fauna of the Great Lakes [14], [15]. For example, a review [16] of the habitat requirements of 113 Great Lakes fishes reported that the vast majority of species used or required coastal habitats. Wetland vegetation of the bays and lagoons includes a high diversity of plants that are adapted to fluctuating water levels [17], [18]. Relative to the open lake, coastal bays and wetlands support high productivity that is enhanced by fluctuations in water levels [19], [20]. Finally, water level variations of Lake Ontario interact with the shoreline features to create complex patterns of coastal habitats [21].

The Study Board [9] identified a set of water management plans (Table 1) as policy alternatives for influencing water levels of Lake Ontario. Lake Ontario was first regulated in 1960 following dam building on the St. Lawrence River and the initial water regulation plan was labeled 1958A. Experience resulted in plan adjustments and the final operational plan used to the present is plan 1958D. Application of plan 1958D showed that adjustments, called deviations, were required at times to accommodate unusual or extreme conditions, and the actual record of actions was termed plan 1958DD (1958D with deviations). Numerous plans were developed and considered by the Study Board and five are used here: 1998, A+, B+, D+, and E. These address a range of interests (Table 1) consistent with the basic expectations of the International Joint Commission to enhance economic and environmental benefits. Finally, a plan was presented by the Study Board as a reference case: Plan E that depicts the unregulated conditions under the present ecosystem configuration and climate.

Table 1. Seven Lake Ontario water management plans defined by the Lake Ontario-St. Lawrence River Study Board [9].

Plan	Description and Purpose
1958D	The original plan used to set weekly outflows of Lake Ontario since 1963. The plan was developed using the 1860–1954 hydrologic record for Lake Ontario.
1958DD	Named for 1958D-with-deviations, this plan uses the decision criteria of actual water management used to set outflows of Lake Ontario since 1960s. Deviations were caused by situations such as winter ice formation and extreme hydrologic conditions outside the design criteria of Plan 1958-D.
1998	A new plan was proposed in 1998 to replace the Plan 1958D but was rejected by the International Joint Commission because it did not address issues on the environment and recreational boating. Plan 1998 was designed using contemporary hydrologic conditions to avoid many of the deviations that were being made from plan 1958D.
A+	This plan was designed to maximize overall economic benefits by striving for stable Lake Ontario water levels in a narrow range that matched desired conditions for a range of businesses.
B+	This plan was designed to simulate a more natural hydrological regime similar to conditions before lake regulation while minimizing impacts to other interests. The plan uses short and long term forecasts of water supplies in conjunction with the pre-project stage-discharge relationship to determine lake releases.
D+	Plan D+ was designed to increase the both economic and environmental benefits relative to plan 1958DD without significant losses to any other interests.
E	This plan was developed as the natural flow option with the current St. Lawrence River channel, dams, and structures. Considered best for the environment, this plan was designed to return the ecosystem to its pre-project state with enhanced condition of the flora and fauna.

Impact Assessment

Our ecosystem impact assessment method was built with a set of established [22], [23] and commonly used impact assessment tools: checklists, classifications, matrices, representative taxa, and performance relations. A checklist was used to identify environmental attributes of interest for the assessment. Classification was used to organize the ecosystem into a series of shoreline classes with different physical properties. A matrix was used to identify the impact-sensitive combinations of environmental attributes and shoreline classes. Representative taxa were used to capture effects of lake level change by environmental attribute. Performance relations linked lake level change to representative taxa by shoreline class. Aside from these standard impact assessment tools, we introduced a somewhat novel concept to cope with the complexity of assessing ecosystem change at the scale of Lake Ontario: shoreline geomorphic classes in archetypal form. By an archetype we mean a model class of shoreline that exhibits defining features, typical characteristics, and distinguishing properties. Archetypes were not perfect representatives of a class, but they were used as a way to simplify the continuous range of variation seen in the biotic and abiotic environment.

Attributes of the environment were selected to address major biotic groups having substantial public and natural resource agency interest within to the coastal zone of Lake Ontario (Table 2). Environmental attributes were assembled from a list of indicator species proposed by the International Lake Ontario-St. Lawrence River Study [24]. Our list is longer and more specific but similar in being focused on prominent animals and plants in the coastal zone of Lake Ontario. Each biotic group was characterized by a representative taxa or assemblage that uses specific habitats; water level management impacts were assessed at this level. The orientation of water management policy-making around a broad array of biotic groups was established over many years of public debate on environmental implications of water level management on Lake Ontario.

A coordinated US and Canadian effort to predict the consequences of varying water levels in the Great Lakes began in the late 1980s following a time when most of the lakes were at historically high levels. At that time a classification was developed [25] to provide a comprehensive description of Great Lakes shorelines that differed in sensitivity to water level changes. This classification system has been refined and applied throughout the Great Lakes [26] making it a standard for mapping and managing shorelines. The classification primarily relies on ten different shoreline types that differ in erosion rates, stability, and capability to adjust position with water level change. The characteristic features of these habitats are summarized in Figure 1 to define archetypes of each class. A quantitative assessment of ecological effects that may occur under different water level scenarios can be made at the full scale of any lake using data on the portion of the lake in each shoreline class.

Table 2. Environmental attributes for assessing impacts of water level regulation on Lake Ontario using management and public interests.

Biotic group	Habitats	Assessment taxa	Rationale for consideration in the assessment
Plants	Protected waters	Submerged aquatic vegetation	These plants stabilize sediments, reduce turbidity, and provide habitat for the spawning and rearing of young fish.
Plants	Protected shores	Emergent vegetation	Water edge plants that create habitat for benthic invertebrates, small forage fishes, juveniles of larger fish species, and nesting birds.
Plants	Land with saturated soil	Wetland vegetation	Plants in hydric soils above the water line form water key habitats for many organisms and are sensitive indicators of water levels and fluctuation regime.
Invertebrates	Littoral zone	Benthic invertebrates	Bottom dwelling organisms that are a major component of the aquatic food web and biodiversity.
Fish	Aquatic vegetation	Bowfin, *Amia calva*	Fish that use vegetated habitats throughout their lives or seasonally for reproduction, feeding, and refuge.
Fish	Wetland and submerged plants	Northern pike, *Esox lucius*	Spawning success often related to submerged and shallow water vegetation for eggs and larvae.
Fish	Tributaries	Rainbow smelt, *Osmerus mordax*	Spawning period water level changes can reduce habitats associated with stream inflows to the lake.
Fish	Littoral zone	Rock bass, *Ambloplites rupestris*	Dependent on shallow shoreline waters along open shores.
Birds	Protected waters	Mallard, *Anas platyhynchos*	Species dependent on submerged aquatic vegetation in still clear waters.
Birds	Emergent plants	Black tern, *Chlidonias niger*	Nesting in emergent vegetation can be disrupted by flooding, wave action, and water level change.
Birds	Marshes	King rail, *Rallus elegans*; Marsh wren, *Cistothorus palustris*	Nesting in marsh vegetation can be disrupted by flooding, wave action, and water level change.
Birds	Open shorelines	Bank swallow, *Riparia riparia*; Piping plover, *Charadrius melodus*; Killdeer, *Charadrius vociferus*	Species using open shoreline features such as sand, cobble beaches, and bluffs.
Mammals	Protected waters	Beaver, *Castor canadensis*	Dependent on protected shoreline waters and shores.

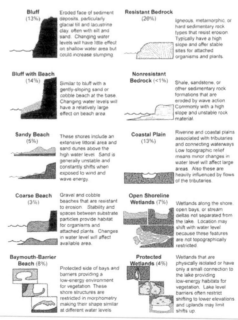

Figure 1. Shoreline classes for assessing lake level effects throughout the Great Lakes.

The classes were identified, characterized, and illustrated by Stewart and Pope [25] and later refined and applied to all shoreline segments in the US and Canada. The percentage of the Lake Ontario shoreline composed of each class and its basic features relative to potential impacts are provided. Not included are artificial shorelines and other minor classes. Drawings included here were made from sketches in Stewart and Pope [25].

A matrix of environmental attributes and shoreline classes was used to identify combinations for detailed assessment of water level change impacts. We considered each attribute and shoreline class in its archetypal form and how water level change would interact with distinguishing properties. This pair by pair review was used to reduce the potential range of impact analyses to those most likely to be important and worthy of further attention. Judgments on the likely significant impacts depended on the nature of water level change, how the shore class would respond to change, and how change could affect environmental attributes. Importance was judged by considering the likelihood that biotic changes symptomatic of ecosystem functional loss [27] would occur: reduction in sensitive species, community shifts toward tolerant organisms, unbalanced taxonomic composition, and food web alteration.

For each environmental attribute, one or more representative taxa were assessed by shoreline class using one or more relations between a water level change and impact magnitude. These models or performance relations were developed from scientific literature accounts of water level effects, and they were parameterized as changes in habitat quantity (e.g., percent change in area) or quality (suitability index; scale 0 for unsuitable and 1 as optimal). Performance relations spanned the range of water level change anticipated under the proposed water management plans.

A 101-year water level sequence for each of the seven Lake Ontario water management plans (Table 1) was simulated using stochastic hydrologic inputs to the lake and the decision rules for each plan [9]. A 3-year portion of the simulated lake levels for each plan is shown in Figure 2. STELLA® version 8 [28] was used to model the effects of each of the seven 101-year simulated water management plans on the representative taxa. The relationship between each taxon and the change in water level caused by each plan was incorporated into the simulation model using step-linearly method and logical function (IF-THEN-ELSE). The 101-year simulation started with month 0 on January 1st and ended at 1,212 months on December 31st. The interval of time between calculations was set to a small value of 0.0625 months to avoid artificial dynamics during the software calculation. The modeling values reflecting the taxa response to the change in water level under each of the water management plans were averaged by time period over the 101 simulated years. The results from the combination of representative taxa and water level change in each shoreline class were then rated by magnitude of expected impact from negative change (adverse impact), minor or equivocal impact, or a desirable change (favorable outcome). These ratings were assembled for all environmental attributes by water management plan to make comparisons of environmental consequences across all plans.

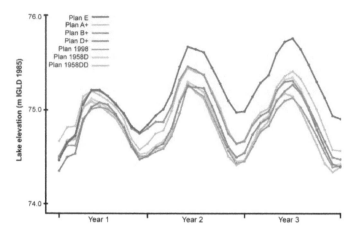

Figure 2. A sample 3-year sequence of monthly water levels under each water management plan.
Simulated data were obtained from the Lake Ontario - St. Lawrence River Study Board [9]. Elevation in meters IGLD 1985 (International Great Lakes Datum of 1985) is the current standard that includes a baseline adjustment for glacial rebound in the earth's crust under the Great Lakes.

Results

The review of potentially important impacts for 130 pairs of environmental attributes (13) and shoreline classes (10) resulted in a matrix (Figure 3) with most entries empty. Expected impacts were concentrated in two shoreline classes that have extensive shallow water and low sloping shores with limited physical response to water level change: baymouth-barrier beach and protected wetland classes. These two shoreline classes were similar in anticipated impacts with some environmental attributes covered in one shoreline class. Impacts on plants were included under both shoreline classes and were found below to be similar in separate analyses. This first analysis step also revealed that birds using shorelines with beaches and bluffs were expected to be affected by water level change across a set of shoreline classes. Finally, four classes of shorelines were not designated to have significant impacts from water level changes at the scale of variability predicted for the water management plans. These shoreline classes were either steep with immobile rock (resistant bedrock, nonresistant bedrock), habitats expected to physically relocate with water levels (open shoreline wetland habitats), or were already highly variable in levels due to tributary flows (riverine and coastal plain). Water level changes in Lake Ontario will have little effect on the areal extent of these habitats or taxa found in them. From the data on amounts of shoreline in each class (Figure 1), the matrix analysis identifies about half the shoreline of Lake Ontario as potentially sensitive to water level change with a minor portion (12%) being highly sensitive.

Environmental attributes	Shoreline class									
	Bluff	Bluff with beach	Sandy beach	Coarse beach	Baymouth - barrier beach	Resistant bedrock	Non-resistant bedrock	Coastal plain	Open shoreline wetlands	Protected wetlands
Submerged aquatic vegetation					✓					✓
Emergent vegetation					✓					✓
Wetland vegetation					✓					✓
Benthic invertebrates				✓	✓					✓
Fish using aquatic vegetation										✓
Fish in wetlands and submerged plants					✓					
Fish using tributaries										✓
Littoral zone fishes				✓	✓					
Birds in protected waters										✓
Emergent plant birds					✓					
Marsh birds					✓					
Birds using beach shores	✓	✓	✓	✓						
Mammal using protected waters										✓

Figure 3. Matrix of environmental attributes and shoreline classes.
Check marks indicate significant impacts were expected; empty cells indicate that no significant effect was anticipated for the range of water level changes being considered.

Bluffs

Coastal bluffs experience erosion and recession from wave action on their face (Figure 1), and changes in water level would alter bluff erosion and recession rates in the short term. A decline in mean lake level would reduce erosion, while a rise in lake level would increase erosion until an equilibrium profile returns [29]. When waves erode the base of a bluff, its slope angle increases and the base becomes unstable, which may result in mass movements of material [30]. Horizontal recession rates of Lake Ontario bluffs range from 0.3–1.5 m/yr [31], and the rate changes with lake level [32]. Consequently, slumping and caving of overlying material would increase as water levels rise and decrease as levels fall.

A rise in water levels and an increase in erosion of the bluff can be detrimental to shoreline birds that nest in bluffs, such as bank swallows (Riparia riparia). Bank swallows are found throughout the Great Lakes region in summer (June to August). They construct nests at the end of a tunnel (60–95 cm long) near the top of coastal bluffs [33]. The length of nest tunnels is equal to the common horizontal recession rate of bluffs, and any increase in bluff recession rate would jeopardize survival of eggs or young. An increase in water level greater than 1 m is expected to elevate bluff recessions and decrease nesting suitability of bank swallows: conditions will become unsuitable when the water level increases 3 m or more (Figure 4A). This relation is very similar to that measured for bank swallow nesting along the Sacramento River, California [34]. Comparing the results among the seven

water management plans indicate plan 1958DD provided the most suitable conditions (suitability = 0.93) for nesting of bank swallows, whereas plan E had the
lowest but still high score (0.82). All other plans were intermediate in suitability
resulting in favorable bank swallow nesting conditions for all plans (Table 3).

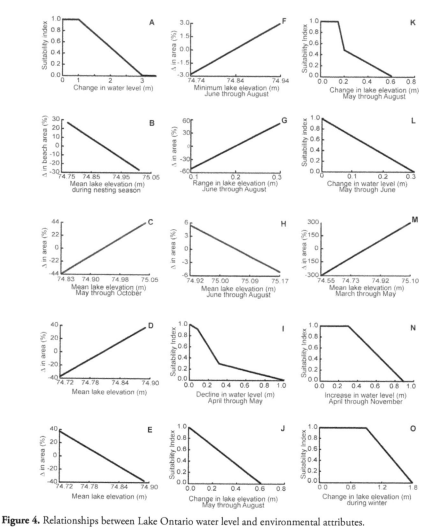

Figure 4. Relationships between Lake Ontario water level and environmental attributes.
A) bank swallow nesting in bluff habitat, B) piping plover nest area on sand beaches, C) benthic invertebrate
habitat along coarse beaches, D) rock bass habitat along coarse beaches, E) killdeer foraging and nesting area on
coarse beaches. Relationships for both baymouth-barrier beach shorelines and protected wetland and backwater
shoreline classes: F) area of submerged aquatic vegetation, G) area of emergent vegetation, and H) area of
wetland vegetation. Baymouth-barrier beach shoreline relations included: I) suitability of habitat for Northern
pike embryos and the earliest fry stages, J) nesting suitability for black tern, and K) nesting suitability for king
rail. Relations only for the protected wetland and backwater shoreline classes were: L) suitability of habitat
for bowfin early life stages, M) area of rainbow smelt adult staging and early life rearing habitat, N) nesting
suitability for marsh wren, and O) suitability of habitat for overwintering beaver.

Table 3. Impact ratings of water regulation plans on each representative taxa in each shoreline class.

Representative taxa	Shoreline class*	1958D	1958DD	1998	A+	B+	D+	E
Bank swallows	Bluff	+	+	+	+	+	+	+
Piping plovers	Bluff with beach, sandy beach	−	o	o	−	o	o	−
Benthic invertebrates	Coarse beach	+	−	−	o	−	−	+
Rock bass	Coarse beach	o	−	o	o	o	o	+
Killdeer	Coarse beach	o	+	o	o	o	o	−
Submerged aquatic vegetation	Baymouth-barrier beach, protected wetlands	+	o	o	+	o	o	+
Emergent vegetation	Baymouth-barrier beach, protected wetlands	o	−	−	−	−	−	+
Wetlands	Baymouth-barrier beach, protected wetlands	o	+	o	o	+	o	−
Northern pike	Baymouth-barrier beach	o	+	+	o	o	o	−
Black tern	Baymouth-barrier beach	+	+	+	+	+	+	+
King rail	Baymouth-barrier beach	+	+	+	+	+	+	−
Bowfin	Protected wetlands	+	+	+	+	+	+	+
Rainbow smelt	Protected wetlands	+	o	o	+	+	+	+
Mallard	Protected wetlands	o	+	+	o	+	+	o
Marsh wren	Protected wetlands	+	+	+	+	+	+	o
Beaver	Protected wetlands	+	+	+	+	+	+	+

Impact rating symbols: − for adverse impact, o for a minor or equivocal impact, + for a desirable change (favorable outcome).
*Note that multiple shoreline classes are shown in four table rows.

Bluffs with Beach and Sandy Beach

Gently sloping sand beaches with or without bluffs (Figure 1) will be inundated or exposed with changes in water levels. We expect that beach area during shorebird nesting and summering can serve as a measure of the nesting habitat extent for some birds. Piping plover (Charadrius melodus) is an endangered species and beach nesting bird [35], [36] found along Lake Ontario from March through August. Nests are maintained for four to five weeks while eggs incubate and several additional days are necessary for the young to fledge [37]. The low slope and constrained width of sand beaches make nesting habitat vulnerable to changes in water level. For example, on a beach with a slope of 0.015 (the average beach slope for Lake Ontario [29]), over 67 m of beach width is lost if water level rises by 1 m. We assume that the width of an archetype beach is about 30 m. Thus, 8.0 m of beach gain or loss will result in the range of change in the beach area for piping plover nesting between −27 and 27% (Figure 4B). The simulated water levels and changes in beach area ranged from a monthly area loss of 81% to a gain of 28%. Again, water management plan 1958DD provided the best conditions (mean loss of 11% beach area) and plan E the worst (59% mean loss). Four plans provided minor beach area losses and were consider equivocal in impact while three plans were predicted to result in large area losses and adverse beach nesting conditions (Table 3).

Coarse Beach

Coarse beaches are composed of rocky material, typically with particle sizes of gravel to cobble, that are resistant to wave action and do not move readily with water level changes (Figure 1). Shallow and turbulent waters with coarse material provide habitat for a variety of invertebrates and fish [38], [39], and the exposed coarse beaches support some birds. Three taxa were used for assessing water level change in coarse beach habitat: benthic invertebrates (e.g., insects, mollusks, and crustaceans), rock bass (Ambloplites rupestris, shallow rocky habitat fish), and killdeer (Charadrius vociferous, shore zone bird). The linear relationship between the change in area of coarse substrate littoral area and exposed beach were developed for each taxon.

Water level changes are expected to affect invertebrates by altering the total amount of submerged habitat available and shifting the location of wave action within this habitat. During the period of rapid growth and reproduction (May through October), high water levels will expand the productive area while lower water levels will contract the available shallow-water area with coarse substrate. Average water level in Lake Ontario from May through October was 74.94±0.1 m with a coarse beach slope of approximated 0.0075 (half the slope of sandy beach [29]). An increase of mean water level by 0.1 m will increase the area available to benthic invertebrates by 13 m, and with a 30-m wide coarse beach zone the usable area would change by 44% (Figure 4C). Under all water plans the area of invertebrate habitat increased in early summer (maximum 142%) and declined (maximum 188%) afterward with falling lake levels. Over the period, water management plans E and 1958D increased invertebrate habitat (42% and 3% respectively) and were considered favorable (Table 3). Plan 1958DD had the most negative (−53%) predicted invertebrate habitat change, and was considered adverse with three other plans of similar outcome. One plan (A+, Table 3) had an expected minor loss of habitat and was considered equivocal in impact. Using the same calculations on an annual basis, a similar magnitude of habitat change was found for fish (Figure 4D) and again plan E provided the largest increase (81%) and plan 1958DD the largest loss (−26%) in habitat. These plans were classified as favorable and adverse (respectively, Table 3) with the other plans displayed minor gains and losses that were considered equivocal in impact. Fish that inhabit the littoral zone of coarse beaches forage among the interstitial spaces of the substrate to find invertebrates. Our representative fish for this habitat was rock bass, a species that commonly inhabits rocky areas in shallow waters of northern lakes year round, and spawns in these habitats during the spring [40].

The killdeer is a shoreline bird that forages and nests on coarse beaches. This species migrates from wintering habitats in southern North America to breeding grounds in the Great Lakes region [41] as early as March, nest in shallow depressions, and feed almost entirely on invertebrates by dabbing the ground or gravel [42]. Their beach habitat area is largely determined by the mean water level. The change in beach area with change in water level was estimated in the same way as for rock bass, but usable habitat varied in an inverse manner (Figure 4E). Exposed beach area for killdeer increased the most with plan 1958DD (26%) and declined the most with plan E (–81%) causing these plans to be classified as favorable and adverse (Table 3). Again, the other plans displayed minor gains and losses that were considered equivocal in impact. Essentially, changing water levels directly tradeoff habitat available to rock bass and killdeer because the beach-littoral zone slope is a constant and both species use habitats either above or below the water surface.

Baymouth-Barrier Beach

The baymouth-barrier beach shoreline class provides protected, low-energy sites supporting abundant submerged aquatic vegetation, emergent plants, and wet-lands. Because of the diversity of vegetated zones, baymouth-barrier habitats are used by a large variety of biotic groups, either permanently or during discrete times of the year. As such, these areas support some of the richest plant and ani-mal communities, including species that are particularly important ecologically and others that are valued by humans. Baymouth-barrier beach shorelines are restricted in morphometry and are unable to change shape or move with changing water levels (Figure 1). A lowering of water level will reduce open water habitat area, reduce the length of land-water margin, and increase intermittently flooded land area because these sites are confined on the protected side. An increase in water level will expand the wetted habitat area, increase the length of margins, and reduce low-lying upslope area.

Lake level fluctuations strongly influence the extent, location, and density of aquatic vegetation [18], [43]–[45]. The areal coverage of submerged and floating-leaved aquatic vegetation (SAV) is determined by the potential habitat space avail-able between the minimum water level and the depth of light penetration (i.e., the photic zone). For baymouth-barrier beach shorelines, the area of SAV declines as water levels fall and increases as water levels rise because gains in habitat upslope exceed downslope loss. SAV cover was mapped in an archetype baymouth-barrier beach site (Blind Sodus Bay [46]) and dimensions on 10 transects across the phot-ic zone showed an average 151-m width of vegetation cover and an average slope of 0.024. This approximates the photic range (up to 4.5 m) predicted for SAV growth [47] for a Lake Ontario waters with a secchi disk transparency of 3.5 m [48]. The maximum SAV depth was also near the maximum available water depth inside a baymouth-barrier beach site. During lake level regulation, the average minimum water level during the SAV growing season (June through August) was 74.84 m. A variation in lake level during this period of 0.1 m would be expected to alter the available area for SAV (Figure 4F). Plan E provided the largest positive change (7%) and was similar to plans 1958D and A+; these plans were considered favorable for SAV coverage (Table 3). Plan 1958DD had the smallest average change in SAV habitat (<2% increase) and was considered equivocal in impacts on SAV coverage. The same rating was applied (Table 3) to three other plans with minor predicted change in SAV cover. No plan was considered adverse to SAV growth because none had a sizable predicted loss in coverage.

Emergent vegetation inhabits the shallow water zone between open water and wetlands. The habitat of emergent vegetation is defined by the variability between mean and minimum water levels [18] during the growing season (June–August). We assume a change in the range of water level variability changes suitable habitat area in a direct linear manner. Since regulation, the average mean to minimum range has been 0.195 m along a slope of 0.024 from the characterization of Blind Sodus Bay. Therefore, the fluctuation range of 0.195 m alternately floods and dewaters an average of 8.12 m of substrate that supports an emergent vegetation fringe between wetlands and open water. With increases in the fluctuation range, the emergent vegetation fringe will expand while decreases in the water level range will reduce the area of this habitat. For a range of 0.3 m, the emergent zone will encompass 12.71 m, an increase of 56%. If the range declines to 0.1 m, emergent vegetation will be limited to a 4.21 m band, a decline of 48% from the present extent (Figure 4G). The two plans bracketing the range of emergent habitat change were again 1958DD and E. Plan 1958DD resulted in a large reduction in area (–75%) and was considered adverse (Table 3) with three other plans with similar outcomes. Plan E was the only plan that provided an increase (26%) in habitat area for emergent vegetation and was considered favorable (Table 3). Two other plans were predicted to result in minor losses in area and were classified as equivocal in impact.

Coastal wetland habitats along the Great Lakes generally develop between the mean and maximum water levels [18]. Wetland plants can survive long periods of flooding, but many require low water levels that expose the substrate for successful germination and seedling establishment [18], [44]. With confined and limited low upslope area at barrier-beach shoreline sites, we assume that increasing the mean water level during the growing season (June–August) would reduce the total wetland area. Prior to regulation of Lake Ontario, the mean water level in a growing season (June–August) was 74.9 m and later the mean for this period increased to 75.0 m. Frenchman's Bay (Ontario), considered here an archetypal barrier-beach site, lost 5.5% of its wetland area [49] under a 0.12 m rise in lake level. Over the water level change of interest, the range of wetland change is expected to be ±5.5% (Figure 4H). Plans 1958DD and B+ show large (6.5%, 5.4%) increases in wetland area and were considered favorable (Table 3). Plan E had a mean loss of 2.5% in area and was classified as adverse. The other plans were predicted to produce minor changes in wetland area and were classified as equivocal.

Great Lakes fish diversity and biomass density are concentrated in nearshore waters, especially in vegetated habitats [50], [51]. The structural complexity of vegetation protects small fish from predation, while abundant invertebrates and high primary productivity provide rich food resources. Thus, a large majority of Great Lakes fishes use protected waters and vegetated habitats for one or more life stages [16]. For baymouth-barrier beach shorelines, we used the northern pike (Esox lucius) because it depends on these habitats, has a well documented biology, supports popular fisheries, and is recognized by the public.

Soon after the winter ice clears, northern pike spawn in shallow vegetated waters (typically submerged terrestrial and wetland plants), eggs usually hatch after 12–14 days, and the young remain attached to vegetation for an additional 6–10 days. Young northern pike feed on zooplankton and aquatic insects; thereafter, fish constitute most of their diet [40], [52]. Although spawning habitat is a limiting factor for northern pike reproduction in many waterways, Casselman and Lewis [53] found no relationship between spring water levels and northern pike spawning success. Instead they reported that early life survival played a dominant role in determining the abundance of northern pike in the Bay of Quinte (Ontario). Mortality rates of small young can reach 99% due to stranding and predation [40]. Field studies [54]–[56] reported that pike spawning and early life occurs in water less than 1.0 m deep over newly inundated vegetation with the highest concentration of embryos and new fry in water less than 0.3 m deep. Figure 4I relates a drop in water level to habitat suitability for northern pike embryos and the earliest fry stages during April and May. Embyros are most sensitive to water level changes because of limited mobility and dispersion across depth ranges. Plans 1958DD and 1998 provided excellent (suitability index 0.60) early life habitat for northern pike, and plan E provided the least suitable (0.33) conditions. These plans were classified as favorable and adverse respectively (Table 3). The other plans were intermediate and considered equivocal in impact on northern pike early survival.

The extensive vegetated waters and shores of baymouth-barrier beach sites support a variety of birds. The abundant invertebrates and fish in these habitats constitute a large portion of the diet of many species [57], [58]. Birds are often associated with specific vegetation types [58], [59] with many species preferring an equal mix of emergent vegetation to open water while others favor nesting sites in or near submerged aquatic vegetation [58]. Birds are affected by water level changes both directly, in terms of nest success, and indirectly, through effects on habitat space and food supplies. Taft et al.. [60] showed that the average water depth and topographic variability affected foraging habitat availability, which influenced waterbird community composition.

Baymouth-barrier beach shorelines are important to birds that nest at the interface of wetlands and open water. Black tern (Chlidonias niger) nest in emergent vegetation and is a New York endangered species [61]. Black terns prefer habitats that have an even mix of open water and vegetation [62] where diverse invertebrates and small fish provide an abundant food supply. Black terns construct their nests on mats of vegetation over water about 0.6 m deep (0.46–1.10 m [41], [63]) from May to August [64]. Eggs are incubated for 17 to 22 days, and young fledge 19 to 25 days after hatching. Rising water levels can result in the loss of nests and young. Figure 4J relates changes in water level during the nesting season with stable water levels being optimal and rises of more than 0.6 m being unsuitable for nesting. Water levels typically increased during the nesting months prior to lake regulation (mean 0.3 m) and the magnitude of increase was greater after lake regulation began (mean 0.4 m). Consequently, all plans resulted in very good nesting suitability (index≥0.66) and considered favorable (Table 3).

Marsh birds construct nests at the interface of open water and plants (emergent, wetland), preferring habitats adjacent to stable shallow water for the nesting season. King rail (Rallus elegans) represented marsh birds in this shoreline class; it is endangered in Canada [65], threatened in New York [61], and declining over its range [66]. King rail are found in the Great Lakes region from April to October [67] when they construct nests of dead grasses or sedges in heavily vegetated waters 0.15 to 0.46 m deep [68], [69]. Most eggs are laid from May to June and incubated for 21 to 24 days. Young roam out of the nest to forage (invertebrates and small fish) soon after hatching but they cannot fly for 9 weeks. Water level increases during the nesting season may flood the nests of king rail because they are fixed in low-lying vegetation in water or on vegetated shorelines. Figure 4K shows a relation between habitat suitability and water level increase. Initial increases in water level to common nest heights do not diminish suitability. Further increases result in sharp declines in habitat suitability and additional increases reduce suitable nesting conditions to zero at magnitudes beyond 0.6 m. Water levels have often risen during rail nesting in Lake Ontario prior to regulation and more so afterward, making this species very sensitive to water management plans. All plans except E provided very good nesting suitability (index≥0.71) and considered favorable (Table 3). Plan E was somewhat lower in suitability (0.57) and considered equivocal in effect on king rail nesting success.

Protected Wetlands

The protected wetlands shoreline class includes tributary mouths, lagoons, and distinct shallow waters with restricted connections to Lake Ontario (Figure 1).

Like baymouth-barrier beach shorelines, this shoreline class provides protected, low-energy sites supporting abundant plant growths. Habitats in protected wetlands shoreline class support a large variety of biotic groups, rich plant and animal communities, and many species important ecologically and valued by humans. Protected wetlands are distinct in morphometry and do not move or change shape with changing water levels. A reduction in water level will reduce habitat area and an increase in water level may increase open water and wetland habitats.

Relations between water level change and change in wetlands, emergent plants, and SAV in protected wetland habitats were found to be very similar to those of the baymouth-barrier beach shoreline class. The change in area for SAV growth was estimated using topography of South Sandy Pond; a protected wetland site. Bathymetric measurements were made as described for Blind Sodus Bay, the archetype baymouth-barrier beach site. The resulting relationship between water level and areal change was not meaningfully different from that shown in Figure 4 (F, G, H) because the difference in average underwater slope was very small (0.017 versus 0.024). Therefore, the relations for change in areas of wetlands, emergent plants, and SAV developed for the baymouth-barrier beach shoreline class were used in characterizing water management impacts in the protected wetlands (Table 3).

Some fish species permanently inhabit shallow vegetated habitats that provide refuge from predators, abundant foods, and nesting sites. Bowfin (Amia calva) is a resident fish in vegetated habitats. Bowfin spawn from May through June when males prepare a nest in shallow (0.30–0.61 m) vegetated areas. After spawning, the eggs and young are guarded by the male for several weeks (Scott and Crossman 1973). Water level declines during the spawning season may cause eggs and young to be stranded. The minimum water depth for nesting of bowfin is about 0.3 m and we estimate (Figure 4L) a decline in spawning and rearing habitat suitability with rapid water level declines. All water management plans were very similar in providing excellent early life habitat for bowfin: suitability index score from 89 to 98. Therefore, all plans were considered favorable for bowfin in protected wetlands.

Rainbow smelt (Osmerus mordax) ascend Lake Ontario tributary streams to spawn in the spring soon after the ice thaws, typically between March and May. Spawning occurs upstream in shallow flowing water locations that tend to be free of vegetation. Eggs adhere to the substrate and hatch after 2 to 3 weeks. Still protected waters associated with tributaries, wetlands, and backwaters provide staging habitats for adults at spawning, and rearing habitat for larvae [40]. The mean water level of Lake Ontario has been 74.81±0.26 m from March through May. With a slope across protected wetlands of 0.017, a 0.26 m rise in mean water level would have flooded approximately 15.24 m at the edge of these habitats. For a small lake tributary with a 1,350 m2 open water area at the stream mouth with wetlands, a 0.26 m rise in lake level would be expected to increase open water habitat to 4,056 m2 – 300% more assuming a roughly circular water body (Figure 4M). The exact change in open water area will depend on the configuration of a protected wetlands site but this scale of habitat change approximates responses for a range of surface areas. From lake level simulations, most plans were found to substantially (>200%) increase rainbow smelt adult staging and early life rearing habitat and were considered favorable to this fish (Table 3). Plan 1958DD and 1998 providing the smaller gains and were classified as equivocal in changing habitats for smelt.

While various water birds live in protected wetlands of Lake Ontario, we focus on mallards (Anas platyrhynchos) to represent effects of water level change. Mallard nests are often constructed in grasses or reeds near water bodies. The mallard dabbles in the shallow water for food, which consists primarily of plants and plant seeds as well as some insects, mollusks, and crustaceans [41], [70]. The loss of wetlands may affect whether mallards are present in an area and will likely influence their nesting success. Both suitable feeding and nesting habitats are needed for supporting mallards. We used a linear combination of SAV and wetland relations (Figure 4F, H) with water level change to capture the cumulative effects on mallards:

% change in mallard support =   (0.5 X % change in SAV area)

                                                         + (0.5 X % change in wetland area).

Water level change affects wetlands and SAV in opposite ways so an optimal water level regime for mallards will be a compromise of the two relations. All water management plans provided at least a minor improvement in support for mallards with four plans providing clearly the best conditions (Table 3).

Marsh wren (Cistothorus palustris) inhabit marshes almost exclusively. During their breeding season (April through November [71]), marsh wrens are found in marshes and swamps across the northern half of the United States [41]. They use reeds to construct dome-shaped nests about 0.3 to 0.9 m above water [41], [72]. Before water regulation of Lake Ontario, water levels rose an average of 0.66 m during the nesting season and following regulation the seasonal rise was 0.74 m. Changes in habitat suitability for nesting of marsh wren respond to water level increases greater than 0.3 m and declines to zero at 0.9 m (Figure 4N). All plans but E were similar in providing excellent nesting conditions for marsh wren: suitability index score from 89 to 98 and considered favorable (Table 3). Plan E was clearly below the others (79) and classified as equivocal in impact on marsh wren nesting.

Beaver (Castor canadensis) is a mammal living in protected wetland habitats and considered is sensitive to lake level fluctuations. Beavers construct lodges at very specific water levels and this species is not normally found in habitats experiencing large fluctuations [73]. Water depth and water level stability are especially important during the winter where ice cover forms. The winter water depth must remain within a range that does not submerge the lodge and remains deep enough for lodge access after ice covers the water surface. Winter water depths between 0.9 and 1.8 m appear necessary for safe beaver lodge sites along the edge of wetlands, tributaries, and land margins. Focusing on water depths for lodges, we use the relation in Figure 4O to estimate effects of water level change on beaver habitat suitability. All water management plans were very similar in providing excellent overwintering conditions for beaver: suitability index score from 93 to 99. Consequently, all plans were rated favorable for beaver habitat Table 3).

Comparison of Water Management Plans

The ratings of each water management plan on the 20 combinations of representative taxa and shoreline classes (Table 3) show some differences in patterns of impact on the Lake Ontario environment. All water management plans were expected to result in favorable conditions for most representative taxa, and the number of favorable outcomes (8 to 10 out of 16) were similar among all plans. Adverse impacts were most common for plan E and this plan generally set the upper extreme of water levels in regulation simulations (Figure 2). Plan E had five adverse impacts affecting almost all (5 of 6, Table 3) of the vulnerable shoreline classes accounting for about a third (34%, Figure 1) of the lake shoreline. Plan 1958DD had three adverse impacts affecting three of the shoreline classes comprising 15% of the shoreline. This plan closely followed the lower extreme of water levels in plan simulations. The other water management plans were consistently intermediate to plans 1958DD and E in impact ratings and simulated water levels. There were differences in the distribution of favorable and adverse impacts across these five plans but there was little to distinguish these plans in terms of overall effect on the lake environment. The five intermediate impact plans (1958D, 1998, A+, B+, D+) often had equivocal impact ratings (Table 3) and very few adverse impacts. Overall, the natural flow plan (E) is expected to result in the most adverse impact on the current Lake Ontario environment, and current water management (plan 1958DD) has many favorable outcomes and few adverse impacts.

This impact assessment was primarily focused on birds, fish, and plant groups and the water level management plans varied in their impacts by group (Table 3). Plan 1958DD had the most favorable ratings for birds while plan E had clearly more adverse impacts on birds. This pattern reflects the higher water levels, reduced beach area, and faster rising waters during the nesting period under plan E. Fish species that gain habitat under high water levels were favored most often by plan E although changing water levels adversely impacted northern pike. Wetlands were least impacted under plans 1958DD and B+ due to low summer water levels, and emergent and submerged plants were favored under plan E with its relatively high summer water levels. Status quo water management (plan 1958DD) tends to favor birds, fish nesting and early survival, and wetlands. In contrast, plan E tends to favor plants and organisms using shallow aquatic habitats. Other plans were intermediate to these contrasting patterns of impact and difficult to distinguish.

The pattern of impact ratings among shoreline classes (Table 3) was mixed; especially for the most complex classes that offered many habitats (baymouth-barrier beach and protected wetlands). Both favorable and adverse impact ranks were seen within each of these shoreline classes under the distinct plans 1958DD and E. There was some trend for favorable ratings in protected wetlands, and adverse ratings in beach habitats. In general, the pattern of impacts is clearer with taxonomic groupings than shoreline classes.

Discussion

This ecosystem impact assessment provided results that show similarities and differences among the water management plans relative to select Lake Ontario environmental attributes. The water management plan that emerged from application experience in the last half century, plan 1958DD, appears to be a good choice for maintaining most environmental resources and harming few. This plan had favorable outcomes most often, adverse impacts for half of the shoreline classes comprising a minor portion of the coast, and a long record of successful application experience. Five plans that were designed as modifications to historical management to better address different social and environmental interests slightly reduced the frequency of adverse impacts. However, these plans commonly provide equivocal impact outcomes suggesting many moderately positive and negative effects. It appears that fine-tuning the actual operational plan results in options with many minor impacts of both directions. With little evidence for clear environmental gains from the five refined water management plans a continuation of the current water management policy seems reasonable for the lake environment.

One clear inconsistency with preconceived plan performance was seen with plan E; intended as the natural water level regime with the greatest environment benefits. Instead this plan was most often the one with adverse impacts, impacts on most of the sensitive shoreline classes, and impacts to the largest portion of the lake coast. It appears this natural water management plan would have broad adverse consequences for the lake environment. The belief that plan E is best for the environment rests on the thought that natural hydrologic variation provides the most natural conditions. Plan E was posed as a reference regime for this purpose. The expected changes under plan E may promote past environmental conditions, but in the context of the current lake setting many of these changes would be considered adverse.

The distribution of favorable and adverse impacts across species and assemblages varied without a clear pattern. There was some trend for the current water management (plan 1958DD) to favor species and assemblages that benefit from low, warm season lake levels: inhabitats of beaches, seasonally inundated lands, and stable land-water margins. Plan E, with its high lake levels much of the time, tends to favor inhabitants of shallow water that gain from increased littoral area. Again, the other plans were intermediate to these contrasting patterns of impact and difficult to distinguish. Comparing patterns of impact across water management plans was our assessment aim and the results provide the clearest conclusions when used across plans. Our methods then appear to provide the information that can inform decisions about water management relative environmental resources of most interest.

The management of water in large aquatic ecosystems is expected to result in varied and complex environmental change. Increasingly the management of ecosystem-scale change is being attempted with planning supported by integrated assessments and a systemwide perspective. However, effective integrated assessment practices have not been established, system scale methods are lacking, and the challenges seem overwhelming. For example, management of the Upper Mississippi River-Illinois Waterway was found [74] to be inadequate to support decision-making because it lacked ecosystem scale understanding and broad information synthesis. Similarly, management of water along the Missouri River was reviewed [75] and judged to be very limited by weak consideration of issues at the ecosystem scale, poor responsiveness to public interests, and lack of an ecosystem context for site-specific actions. Ecosystem management requires organizing and synthesizing information at the ecosystem scale while simultaneously addressing public interests, continua and dynamics of physicochemical conditions, and diverse ecological properties. No approach and set of methods will perfectly accomplish integrated ecosystem assessment making the challenge one of selecting a set of tools to present broad, relevant, unbiased, and accessible information to guide decision making. Thus, the product of the assessment needs to be concise but revealing of tradeoffs among policy options.

In structuring our assessment, we made a series of decisions on what to consider, how to encompass the ecosystem, how to consolidate information, and what precision of results are needed for management decisions. Binational government policies, management history, and treaties and laws determined some important attributes of our impact assessment: the management options (water regulation plans), important environmental attributes (species, assemblages), and water level reference data. All large scale ecosystem management cases will have history, law, and precedent that constrain assessment methods and analysis capabilities. We chose a set of impact assessment techniques to fit the situation and meet the need of broad, relevant, unbiased, and accessible information to guide decision-making.

Classification of the Lake Ontario shoreline allowed inferences to be made on how water level change would interact with the biophysical environment. This method provided a whole ecosystem context for analyzing potentially significant water level impacts. A checklist of biotic resources of public interest provided the assignment of what information was relevant in the policy arena. A matrix of environmental attributes and shoreline classes allowed us to infer at a manageable scale which biotic and physical properties needed more detailed analysis. Performance relations integrating water levels and taxon biology enabled us to capture the basic effects of water level management. Analyses were made tractable by working within the concept of representative organisms and archetypal physical settings. Finally, rating impact direction and magnitude diminished the importance of specific numeric predictions (e.g., change in areas) while retaining the capability to compare management options.

Many details of this assessment can be challenged, debated, and improved but as a whole our mix of methods yielded useful findings on the relative merit of seven water management plans for the Lake Ontario ecosystem. The overall outcome was found to be roughly consistent with the ideas used to generate the management plans although they were developed independently of past decisions and management issues. We also draw into question some widely held beliefs: a natural water regime will be best for the environment today, and there are good and bad options for the environment. We do not provide an answer on what is best for the environment although we make a recommendation that one plan appears a reasonable and safe choice. More important, we provide a one page synthesis (Table 3) of information that can inform government officials what environmental tradeoffs are at stake when a plan is considered with other social needs like economic development, transportation efficiency, hydropower, recreation, and shoreline property security.

Acknowledgements

We thank John Barko (US Army Engineer Research and Development Center, Vicksburg, MI) for organizing a workshop on assessing Great Lakes hydrologic change where the shoreline classification was introduced. We are grateful to the Lake Ontario - St. Lawrence River Study Board for freely providing extensive and detailed information on their water management plans and simulated hydrologic data by way of an open access web site. Glenn Suter and Reuben Goforth provided important suggestions that improved the assessment and paper.

Author Contributions

Conceived and designed the experiments: MBB KEM. Analyzed the data: MBB NS. Wrote the paper: MBB NS. Assembled data and relations: MBB KEM.

References

1. Curnutt JL, Comiskey J, Philip Nott M, Gross LJ (2000) Landscape-based spatially explicit species index models for Everglades restoration. Ecol Appl 10: 1849–1860.

2. Fitz C, Sklar F, Waring T, Voinov A, Costanza R, et al.. (2004) Development and application of the Everglades landscape model. In: Costanza R, Voinov A, editors. Landscape simulation modeling: A spatially explicit, dynamic approach. New York: Springer. pp. 143–171.

3. Chesapeake Bay Program (2007) Chesapeake Bay: 2006 health and restoration assessment. Annapolis, MD: US Environmental Protection Agency, EPA 903R-07001.

4. Lund J, Hanak E, Fleenor W, Howitt R, Mount J, et al.. (2007) Envisioning futures for the Sacramento-San Joaquin Delta. San Francisco: Public Policy Institute of California.

5. US Army Corps of Engineers (2004) 2004 Report to Congress. Rock Island, IL: U.S. Army Corps of Engineers, Upper Mississippi River System Environmental Management Program.

6. Clark JS, Carpenter SR, Barber M, Collins S, Dobson A, et al.. (2001) Ecological forecasts: an emerging imperative. Science 293: 657–660.

7. Carpenter SR (2002) Ecological futures: building an ecology of the long now. Ecology 83: 2069–2083.

8. Haug PT, Burwell RW, Stein A, Bandurski BL (1984) Determining the significance of environmental issues under the National Environmental Policy Act. J Environ Manage 18: 15–24.

9. Lake Ontario - St. Lawrence River Study Board (2006) Options for managing Lake Ontario and St. Lawrence River water levels and flows. Washington DC/Ottawa: International Joint Commission.

10. St. Lawrence River - Lake Ontario Plan of Study Team (1999) Plan of study for the criteria review in the orders of approval for regulation of Lake Ontario – St. Lawrence River levels and flows. Washington DC/Ottawa: International Joint Commission.

11. US Environmental Protection Agency (1987) The Great Lakes: An environmental atlas and resource book. Chicago: US Environmental Protection Agency.

12. Wilcox DA, Thompson TA, Booth RK, Nicholas JR (2007) Lake-level variability and water availability in the Great Lakes. U.S. Geological Survey, Circular 1311.

13. Great Lakes Commission (2003) Toward a water resources management decision support system for the Great Lakes-St. Lawrence River Basin. Ann Arbor, MI: Great Lakes Commission. Available: http://glc.org/wateruse/wrmdss/final-report.html. Accessed 2008 Mar 16.

14. Herdendorf CE (1992) Lake Erie coastal wetlands: an overview. J Great Lakes Res 18: 533–551.

15. Krieger KA, Klarer DM, Heath RT, Herdendorf CE (1992) A call for research on Great Lakes coastal wetlands. J Great Lakes Res 18: 525–528.

16. Jude JD, Pappas J (1992) Fish utilization of Great Lakes coastal wetlands. J Great Lakes Res 18: 651–672.

17. Keddy PA, Reznicek AA (1982) The role of seed banks in the persistence of Ontario's coastal plain flora. Amer J Botany 69: 13–22.

18. Keddy PA, Reznicek AA (1985) Vegetation dynamics, buried seeds, and water-level fluctuations on the shorelines of the Great Lakes. In: Prince HH, D'Itri FM, editors. Coastal Wetlands. Chelsea, MI: Lewis Publishers, Inc. pp. 38–58.

19. Wilcox DA (1995) Wetland and aquatic macrophytes as indicators of anthropogenic hydrologic disturbance. Natural Areas J 15: 240–248.

20. Mortsch LD (1998) Assessing the impact of climate change on the Great Lakes shoreline wetlands. Climatic Change 40: 391–416.

21. Bedford KW (1992) The physical effects of the Great lakes on tributaries and wetlands. J Great Lakes Res 18: 571–589.

22. Shopley JB, Fuggle RF (1984) A comprehensive review of current environmental impact assessment methods and techniques. J Environ Manage 18: 25–47.

23. Westman WE (1985) Ecology, impact assessment and environmental planning. New York: John Wiley and Sons.

24. Parker B, Tracy M (2003) Environmental technical work group update. Buffalo, NY: International Joint Commission, International Lake Ontario - St. Lawrence River Study Office, Ripple Effects 6.

25. Stewart CJ, Pope J (1993) Erosion processes task group: land use and management. Washington DC/Ottawa: International Joint Commission, Great Lakes Water Level Reference Study, Erosion Processes Task Group.

26. Christian J. Stewart Consulting (2001) A review of potential data sources for use in evaluating coastal process and riparian impacts associated with changing water levels on Lake Ontario and the St. Lawrence River. Washington DC/Ottawa: International Joint Commission, Great Lakes Water Level Regulation Study, Coastal Processes Working Group.

27. Davies SP, Jackson SK (2006) The biological condition gradient: a descriptive model for interpreting change in aquatic ecosystems. Ecol Appl 16: 1251–1266.

28. High Performance Systems, Inc. (2003) STELLA @ Research, Version 8.0 for Windows. Lebanon, NH: High Performance Systems, Inc.

29. Functional Group 2 (1989) Living with the lakes: challenges and opportunities, Annex B - Environmental features, processes, and impacts: an ecosystem perspective on the Great Lakes-St. Lawrence River system. Washington DC/Ottawa: International Joint Commission, Water Levels Reference Study.

30. Lawrence PL (1994) Natural hazards of shoreline bluff erosion: a case study of Horizon View, Lake Huron. Geomorphology 10: 65–81.

31. Davidson-Arnott RGD, Ollerhead J (1995) Nearshore erosion on a cohesive shoreline. Marine Geol 122: 349–365.

32. Amin SMN, Davidson-Arnott RGD (1997) A statistical analysis of the controls on shoreline erosion rates, Lake Ontario. J Coastal Res 13: 1093–1101.

33. Tufts RW (1986) Birds of Nova Scotia. Halifax: Nimbus Publishing Limited and the Nova Scotia Museum.

34. Stillwater Sciences (2006) Sacramento River ecological flows study: state of the system report. Chico, CA: The Nature Conservancy.

35. Environment Canada (2008) The piping plover in Eastern Canada. Dartmouth, NS: Environment Canada, Atlantic Region. Available: http://www.ns.ec.gc.ca/wildlife/plover/brochure/page2.html. Accessed 2008 Mar 16.

36. US Fish and Wildlife Service (2008) Piping plover. Washington DC: US Fish and Wildlife Service, Environmental Conservation Online System. Available: https://ecos.fws.gov/species_profile/SpeciesProfile?spcodeB079. Accessed 2008 Mar 16.

37. New York Department of Environmental Conservation (2008) Piping Plover Fact Sheet. Albany, NY: New York Department of Environmental Conservation. Available: http://www.dec.ny.gov/animals/7086.html. Accessed 2008 Mar 16.

38. Barton DR, Hynes HBN (1978) Wave-zone macrobenthos of the exposes Canadian shores of the St. Lawrence Great Lakes. J Great Lakes Research 4: 27–45.

39. Barton DR, Griffiths M (1984) Benthic invertebrates of the nearshore zone of eastern Lake Huron, Georgian Bay, and North Channel. J Great Lakes Res 10: 407–416.

40. Scott WB, Crossman EJ (1973) Freshwater fishes of Canada. Ottawa: Fisheries Research Board of Canada.

41. Alsop FJ, III (2001) Birds of North America. New York: D.K. Publishing.

42. Canadian Wildlife Service (2008) Bird fact sheets: killdeer. Ottawa: Canadian Wildlife Service. Available: http://www.hww.ca/hww2.asp?id50. Accessed 2008 Mar 16.

43. Robel RJ (1962) Changes in submersed vegetation following a change in water level. J Wildl Manage 26: 221–224.

44. Harris SW, Marshall WH (1963) Ecology of water-level manipulations on a northern marsh. Ecology 44: 331–343.

45. Hudon C, Lalonde S, Gagnon P (2000) Ranking the effects of site exposure, plant growth form, water depth, and transparency on aquatic plant biomass. Can J Fish Aqu Sci 57: Suppl 131–42.

46. Lake Ontario Biocomplexity Project (2007) Lake Ontario biocomplexity study: The physical, biological, and human interactions shaping the ecosystems of freshwater bays and lagoons. Ithaca, NY: Cornell University. Available: http://ontario.cfe.cornell.edu/. Accessed 2008 May 9.

47. Chambers PA, Kalff J (1985) Depth distribution and biomass of submersed aquatic macrophyte communities in relation to Secchi depth. Can J Fish Aqu Sci 42: 701–709.

48. Hall SR, Pauliukonis NK, Mills EL, Rudstam LG, Schneider CP, et al.. (2003) A comparison of total phosphorus, chlorophyll a, and zooplankton in embayment, nearshore, and offshore habitats of Lake Ontario. J Great Lakes Res 29: 54–69.

49. Williams DC (1995) Dynamics of area changes in Great Lakes coastal wetlands influenced by long term fluctuations in water levels. Ann Arbor, MI: University of Michigan, School of Natural Resources and Environment, Doctoral Dissertation.

50. Keast A, Harker J (1977) Fish distribution and benthic invertebrate biomass relative to depth in an Ontario lake. Environ Bio Fish 2: 235–240.

51. Randall RG, Minns CK, Cairns VW, Moore JE (1996) The relationship between an index of fish production and submerged macrophytes and other habitat features at three littoral areas in the Great Lakes. Can J Fish Aqu Sci 53: Suppl 135–44.

52. Smith CL (1985) The inland fishes of New York State. Albany, NY: New York State Department of Environmental Conservation.

53. Casselman JM, Lewis CA (1996) Habitat requirements of northern pike (Esox lucius). Can J Fish Aqu Sci 53: Suppl 1161–174.

54. Williamson LQ (1942) Spawning habitat of muskellunge, northern pike. Wisc Conserv Bull 7: 10–11.

55. Clark CF (1950) Observations of the spawning habits of northern pike, Esox lucius, in northwestern Ohio. Copeia 1950: 258–288.

56. Fabricius E (1950) Heterogeneous stimulus summation in the release of spawning activities in fish. Fisheries Board Sweden, Institute of Freshwater Research Drottningholm 29: 57–99.

57. Palmer RS (1962) Handbook of North American birds. New Haven, CT: Yale University Press.

58. Wilcox DA, Meeker JE (1992) Implications for faunal habitat related to altered macrophyte structure in regulated lakes in northern Minnesota. Wetlands 12: 192–203.

59. Johnsgard PA (1956) Effects of water fluctuation and vegetation change on bird populations, particularly waterfowl. Ecology 37: 689–701.

60. Taft OW, Colwell MA, Isola CR, Safran RJ (2002) Waterbird responses to experimental drawdown: implications for the multispecies management of wetland mosaics. J Appl Ecology 39: 987–1001.

61. New York Department of Environmental Conservation (2007) List of Endangered, Threatened, and Special Concern Fish and Wildlife Species of New York State. Albany, NY: New York Department of Environmental Conservation. Available: http://www.dec.ny.gov/animals/7494.html. Accessed 2008 May 9.

62. Hickey JM, Malecki RA (1997) Nest site selection of the black tern in western New York. Colonial Waterbirds 20: 582–595.

63. Mazzocchi IM, Hickey JM, Miller RL (1997) Productivity and nesting habitat characteristics of the black tern in western New York. Colonial Waterbirds 20: 596–603.

64. Currier CL (2000) Special animal abstract for Chlidonias niger (black tern). Lansing, MI: Michigan Natural Features Inventory. Available: http://www. msue.msu.edu/mnfi/abstracts/zoology/chlidonias_niger.pdf. Accessed 2007 December 12.

65. Environment Canada (2007) Species at risk public registry. Gatineau Québec: Environment Canada. Available: http://www.sararegistry.gc.ca/default_e.cfm. Accessed 2008 May 9.

66. National Biological Information Infrastructure (2006) US Fish and Wildlife Service focal species: King rail (Rallus elegans). Reston, VA: US Geological Survey, Biological Informatics Office. Available: http://www.nbii.gov/portal/community/Communities/Ecological_Topics/Bird_Conservation/USFWS_Focal_Species/. Accessed 2008 May 9.

67. Brewer R, McPeek GA, Adams RJ, Jr (1991) The Atlas of Breeding Birds of Michigan. East Lansing, MI: Michigan State University Press.

68. Meanley B (1969) Natural history of the king rail. Washington DC: Bureau of Sport Fisheries and Wildlife, North American Fauna 67.

69. Harrison HH (1979) A field guide to western bird's nests. Boston: Houghton Mifflin Company.

70. Lister R (2008) Bird fact sheets: mallard. Ottawa: Canadian Wildlife Service. Available: http://www.hww.ca/hww2.asp?id54 .Accessed 2008 Mar 19.

71. Mumford RE, Keller CE (2000) The Birds of Indiana. Bloomington, IN: Indiana University Press. Available: http://www.ulib.iupui.edu/collections/birds/bird227_1.html. Accessed 2008 May 19.

72. Manitoba Museum of Man and Nature (2008) Marsh Wren (Cistothorus palustris). Winnipeg, Manitoba: The birds of Manitoba Online. Available: http://www.virtualmuseum.ca/Exhibitions/Birds/MMMN/English/a_marshwren.html. Accessed 2008 May 19.

73. Allen AW (1983) Habitat suitability index models: beaver. US Fish and Wildlife Service FWS/OBS-82/10.30(rev).

74. National Research Council (2001) Inland Navigation System Planning: The Upper Mississippi River-Illinois Waterway. Washington DC: National Academy Press.

75. National Research Council (2002) The Missouri River Ecosystem: Exploring the Prospects for Recovery. Washington DC: National Academy Press.

Copyrights

Index

T

U

V

W

9 781774 632444